ARMS CONTROL

The International Peace Research Institute, Oslo (PRIO) is an independent, international institute for peace and conflict research, founded in 1959. It was one of the first centres of peace research in the world and is Norway's only peace research centre. Its foundation and early influence were instrumental in projecting the idea of peace research.

PRIO's publications include the *Journal of Peace Research*, published bi-monthly, and the quarterly *Security Dialogue*. PRIO's scholarly work is published in peer-reviewed journals, as well as in books, reports and conference papers, some of which can be accessed at **www.prio.no**

PRIO's overall research is organized into four Strategic Institute Programmes:

- Conditions of War and Peace
- Foreign and Security Policies
- Ethics, Norms and Identities
- Conflict Resolution and Peacebuilding

 PRIO

International Peace Research Institute, Oslo
Fuglehauggata 11, N-0260 Oslo, Norway
Telephone: +47 22 54 77 00 Telefax: +47 22 54 77 01
Email: **info@prio.no**
www.prio.no

SIPRI is an independent international institute for research into problems of peace and conflict, especially for those of arms control and disarmament. It was established in 1966 to commemorate Sweden's 150 years of unbroken peace.

The Institute is financed mainly by the Swedish Parliament. The staff and the Governing Board are international. The Institute also has an Advisory Committee as an international consultative body.

SIPRI's major research projects are:

- Armed conflicts and prevention
- Arms production
- Arms transfers
- Chemical and biological weapons
- European security
- Export controls
- Military expenditure
- Military technology
- Integrating Fact Databases in the Field of International Relations and Security

SIPRI publishes its research findings in books and on the Internet at **www.sipri.org**

sipri

Stockholm International Peace Research Institute
Signalistgatan 9, SE-169 70 Solna, Sweden
Telephone: +46 8/655 97 00 Telefax: +46 8/655 97 33
Email: **sipri@sipri.org**
www.sipri.org

Jozef Goldblat

ARMS CONTROL

The New Guide to
Negotiations and Agreements

Fully Revised and Updated Second Edition
with
New CD-ROM Documentation Supplement

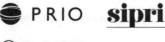

Los Angeles | London | New Delhi
Singapore | Washington DC

 SAGE Publications Ltd
1 Oliver's Yard
55 City Road
London EC1Y 1SP

SAGE Publications Inc.
2455 Teller Road
Thousand Oaks, California 91320

SAGE Publications India Pvt Ltd
B 1/I 1 Mohan Cooperative Industrial Area
Mathura Road
New Delhi 110 044

SAGE Publications Asia-Pacific Pte Ltd
33 Pekin Street #02-01
Far East Square
Singapore 048763

British Library Cataloguing in Publication data

A catalogue record for this book is available from the British Library

ISBN 978-0-7619-4015-9 (C)
ISBN 978-0-7619-4016-6 (P)

Library of Congress Control Number 2002102784

Printed in India by Replika Press Pvt. Ltd.

Contents

Tables and Figures

On the accompanying CD-ROM

PART II. AGREEMENTS AND PARTIES

Introductory Note

1989   Agreement between the USA and the USSR on reciprocal advance
        notification of major strategic exercises

1990   Agreement between the USA and the USSR on destruction and non-
        production of chemical weapons and on measures to facilitate the
        multilateral convention on banning chemical weapons (US–Soviet
        Chemical Weapons Agreement), Agreed Statement

1990   Treaty on Conventional Armed Forces in Europe (CFE Treaty), as
        amended by the 1999 Agreement on Adaptation, Protocols

1990   Charter of Paris for a New Europe, Supplementary Document

1991   United Nations Security Council Resolution 687 imposing arms
        restrictions on Iraq and establishing the UN Special Commission on
        Iraq (UNSCOM)

1991   Treaty between the USA and the USSR on the reduction and
        limitation of strategic offensive arms (START I Treaty), Annexes,
        Protocols, Memorandum of Understanding, Agreements,
        Statements, Declarations

        See also 1992 Lisbon Protocol to the START I Treaty

1991   US President's announcement regarding unilateral reductions of
        nuclear weapons

1991   Protocol on Environmental Protection to the 1959 Antarctic Treaty
        (Madrid Protocol)

1991   Soviet President's announcement regarding unilateral reductions of
        nuclear weapons

1991   United Nations Security Council Resolution 715 on the monitoring
        of Iraqi compliance with arms restrictions

1991   Guidelines for conventional arms transfers agreed by the permanent
        members of the UN Security Council

1991   Cartagena Declaration on renunciation of weapons of mass
        destruction

1991   UN General Assembly Resolution 46/36 on transparency in
        armaments, Annex: Register of conventional arms

1992   Joint Declaration by South and North Korea on the denuclearization
        of the Korean Peninsula

1992   Treaty on Open Skies, Annexes

1992   Tashkent Document: Joint Declaration and Agreement on the
        principles and procedures for implementing the 1990 CFE Treaty,
        Protocols

Foreword

This volume describes and analyses the international arms control agreements concluded from the Hague Declarations and Conventions of the 1800s to those concluded in the first years of this century. Together with the documents reproduced on the accompanying CD-ROM, it provides the most comprehensive survey of arms control available anywhere.

Jozef Goldblat, an acknowledged expert with long experience in the field, places arms control in the political context of the times. The history of arms control negotiations – both those that had led to agreements and those that failed to do so – is itself instructive. However, the applicability of the established norms and the lessons learned from their implementation must be assessed in the light of the challenges faced today and predictably on the horizon. Jozef Goldblat does this and provides pointers for action to be taken to further develop the law of arms control.

While major accords were reached in the 1990s owing to the new atmosphere of mutual trust immediately after the end of the Cold War, the weapons of mass destruction have not been abolished. The world continues to live in the fear that they will be used, by state or non-state actors. In the situation of today, characterized by the prevalence of intra-state wars coupled with international terrorism and by unilateral action taken in the field of armaments by militarily powerful states, arms control is needed more than ever. Hence the timeliness of the present book.

Jozef Goldblat has published comprehensive surveys of arms control negotiations and agreements in 1978 and 1982 (for SIPRI) and 1994 (for PRIO). The present study was supported by two grants. PRIO received a grant from the Ford Foundation, and SIPRI received a grant from the Volkswagen Foundation through the Centre for European Security Studies (CESS). We are grateful for this support.

Stein Tønnesson
Director, PRIO

Adam Daniel Rotfeld
Director, SIPRI

June 2002

Preface

Both progress and reversals have characterized the field of arms control since 1994, when the last edition of this book appeared. New agreements on weapons of mass destruction and conventional arms have been concluded, and the humanitarian law of armed conflict has been strengthened by new restrictions on the use of the means of warfare. Non-governmental organizations have played an important role in generating popular support for the reduction and elimination of arms and have engaged in monitoring states' compliance with the assumed obligations; a new type of multi-lateral diplomacy is emerging, based on a partnership between state authorities and the civil society.

At the beginning of this century, disarmament efforts began to lose momentum. The need for negotiated agreements, subject to strict verification, began to be questioned, mainly in the United States, a major party to most of the agreements concluded in the past. For the first time, an arms control treaty was abrogated by the withdrawal of a party. This action may undermine the validity of several other treaties and create an international climate of mistrust that would be disastrous for arms control. By 2002, multilateral negotiations on further arms control measures had come to a standstill.

The fight against international terrorism should not divert the attention of governments from the dangers of the arms races between states. The setback in arms control may be taken advantage of by terrorists.

Arms control remains an essential building-block of the edifice of peace that the United Nations has endeavoured to construct ever since it undertook to free the world from the scourge of war. It is therefore imperative that the arms control process be put back on track. The present publication is intended to facilitate the achievement of this goal by providing a comprehensive guide for politicians, diplomats, military officers, teachers, students, journalists and non-governmental organizations.

The achievements and failures of arms control are described, analysed and assessed in Part I of this book; tables, graphs and maps illustrate the intricate issues discussed. Part I also contains a glossary and select bibliography.

Part II, presented on the accompanying CD-ROM, contains full texts or excerpts of over 150 documents adopted since the 1800s as well as lists of the signatories and parties to arms control agreements as of early 2002. The electronic search function provides a useful tool to the reader. In addition to the entries for Part I, the table of contents and the detailed index include entries for the titles of the documents reproduced in Part II.

Acknowledgements

I am very grateful for the invaluable research assistance of Ragnhild Ferm Hellgren and editorial assistance of Connie Wall, both at SIPRI. I also thank SIPRI librarians Christer Berggren and Christine-Charlotte Bodell for their expert help in preparing the document collection. I am indebted to both PRIO and SIPRI for their support.

Adam Daniel Rotfeld, SIPRI Director 1991–2001, encouraged and supported the project in many important ways, as did Dan Smith, the previous PRIO Director, and Stein Tønnesson, current Director. Ingeborg K. Haavardsson, PRIO Information Director, coordinated production of the book with SAGE Publications. Several chapters benefited from review by prominent individuals in the arms control community.

Jozef Goldblat
July 2002

Typeset by Connie Wall, SIPRI; index by Peter Rea, UK.

Acronyms

ABACC	Argentine–Brazilian Agency for Accounting and Control of Nuclear Materials	**CBM**	Confidence-building measure
ABM	Anti-ballistic missile	**CBW**	Chemical and biological weapons
ACA	Agency for the Control of Armaments	**CCD**	Conference of the Committee on Disarmament
ACV	Armoured combat vehicle	**CD**	Conference on Disarmament
AIFV	Armoured infantry fighting vehicle	**CFE**	Conventional Armed Forces in Europe (Treaty)
ALCM	Air-launched cruise missile	**CIS**	Commonwealth of Independent States
ANZUS	Australia–New Zealand–United States (Treaty)	**CISAC**	Committee on International Security and Arms Control
APM	Anti-personnel mine		
ASAT	Anti-satellite	**COCOM**	Co-ordinating Committee for Multilateral Export Controls
ASBM	Air-to-surface ballistic missile		
ASEAN	Association of South-East Asian Nations	**COPREDAL**	Preparatory Commission for the Denuclearization of Latin America
ATBM	Anti-tactical ballistic missile		
ATTU	Atlantic-to-the-Urals (zone)	**CORRTEX**	Continuous reflectometry for radius versus time experiments
BCC	Bilateral Consultative Commission	**CRAMRA**	Convention on the Regulation of Antarctic Mineral Resource Activities
BIC	Bilateral Implementation Commission		
BMD	Ballistic missile defence	**CSBM**	Confidence- and security-building measure
BW	Biological weapon		
BWC	Biological Weapons Convention		
CANWFZ	Central Asian Nuclear-Weapon-Free Zone	**CSCE**	Conference on Security and Co-operation in Europe
CAT	Conventional arms transfers (talks)	**CTBT**	Comprehensive Nuclear Test-Ban Treaty

CTBTO	Comprehensive Nuclear Test-Ban Treaty Organization	**GPALS**	Global Protection Against Limited Strikes
CW	Chemical weapon	**GZA**	Geographical zone of application
CWC	Chemical Weapons Convention	**HNEs**	Hydronuclear experiments
DC	Disarmament Commission	**IAEA**	International Atomic Energy Agency
DEW	Directed-energy weapon	**ICBL**	International Campaign to Ban Landmines
DMA	Dangerous Military Activities (Agreement)	**ICBM**	Intercontinental ballistic missile
ECOSOC	Economic and Social Council (of the United Nations)	**ICJ**	International Court of Justice
ECOWAS	Economic Community of West African States	**ICRC**	International Committee of the Red Cross
EEZ	Exclusive economic zone	**ILO**	International Labour Organization
EMP	Electromagnetic pulse	**IMO**	International Maritime Organization
ENDC	Eighteen-Nation Committee on Disarmament	**IMS**	International Monitoring System
Enmod	Environmental modification	**INESAP**	International Network of Engineers and Scientists Against Proliferation
EU	European Union		
Euratom	European Atomic Energy Community	**INF**	Intermediate-range nuclear forces
FOBS	Fractional orbital bombardment system	**Interpol**	International Criminal Police Organization
FRG	Federal Republic of Germany	**IPPAS**	International Physical Protection Advisory Service
FRY	Federal Republic of Yugoslavia	**IPS**	International plutonium storage
FSC	Forum for Security Co-operation	**IRM**	Intermediate-range missile
FYROM	Former Yugoslav Republic of Macedonia	**ISMA**	International satellite monitoring agency
GCS	Global Control System	**JCC**	Joint Consultative Commission
GDR	German Democratic Republic		
GLCM	Ground-launched cruise missile	**JCG**	Joint Consultative Group

JCIC	Joint Compliance and Inspection Commission	**NSG**	Nuclear Suppliers Group
		NSS	National seismic station
JDEC	Joint Data Exchange Center	**NTM**	National technical means (of verification)
JMC	Joint Military Commission	**NWC**	Nuclear weapons convention
JNCC	Joint Nuclear Control Commission	**NWFZ**	Nuclear-weapon-free zone
KEDO	Korean Peninsula Energy Development Organization	**OAS**	Organization of American States
		OAU	Organization of African Unity
LPAR	Large phased-array radar		
LTBT	Limited Test Ban Treaty	**OPANAL**	Agency for the Prohibition of Nuclear Weapons in Latin America and the Caribbean
LWR	Light-water reactor		
MAD	Mutual assured destruction		
MBFR	Mutual and Balanced Force Reduction	**OPCW**	Organisation for the Prohibition of Chemical Weapons
MIRV	Multiple independently targetable re-entry vehicle	**OPP**	Other physical principles
MNLH	Maximum national levels for holdings	**OSCE**	Organization for Security and Co-operation in Europe
MOU	Memorandum of Understanding		
		OSI	On-site inspection
MOUS	Memorandum of Understanding on Succession	**PAL**	Permissive Action Link
		PAROS	Prevention of an arms race in outer space
MOX	Mixed-oxide (fuel)		
MTCR	Missile Technology Control Regime	**PCASED**	Programme for Co-ordination and Assistance for Security and Development
MW(e)	Megawatt-electric		
NATO	North Atlantic Treaty Organization	**PLNS**	Pre- and post-missile launch notification system
NGO	Non-governmental organization		
		PMDA	Plutonium Management and Disposition Agreement
NMD	National missile defence		
NPT	Non-Proliferation Treaty		
NRC	NATO–Russia Council	**PNET**	Peaceful Nuclear Explosions Treaty
NRRC	Nuclear Risk Reduction Center		
		PRC	People's Republic of China

PTBT	Partial Test Ban Treaty	**UNESCO**	UN Educational, Scientific and Cultural Organization
R&D	Research and development		
RV	Re-entry vehicle	**UNGA**	UN General Assembly
SALT	Strategic Arms Limitation Talks/Treaty	**UNIDIR**	UN Institute for Disarmament Research
SALW	Small arms and light weapons	**UNMOVIC**	UN Monitoring, Verification and Inspection Commission
SAM	Surface-to-air missile		
SCC	Standing Consultative Commission	**UNSC**	UN Security Council
		UNSCEAR	UN Scientific Committee on the Effects of Atomic Radiation
SCCC	Common System of Accounting and Control of Nuclear Materials		
SDF	Self-defence Forces	**UNSCOM**	UN Special Commission on Iraq
SDI	Strategic Defense Initiative	**VEREX**	Verification experts (group)
SFRY	Socialist Federal Republic of Yugoslavia	**WEU**	Western European Union
SLBM	Submarine-launched ballistic missile	**WHO**	World Health Organization
SRAM	Short-range attack missile	**WMO**	World Meteorological Organization
SRM	Shorter-range missile	**WTO**	Warsaw Treaty Organization
START	Strategic Arms Reduction Talks/Treaty		
		ZOPAN	Zone of Peace, Freedom and Neutrality
SVC	Special Verification Commission		
THAAD	Theater High-Altitude Area Defense		
TLE	Treaty-limited equipment		
TMD	Theatre missile defence		
TNT	Trinitrotoluene		
TTBT	Threshold Test Ban Treaty		
UN	United Nations		
UNCLOS	UN Convention on the Law of the Sea		
UNEP	UN Environment Programme		

Glossary

Anti-ballistic missile (ABM) system

See Ballistic missile defence and National missile defence.

Anti-personnel mine (APM)

Landmine designed to be exploded by the presence, proximity or contact of a person and that can incapacitate, injure or kill one or more persons.

Anti-satellite (ASAT) weapon

Weapon designed to interfere with, damage or destroy earth satellites in orbit.

Anti-submarine warfare (ASW)

Activities involved in the detection, identification, tracking and destruction of hostile submarines.

Association of South-East Asian Nations (ASEAN)

Established in 1967 to promote economic, social and cultural development as well as regional peace and security in South-East Asia.

Atlantic-to-the-Urals (ATTU) zone

Zone of application of the 1990 CFE Treaty, the 1992 CFE-1A Agreement and the 1999 Agreement on Adaptation of the CFE Treaty, stretching from the Atlantic Ocean to the Ural Mountains. It covers the entire land territory of the European states parties (excluding part of Turkey) and the territory of Russia and Kazakhstan west of the Ural River.

Australia Group

Group of states formed in 1985 to discuss chemical and biological weapon-related items which should be subject to national regulatory measures.

Ballistic missile

Missile that is lifted into space by a booster rocket and then descends towards its target in a free-falling ballistic trajectory.

Ballistic missile defence (BMD)

Weapon system designed to defend against a ballistic missile attack by intercepting and destroying ballistic missiles or their warheads in flight.

Binary chemical weapon

Shell or other device filled with two chemicals of relatively low toxicity which mix and react while the device is being delivered to the target, the reaction product being a super-toxic chemical warfare agent, such as a nerve agent.

Biological weapon (BW)

Weapon containing living organisms (as well as the means of their delivery) which are intended to cause disease or death in humans, animals or plants, and which for their effect depend on the ability to multiply within the target organism. *See also* Toxins.

Boost phase

First phase of a ballistic missile flight.

Breakout   Sudden abrogation or massive violation of an arms control agreement.

Breeder reactor   Nuclear reactor that produces more fissile material than it consumes while generating power.

Chain reaction   Continuing process of nuclear fissioning in which the neutrons released from a fission trigger another nuclear fission.

Chemical weapon (CW)   Chemical substance – whether gaseous, liquid or solid – as well as the means of its delivery, intended for use in war because of its direct toxic effects.

Commonwealth of Independent States (CIS)   Established in 1991 as a framework for multilateral cooperation among former Soviet republics.

Comprehensive Nuclear Test-Ban Treaty Organization (CTBTO)   Established by the 1996 CTBT to deal with questions of compliance with the Treaty and as a forum for consultation and cooperation among the parties.

Conference on Disarmament (CD)   Multilateral arms control negotiating body.

Confidence- and security-building measure (CSBM)   Measure undertaken to promote confidence and security through military transparency, openness and demonstration of a nation's lack of hostile intent, as distinguished from a measure actually reducing military capabilities.

Confidence-building measures (CBM)   Measure undertaken to help reduce the danger of armed conflict and of misunderstanding or miscalculation of military activities.

Conventional weapon   Weapon not having mass destruction effects. *See also* Weapon of mass destruction.

Conversion   Term used to describe the reallocation of resources from military to civilian use.

Counter-proliferation   Measures or policies to enforce the non-proliferation of weapons of mass destruction.

Cruise missile   Pilotless, guided weapon-delivery vehicle which sustains flight at subsonic or supersonic speeds through aerodynamic lift, generally flying at very low altitudes to avoid radar detection, sometimes following the contours of the terrain. It can carry a conventional or non-conventional warhead.

Decoy   Facsimile of a weapon system or component (such as a missile warhead) designed to complicate attempts to destroy or disable the actual weapon.

Deterrence strategy	Strategy to prevent war by confronting a potential aggressor, with the prospect that the response to his attack would bring unacceptable damage upon himself. *See also* Mutual assured destruction (MAD).
Directed-energy weapon (DEW)	Weapon system based on the delivery on the target of destructive energy in the form of a beam of light or of particles with nearly the speed of light.
Dual-capable	Term that refers to a weapon system that can carry either conventional or non-conventional explosives.
Economic Community of West African States (ECOWAS)	Established in 1975 to promote trade and cooperation and contribute to development in West Africa.
Electromagnetic pulse (EMP)	Burst of electromagnetic energy produced by a nuclear explosion which may damage electrical and electronic equipment at great distances.
European Atomic Energy Community (Euratom or EAEC)	Established in 1957 to promote the development of nuclear energy for peaceful purposes and to administer the nuclear safeguards system covering the European Union member states.
European Union (EU)	Established in 1957 as the European Community, it deals with economic cooperation and elaborates a common foreign and security policy for EU member-states.
Fall-out	Spread of radioactive particles from clouds of debris produced by a nuclear explosion.
Fertile material	Material composed of atoms that readily absorb neutrons. *See also* Chain reaction.
First-strike capability	Capability to launch an attack on an adversary's strategic nuclear forces that would eliminate the retaliatory, second-strike capability of the adversary.
Fissile material	Material composed of atoms which can be split by neutrons. Uranium-235 and plutonium-239 are the most common fissile materials.
Fission	Process whereby the nucleus of a heavy atom splits into lighter nuclei with the release of substantial amounts of energy, as in a fission-type nuclear weapon (atomic weapon).
Fusion	Process whereby light atoms, especially those of the isotopes of hydrogen, combine to form a heavy atom with the release of very substantial amounts of energy, as in a thermonuclear weapon (hydrogen weapon).

Global Protection Against Limited Strikes (GPALS)

See Strategic Defense Initiative (SDI).

Group of Seven/Eight (G7/G8)

Group of the seven leading industrialized nations which have met informally at the level of heads of state or government since the 1970s; from 1997 Russia has participated with the G7 in meetings of the G8.

Heavy water

Isotope of hydrogen. Serves as a moderator and coolant in a heavy water reactor (HWR).

Herbicides

Chemical (or biological) agents that destroy plants. *See also* Chemical weapon and Biological weapon.

Intercontinental ballistic missile (ICBM)

Ground-launched ballistic missile with a range longer than 5,500 kilometres.

Intermediate-range nuclear forces (INF)

Nuclear forces with a range of from 1,000 kilometres up to and including 5,500 kilometres. *See also* Theatre nuclear forces.

International Atomic Energy Agency (IAEA)

Established in 1957 to promote the peaceful uses of atomic energy and ensure that such uses do not further military purposes.

International Court of Justice (ICJ)

Principal judicial organ of the United Nations, set up in 1945.

Isotopes

Nuclides with the same atomic number but different mass numbers.

Joint Consultative Group (JCG)

Established by the 1990 CFE Treaty to promote the objectives and implementation of the Treaty.

Joint Compliance and Inspection Commission (JCIC)

Forum to resolve questions of compliance, clarify ambiguities and discuss ways to improve implementation of the 1991 START I and 1993 START II treaties.

Kiloton (kt)

Measure of the explosive yield of a nuclear weapon equivalent to 1,000 tons of trinitrotoluene (TNT) high explosive. (The bomb detonated at Hiroshima in World War II had a yield of about 12–15 kilotons.)

Landmine

Anti-personnel or anti-vehicle mine, emplaced on land.

Laser

Acronym for a device which operates by the principle of 'light amplification by stimulated emission of radiation'. Lasers use narrow focused light beams to provide powerful directed force for a variety of applications.

Launcher	Equipment which launches a missile. ICBM launchers are land-based launchers which can be either fixed or mobile. SLBM launchers are missile tubes on submarines.
Light water	Ordinary water which serves as a moderator and coolant in a light water reactor (LWR).
Megaton (Mt)	Measure of the explosive yield of a nuclear weapon equivalent to 1 million tons (1,000 kt) of trinitrotoluene (TNT) high explosive.
Mid-course phase	Flight of ballistic missile warhead through space after the boost phase but before re-entry.
Mine	Munition placed under, on or near the ground or other surface area, designed to be detonated or exploded by the presence, proximity or contact of a person or vehicle.
Missile Technology Control Regime (MTCR)	Informal international military-related export control regime, established in 1987 to limit the spread of weapons of mass destruction by controlling missile delivery systems.
Moderator	Component of a nuclear reactor that slows neutrons, thereby increasing their chances of fissioning fertile material.
Multiple independently targetable re-entry vehicles (MIRVs)	Several re-entry vehicles, carried by a single ballistic missile, which can be directed to separate targets along separate trajectories.
Mutual assured destruction (MAD)	Concept of reciprocal deterrence which rests on the ability of the nuclear weapon powers to inflict intolerable damage on one another after suffering a nuclear attack. *See also* Deterrence strategy.
National missile defence (NMD)	Anti-ballistic missile system – prohibited under the 1972 ABM Treaty – capable of defending the national territory of a state against an attack from strategic ballistic missiles.
National technical means (NTM) of verification	Technical means of intelligence, under the control of a state, which are used to monitor compliance with arms control agreements.
Nerve agent	Chemical warfare agent that interferes with or inhibits the transmittal of nerve impulses by disrupting the enzyme reactions in the nervous system; it carries a degree of lethality considerably greater than that of the agents used in World War I. *See also* Chemical weapon.

Neutron	Elementary particle slightly heavier than a proton, with no electric charge. *See also* Proton.
Neutron bomb	Enhanced radiation nuclear warhead which kills by radiation rather than by blast.
Non-Aligned Movement (NAM)	Established in 1961 as a forum for consultations and coordination of positions on political, economic and arms control issues among non-aligned states.
North Atlantic Treaty Organization (NATO)	Established in 1949 by the North Atlantic Treaty (Washington Treaty) as a Western defence alliance. Article 5 of the treaty defines the member-states' commitment to respond to an armed attack on any party.
Nuclear fuel cycle	System of nuclear installations consisting of uranium mines, ore processing, conversion, enrichment and fuel fabrication plants, reactors, spent fuel storages, reprocessing plants and associated storage.
Nuclear Suppliers Group (NSG)	Coordinates export controls on nuclear materials; also known as the London Club.
Nuclear weapon-grade material	Material with a sufficiently high concentration either of uranium-233, uranium-235 or plutonium-239 to make it suitable for a nuclear weapon.
Nuclear weapons	Collective term for atomic and hydrogen weapons of all types and their delivery systems. *See also* Fission and Fusion.
Nuclear silo	Hardened underground facility for a fixed ballistic missile, designed to provide protection and to act as a launching platform.
Nuclide	Nuclear species characterized by the number of protons (atomic number) and number of neutrons. The total number of protons and neutrons is called the mass number of the nuclide.
Open Skies Consultative Commission (OSCC)	Established by the 1992 Open Skies Treaty to resolve questions of compliance with the Treaty.
Organisation for the Prohibition of Chemical Weapons (OPCW)	Established by the 1993 Chemical Weapons Convention to oversee implementation of the Convention and resolve questions of compliance.
Organization for Security and Co-operation in Europe (OSCE)	Initiated in 1973 as the Conference on Security and Co-operation in Europe, the CSCE was renamed the OSCE in 1995. Its Forum for Security Co-operation (FSC) deals with CSBMs and arms control.

Organization of African Unity (OAU)	Established in 1963 to promote African international cooperation and harmonization of, *inter alia*, defence policies. In 2001 the OAU member-states decided that the Organization would be replaced by the African Union.
Organization of American States (OAS)	Group of states in the Americas, which adopted a charter in 1948 with the objective of strengthening peace and security in the western hemisphere.
Payload	Weapon and penetration aids carried by a delivery vehicle.
Peaceful nuclear explosion (PNE)	Nuclear explosion for non-military purposes, such as digging canals or harbours or creating underground cavities.
Penetration aids	Techniques and/or devices employed to increase the probability of penetrating the opponent's defences and reaching the intended target.
Permissive Action Link (PAL)	Locks which prevent a nuclear weapon from being used without authorization.
Plutonium	Radioactive element which occurs only in trace amounts in nature, with atomic number 94 and symbol 'Pu'. As produced by irradiating uranium fuels, plutonium contains varying percentages of the isotopes 238, 239, 240, 241 and 242.
Precursor	Chemical reagent which takes part in the production of a toxic chemical.
Proton	Elementary particle with a positive electric charge.
Radar	Acronym for 'radio detection and ranging', referring to a device that uses the emission of electromagnetic energy for the detection and location of objects.
Reprocessing plant	Facility separating the plutonium and uranium present in spent reactor fuel.
Re-entry vehicle (RV)	Part of a ballistic missile which carries a nuclear warhead and penetration aids to the target. It re-enters the earth's atmosphere and is destroyed in the final phase of the missile's trajectory. A missile can have one or several RVs and each RV contains a warhead.
Safeguards system	System under which the IAEA checks nuclear activities of states to guard against attempts to divert nuclear material and equipment intended for peaceful use to not-permitted military purposes.

Second-strike capability   Ability to launch, in response to a nuclear attack, a retaliatory strike large enough to inflict intolerable damage on the opponent. *See also* Mutual assured destruction.

Small arms and light weapons (SALW)   According to the 1997 UN experts' report, small arms are those designed for personal use, and light weapons are those designed for use by several persons serving as a crew. (These definitions are not yet internationally agreed.)

Short-range nuclear forces (SNF)   *See* Theatre nuclear forces.

South Pacific Forum   Group of South Pacific states, which proposed the South Pacific Nuclear Free Zone established by the 1985 Treaty of Rarotonga.

Spent nuclear fuel   Fuel removed from a nuclear reactor after use. *See also* Nuclear fuel cycle.

Standing Consultative Commission (SCC)   Established by the 1972 ABM Treaty as the body to which parties could refer issues regarding implementation of the Treaty.

Strategic Defense Initiative (SDI)   Programme of research and development of systems capable of intercepting and destroying nuclear weapons in flight and thus protecting the whole territory of the USA against a massive Soviet nuclear missile attack. The programme was pursued by the United States in the 1980s.

Strategic nuclear weapons   ICBMs and SLBMs as well as bombs and missiles carried on aircraft of intercontinental range (over 5,500 kilometres).

Strategic stability   Situation in the relation of forces between potential adversaries which leads them to conclude that an attempt to settle their disputes by military means would constitute a risk of unacceptable proportions.

Subcritical experiments   Experiments in which the configuration and quantities of explosives and nuclear materials used do not produce a critical mass, i.e., there is no self-sustaining nuclear fission chain reaction.

Submarine-launched ballistic missile (SLBM)   Ballistic missile launched from a submarine, with a range in excess of 600 kilometres (as defined in the 2000 US–Russian MOU on notifications of missile launches).

Sub-Regional Consultative Commission (SRCC)
Established by the 1996 Agreement on Sub-Regional Arms Control (Florence Agreement), it is the forum for the parties to resolve questions of compliance with the agreement.

Tactical nuclear weapon
Low-yield, short-range nuclear weapon deployed with general-purpose forces along with conventional weapons; sometimes referred to as battlefield nuclear weapon. *See also* Theatre nuclear forces.

Telemetry
Transmission of electronic signals by missiles to earth; monitoring these signals helps to evaluate the missile's performance.

Theatre missile defence (TMD)
Weapon systems designed to defend against non-strategic nuclear missiles by intercepting and destroying them in flight.

Theatre nuclear forces (TNF)
Nuclear weapons with ranges of up to 5,500 kilometres. In the 1987 INF Treaty, nuclear missiles were divided into intermediate-range (1,000–5,500 kilometres) and shorter-range (500–1,000 kilometres). Also called non-strategic nuclear forces. Nuclear weapons with ranges of up to 500 kilometres fall in the category of short-range nuclear forces.

Throw-weight
Total weight of a ballistic missile's re-entry vehicle(s), penetration aids and targeting devices, that is, the militarily significant portions of the missile sent towards the target, as distinct from launch-weight, which is the weight of a fully loaded ballistic missile at the time of launch.

Toxins
Poisonous substances which are products of organisms but are inanimate, not capable of reproducing themselves, as well as chemically created variants of such substances.

Treaty-limited equipment
Categories of equipment on which numerical limits are established by the 1990 CFE Treaty and the 1999 Agreement on Adaptation of the CFE Treaty.

Tritium
Radioactive isotope of hydrogen; an essential ingredient of thermonuclear weapons.

United Nations (UN)
World intergovernmental organization founded in 1945 'to save succeeding generations from the scourge of war'.

United Nations Monitoring, Verification and Inspection Commission (UNMOVIC)	Body established by the UN Security Council to undertake responsibilities previously mandated to the United Nations Special Commission on Iraq (UNSCOM) with regard to the verification of compliance by Iraq with ceasefire Resolution 687 (1991).
United Nations Register of Conventional Arms	Voluntary reporting mechanism set up in 1992 for UN member states to report annually their imports and exports of seven categories of conventional arms.
Uranium	Naturally occurring radioactive element with atomic number 92 and symbol 'U'.
Uranium enrichment plant	Installation for increasing the concentration of uranium-235 in uranium through isotope separation processes. Highly enriched uranium is used for nuclear fission weapons.
Warhead	Part of a weapon which contains the explosive or other material intended to inflict damage.
Warsaw Treaty Organization (WTO)	The WTO, or Warsaw Pact, was established in 1955 by the Treaty of Friendship, Cooperation and Mutual Assistance between the USSR and seven East–Central European countries: Albania (withdrew in 1968), Bulgaria, Czechoslovakia, the German Democratic Republic, Hungary, Poland and Romania. It was dissolved in 1991.
Wassenaar Arrangement	Arrangement on Export Controls for Conventional Arms and Dual-Use Goods and Technologies, formally established in Wassenaar, the Netherlands, in 1996, which aims to prevent the acquisition of armaments and sensitive dual-use goods and technologies for military uses by states whose behaviour is cause for concern to the member states.
Weapon of mass destruction	As defined in 1948 by the Commission for Conventional Armaments, these weapons include atomic explosive weapons, radioactive material weapons, lethal chemical and biological weapons, and any weapons developed in the future which have characteristics comparable in destructive effect to those of the atomic bomb or other weapons mentioned above.
Western European Union (WEU)	Established by the 1954 Protocols to the 1948 Brussels Treaty and the 1954 Modified Brussels Treaty, it is at present essentially intended to ensure the respect of obligations stemming from Article V (mutual assistance in case of aggression) of the Modified Brussels Treaty.

Yield   Energy released in an explosion. *See also* Kiloton and
                                 Megaton.

Zangger Committee   Established in 1972, the Nuclear Exporters Commit-
                                 tee (called the Zangger Committee after its first chair-
                                 man) is a group of nuclear supplier countries that
                                 meets informally to coordinate export controls on
                                 nuclear materials.

Part I. Analytical Survey

1

Basic Concepts

Over the years, proposals have been made in various forums for the universal and complete elimination of armed forces and armaments. However, for several reasons, the idea of total and general disarmament has proved unacceptable to many nations. So far, only *arms control* measures have been agreed on.

Originally, 'arms control' was meant to denote rules for limiting arms competition (mainly nuclear) rather than reversing it. This term had a connotation distinct from 'regulation of armaments' or 'disarmament', the terms used in the United Nations Charter. Subsequently, however, a wide range of measures have come to be included under the rubric of arms control, in particular those intended to: (a) freeze, limit, reduce or abolish certain categories of weapons; (b) ban the testing of certain weapons; (c) prevent certain military activities; (d) regulate the deployment of armed forces; (e) proscribe transfers of some militarily important items; (f) reduce the risk of accidental war; (g) constrain or prohibit the use of certain weapons or methods of war; and (h) build up confidence among states through greater openness in military matters. Today, 'arms control' is often used interchangeably with 'arms regulation', 'arms limitation', 'arms reduction' or even 'disarmament'.

1.1 Arms Control Agreements

Arms control can take various forms. It can be part of interstate ceasefire or armistice arrangements, as was the case after the 1950–53 war in Korea and the 1946–54 war in Indo-China. It can be imposed upon defeated countries by peace treaties, such as those concluded after World War I and World War II. It can follow the termination of intra-state conflicts, as in the case of Bosnia and Herzegovina in 1995 and Kosovo in 1999. Finally, it can take the form of sanctions applied in accordance with the UN Charter against aggressor states, as in the case of Iraq after the 1991 Gulf War. However, an 'arms control agreement' is an agreement among sovereign states, freely arrived at in time of peace through a process of formal inter-governmental negotiation.

Arms control agreements may be bilateral or multilateral. In the latter category, many agreements are of a regional nature, valid for a specific geographical zone or continent. Agreements vary in form – from treaties, conventions, protocols and documents, to guidelines, memoranda, declarations or common understandings, to statutes, charters and final acts of international conferences, to joint or simultaneous statements by governments or exchanges of letters or notes among the states concerned.

In recent years the conclusion of so-called framework agreements has become an acceptable practice. Their characteristic is that the basic instrument, 'the framework', sets out the objective pursued but specifies few substantive obligations of the parties. However, a mechanism included in the framework agreement provides for the adoption of protocols which contain substantive obligations and to which the parties to the agreement are expected, but not obliged, to adhere.

Succession

The documents reproduced on the CD-ROM with this book cover a long period of time during which some parties have ceased to exist as independent states because of voluntary or forced mergers with other states, while others, such as the Union of Soviet Socialist Republics, the Socialist Federal Republic of Yugoslavia (SFRY) and Czechoslovakia, have broken up into several sovereign states. These developments, as well as the disintegration of the colonial empires, have given rise to new political entities (under old or new names) whose status vis-à-vis the existing treaties is uncertain because the international law dealing with the succession of states in respect of international treaties is somewhat vague.

According to the 1978 Vienna Convention on Succession of States in Respect of Treaties (not yet widely adhered to but regarded by many as stating customary law), states emerging from colonial domination are entitled to a clean slate enabling them to choose freely whether or not to succeed to the treaties by which they were formerly bound. Other new states may be subject to certain restraints in this respect. Thus, in respect of a state that splits off from another state, any treaty in force for its territory at the time of separation generally continues in force after independence. (The 1978 Vienna Convention as such did not apply to the Soviet Union or Yugoslavia because it entered into force only in 1996, after the dissolution of both states.) Consideration must also be given to the nature of the treaty. With arms control treaties, a specific declaration of succession may be required. Regarding some other treaties, such as human rights or humanitarian law treaties, succession is almost automatic, and a general declaration by the new state of a wish to succeed to all such treaties may suffice.

In 1992, consequent on the decision of the UN Security Council (UNSC), endorsed by the UN General Assembly (UNGA), that the Federal Republic of Yugoslavia (FRY) was not the successor of the SFRY, which the UNSC said had ceased to exist, the status of the FRY in the United Nations became ambiguous. The FRY was barred from participation in the UNGA and the UN Economic and Social Council (ECOSOC) and in their subsidiary organs. However, its name remained on membership lists, and its messages to UN organs were published in the same way as those of a UN member. In November 2000, after a change of government, the FRY gave up its claim to be the successor of the SFRY, applied for membership of the United Nations and was admitted as a 'new' state. It then began to notify the depositaries of its succession or accession to the arms control and other treaties to which the SFRY had been a party, treating each treaty individually. (Bosnia and Herzegovina, Croatia, Slovenia and the Former Yugoslav Republic of Macedonia chose to make a general statement of succession to the treaties to which the SFRY had been a party.)

Normally, it is the depositary or depositaries of a treaty that have the authority to determine which states are parties to it. Sometimes such determination is complicated. The Russian Federation declared itself, as from 24 December 1991, the legal successor of the Soviet Union as regards the fulfilment of obligations under all arms control agreements. Although the world community took note of this declaration and it went unchallenged by the non-Russian republics at the time it was made, Belarus, Kazakhstan and Ukraine were in 1992 recognized by the United States as successor states of the Soviet Union – on terms of equality with Russia – with regard to the US–Soviet 1991 START I Treaty on the Reduction and Limitation of Strategic Offensive Arms. This decision reflected the fact that a significant portion of the total Soviet inventory of strategic nuclear weapons, subject to reduction or limitation under the above treaty, was stationed on the territories of these three non-Russian republics. Subsequently, in 1997, a memorandum of understanding was signed establishing that Belarus, Kazakhstan and Ukraine are to be considered as successor states of the Soviet Union with regard to the US–Soviet 1972 ABM Treaty on the Limitation of Anti-Ballistic Missile Systems, as a number of former Soviet early-warning radars and a former Soviet ABM test range were located on the territories of these states. Another case in which the continuity rule was applied to non-Russian republics was the multilateral 1990 Treaty on Conventional Armed Forces in Europe (CFE Treaty); there was no other way to render this treaty effective. As regards other multilateral arms control agreements, the former Soviet republics follow the procedure of accession. In any event, the continuity rule was inapplicable to the 1968 Non-Proliferation Treaty (NPT) because, if the former Soviet republics inherited the Soviet Union's nuclear-weapon-state status, the fundamental purpose of this treaty – to prevent the number of nuclear-weapon states from increasing – would be defeated.

The People's Republic of China (PRC) declared that, as regards the multilateral treaties to which China was a party before the establishment of the People's Republic, its government would decide in light of the circumstances whether it should recognize them. As to the treaties concluded by the Republic of China (Taiwan) after 1 October 1949, the PRC stated that it considered Taiwan's actions as null and void. However, as Taiwan is still recognized by several states (although it is excluded from the United Nations), it is listed in Part II of this book as a party to the arms control agreements which it joined after World War II.

Parties

As a rule, multilateral arms control agreements, with the exception of regional agreements, are open for participation by all states. This is an acknowledgement of the principle that, by its very nature, arms control ought to have universal application. The question has arisen whether, by subscribing to a treaty, a political entity or a regime can gain recognition as a state or a government by other parties which do not formally recognize it. To guard against such implications, some countries have found it expedient to issue special declarations. Most of these declarations relate to Israel or Taiwan. (Until the unification of Germany, many also related to the German Democratic Republic or to West Berlin.)

It is, however, generally understood that neither the signature of nor the deposit of any instrument in relation to a multilateral treaty brings about recognition between parties to the treaty that do not recognize each other. Indeed, within the framework of multilateral treaties open for general adherence, states could even have dealings with a non-recognized regime without thereby recognizing it. Nevertheless, Taiwan has been barred from participating in conferences that review the treaties it has signed and ratified. Yugoslavia, a party to the NPT, was not invited to participate in the 1995 NPT Review and Extension Conference; it formally protested against this exclusion.

Another anomaly arose with regard to Cambodia, when for several years two governments claiming to represent the country were listed as parties to the NPT under two different names: Democratic Kampuchea and the People's Republic of Kampuchea. The situation became normalized in 1993 with the establishment of a Cambodian Government of National Unity.

Depositaries

For bilateral and some very restricted multilateral treaties, all the parties sign copies of the treaty for every other party and submit instruments of ratification to each of them. This is hardly practical for most multilateral treaties. Therefore, a depositary is designated whose duties include accepting signatures; receiving instruments of ratification, acceptance, approval or accession; informing the signatories of the date of each signature, of the deposit of each instrument and of the entry into force of the treaty; as well as receiving and circulating other notices, which may include notifications of succession to the treaty, denunciation or withdrawal and proposals for amendment. The depositary makes arrangements for registering treaties with the United Nations pursuant to Article 102 of the UN Charter.

Formerly, when treaty-making conferences were convened by states, the host state normally acted as depositary for the treaty that was concluded. However, for some time now, many treaties have been formulated under the aegis of international organizations – especially the United Nations – which then normally serve as depositaries of these treaties and even of those produced at some state-convened conferences. At the height of the Cold War, it was necessary to make an exception for certain arms control agreements where universal participation was considered desirable, so as to include states (such as the German Democratic Republic, North Korea, North Viet Nam, and originally the People's Republic of China and later Taiwan) which were not recognized by most states and with which international intergovernmental organizations maintained no formal contacts. The practice was then developed of naming the Soviet Union, the United Kingdom and the United States as co-depositaries. (The Russian Federation now performs the depositary functions formerly performed by the Soviet Union.) This was done for the 1963 Partial Test Ban Treaty (PTBT), the 1967 Outer Space Treaty, the 1968 NPT, the 1971 Seabed Treaty and the 1972 Biological Weapons Convention (BW Convention). It is sufficient for a state to sign a treaty or to deposit its instrument of ratification or accession in the capital of one of the three depositaries to become formally committed. If a state takes the same action in different capitals on different dates, the earliest date is considered to be the effective one. This device facilitated wider adherence to agreements without embarrassing any of the depositaries. As states do not present signatures or instruments of ratification to depositaries with which they have no diplomatic relations, the records of signatories and parties kept by the depositary governments differ. Since the number of countries not universally recognized is rather small today, the cumbersome practice of dealing with three depositaries has lost its justification. The task of depositary is now often assigned to the UN Secretary-General or – for agreements related to nuclear arms control – to the Director General of the International Atomic Energy Agency (IAEA). The 1992

Treaty on Open Skies, signed within the framework of the Conference on Security and Co-operation in Europe, has two depositaries – Canada and Hungary.

Entry into Force

The way in which a treaty enters into force is usually specified in its final clauses. Some agreements enter into force on signature. More frequently, depending on the constitutional requirements of the potential parties, what is required is ratification. This may involve securing the approval of a national legislative body. After such approval has been secured, an instrument of ratification, acceptance or approval is deposited with the depositary in respect of a treaty that has been signed. For a treaty that has not been signed, it is the instrument of accession (or succession) that is deposited. All these procedures are equivalent to and are normally referred to as 'ratification'.

The conditions for entry into force are normally specified in terms of a certain minimum number of ratifications, and it is sometimes required that particular states participate. For example, the BW Convention entered into force after the deposit of the instruments of ratification by 22 signatory governments, but this number was to include the governments of the United Kingdom, the United States and the Soviet Union. Still more restrictive is the provision for entry into force of the 1996 Comprehensive Nuclear Test-Ban Treaty (CTBT), which requires the deposit of the instruments of ratification by 44 states – those which were members of the Conference on Disarmament as at 18 June 1996 and participated in the work of the 1996 session of the Conference, and which possessed nuclear power or research reactors.

When signatures may be affixed is also specified: sometimes a treaty is open for signature during a limited period of time, sometimes until the treaty's entry into force, and sometimes indefinitely. Accession – which is resorted to by states that either prefer not to sign or are unable to do so because the deadline for signing has passed or for other reasons – may be possible from the date a treaty is opened for signature, as allowed for the 1997 Convention Prohibiting Anti-Personnel Mines (APM Convention), or only after it is no longer open for signature. Having signed but not yet exchanged or deposited the instruments of ratification, acceptance or approval of a treaty requiring such action, a state is considered obligated to refrain from acts which would defeat the object or purpose of the treaty until such time as it has made its intention clear not to become party to it.

The 1967 Treaty of Tlatelolco, which established a nuclear-weapon-free zone in Latin America and the Caribbean, contains an unusual clause: it may enter into force among states that have ratified it only when several conditions, specified in the treaty, have been met. However, these conditions may be waived at the time of ratification or later.

Certain agreements, whether signed or not signed, are not intended to be legally binding; they cannot be registered with the United Nations. This is true of many of the documents of the Conference on Security and Co-operation in Europe (CSCE), since 1995 called the Organization for Security and Co-operation in Europe (OSCE). These are only politically binding.

After the required number of ratifications has been deposited, a period of delay may be specified before entry into force. When a treaty formally enters into force, it does so only for those states that have ratified it. For states whose instruments of ratification are deposited after entry into force of the treaty, the treaty enters into force either immediately or after a specified period of time, which varies depending on the type of the treaty.

Arms control treaties may be modified by various procedures. Agreed amendments often enter into force in accordance with the procedures that govern the entry into force of the treaty concerned. In certain cases, a subsidiary agreement must be concluded, within defined time limits, after the treaty comes into force. One example is the NPT, which requires that non-nuclear-weapon parties conclude 'safeguards agreements' with the IAEA. Another is the 1993 Chemical Weapons Convention (CW Convention), which requires that all parties conclude 'facility agreements' with the Organisation for the Prohibition of Chemical Weapons (OPCW).

Once an agreement has entered into force in respect of a state, the state must comply with it in good faith. No party may invoke a provision of its internal law (including its constitution) as justification for failure to observe an agreement. A treaty or a part of it may also be applied provisionally, pending its entry into force, if the treaty itself so provides or if the negotiating states have in some other manner so agreed, as was the case with the implementation of the CFE Treaty.

Duration and Denunciation

Arms control agreements may remain in force indefinitely or for a limited period of time. Many agreements contain a clause permitting unilateral withdrawal in cases when extraordinary events relating to the subject matter of the treaty have jeopardized the supreme interests of the withdrawing state. However, only rarely have states spelled out – upon signing or ratifying a treaty – what kind of event they would consider 'extraordinary'.

Even in the absence of the withdrawal clause, a material breach of a bilateral treaty by one of the parties – which may involve a violation of a provision essential to the accomplishment of its object or purpose – entitles the other party to invoke the breach as grounds for terminating the treaty or suspending its operation. In an unprecedented move, Russia stated several times that it would withdraw from all US–Russian nuclear arms control agreements in response to the denunciation by the United States of another (but related) agreement, namely, the 1972 ABM Treaty.

According to the 1969 Vienna Convention on the Law of Treaties (widely adhered to), a material breach of a multilateral treaty by a party entitles other parties by unanimous agreement to suspend the operation of the treaty, in whole or in part, or to terminate it, either in relations between themselves and the defaulting state or as between all parties. A party specially affected by the breach may invoke it as grounds for suspending the operation of the treaty, in whole or in part, in relations between itself and the defaulting state. Any party other than the defaulting state has the right to invoke the breach to suspend the operation of the treaty, again in whole or in part, with respect to itself, if the treaty is of such a character that a material breach of its provisions by one party radically changes the position of every party with respect to the further performance of its obligation under the treaty.

These rules do not apply to provisions for the protection of human beings, contained in treaties of a humanitarian nature, in particular provisions prohibiting any form of reprisal against persons protected by such treaties.

Reservations

When signing, ratifying, accepting, approving or acceding to a multilateral treaty, or when making a notification of succession establishing its status as a party, a state may formulate a reservation whereby it unilaterally excludes or modifies the legal effect of certain provisions of the treaty in their application to that state. A reservation expressed by the predecessor state is deemed to be maintained by the successor state if the latter remains silent on that point. In general, reservations may be made only if they are of a type explicitly allowed by the treaty or if the treaty neither allows nor prohibits reservations, and if they are not incompatible with the object and purpose of the treaty. The depositary may determine, at least initially, whether a particular reservation is admissible, or it may just circulate it to all states concerned and record their reactions.

Reservations may be objected to. A state that has made and maintained a reservation which has been objected to by one or more of the parties to a treaty, but not by others, can be regarded as still being a party to the treaty if the reservation is compatible with the object and purpose of the treaty. If a party objects to a reservation that it finds incompatible with the object and purpose of the treaty, it may consider the reserving state as not being a party to the treaty. A reservation or an objection to it may be withdrawn at any time. This has been practised, for example, with regard to the 1925 Geneva Protocol prohibiting the use of chemical and bacteriological weapons.

Certain multilateral treaties explicitly rule out reservations. Nevertheless, states sometimes make statements of understanding containing unilateral interpretations of some key provisions.

1.2   Multilateral vs Unilateral Arms Control

The value of negotiated arms control agreements has often been disputed in both governmental and academic circles, especially in the United States. In 2000 the US administration formally declared its preference for unilateral action.

Unilateral measures which were carried into effect during the past decades include reductions of military expenditures; reductions in the strength of troops and changes in their deployment; cuts in the number of certain weapons or even the elimination of an entire category of weapon; cessation of the production of nuclear-weapon-usable material; moratoria on nuclear-weapon testing; freezes on weapon development; undertakings not to use certain means of warfare, including commitments of no first use of weapons of mass destruction; the establishment of nuclear-weapon-free areas; and a variety of other restraints on military programmes. A country embarking on unilateral arms control usually expects similar (although not necessarily immediate and identical) action on the part of other countries, especially its potential adversaries.

Unilateral measures may reduce threat perceptions, thereby helping to set aside problems connected with asymmetries in geography, strategies and components of the military establishments of the parties, as well as the sensitive issues of verification of compliance. Such measures are less subject to bureaucratic opposition within the countries concerned than are interstate agreements, which often require parliamentary approval. Another advantage of unilateral measures is that they avoid situations, so common in a negotiation, in which each side tries to improve its bargaining position by developing or deploying weapons it would otherwise not have developed or deployed, thus stimulating arms competition instead of abating it. Unilateral measures may be especially useful when some urgent problems have to be dealt with and there is no time for a formal process of negotiation.

Reciprocal restraints assumed without legally binding commitments may supplement the conventional means of achieving arms control, but they cannot replace them. To become durable, verifiable and enforceable, limitations resulting from unilateral moves or from merely politically binding agreements – especially those significantly affecting the military potential of states – need to be codified so as to define the range of prohibited activities and give the force of law to the prohibition. A formal treaty may include incentives that increase the likelihood of compliance and provide means for the resolution of disputes. It may also neutralize forces within each state which would otherwise urge new arms acquisitions; abrogation of a contractual commitment is more complicated and politically more hazardous than reversing a unilateral one. Finally, once a treaty gains widespread acceptance, it sets a standard of international behaviour which even non-parties must take into account.

1.3 Confidence-Building Measures

Although confidence building among nations has been practised for many years, the term 'confidence-building measure' (CBM) entered the vocabulary of international relations only in the early 1970s. Since the CBMs subsequently discussed and agreed upon have come to accentuate security aspects, they are also referred to as 'confidence- and security-building measures' (CSBMs).

The objective of CSBMs is to translate certain principles of international law into positive action so as to provide credibility to states' affirmations of their peaceful intentions. Such action means implementing measures aimed at: (a) reassuring states of the non-aggressive intentions of their potential adversaries and reducing the possibility of misrepresentation of certain activities; (b) narrowing the scope of political intimidation by the forces of stronger powers; and (c) minimizing the likelihood of inadvertent escalation of hostile acts in a crisis situation.

In general, CSBMs do not directly affect the strength of armed forces or arms inventories, but in facilitating progress towards disarmament they constitute a separate category of arms control measures. They also make less likely the use of force for settling disputes. To have the intended effect, CSBMs must be significant in scope and binding. A mere exchange of solemn declarations is rarely sufficient.

For a great majority of states, threats to national security arise from conditions within their own region. Hence attention is most often devoted to regional approaches. For confidence-building purposes, a region could embrace states which do not meet the geographical criteria of a 'region' but are linked economically or politically. Arrangements initiated by neighbouring states may subsequently attract more distant states as well. Regional confidence-building measures cannot be imposed by outsiders; they must be freely negotiated and agreed to by states in the region. It is only these states that can address the causes of their specific security problems and determine the type, scope and area of application of the required undertakings. In one region, distrust and tension could be generated by a lack of reliable information about the military activities of neighbouring states and the inadequacy of channels of communication among political decision-makers. In another region, distrust and tension could be generated by the absence of agreed restraints on the behaviour of the armed forces and uncertainty about compliance with international obligations.

Confidence building to promote better communication and understanding among the parties may include: (a) exchange of information about military expenditures, strength of armed forces, arms production and arms transfers; (b) open presentation and clarification of defence doctrines; (c) prior notification of military manoeuvres and major military movements, including their scope and extent; (d) the establishment of a mechanism to check the accuracy of the data provided; (e) the presence of foreign observers at military exercises; (f) exchanges of visits by military officers; (g) exchanges of cadets between military academies; and (h) the establishment of direct, rapid communication links – 'hotlines' – for crisis management.

Confidence-building measures that impose military constraints may include: (a) abstaining from certain specified military activities in border areas; (b) disengagement of armed forces by establishing zones between neighbouring countries that are partly or fully demilitarized; (c) voluntary submission to inspections to demonstrate compliance with agreed standards of behaviour; and (d) formalized commitment to the peaceful settlement of disputes.

Security cannot be obtained by promoting measures solely in the field of military affairs; it embraces economic and social factors as well. However, the military factor is of prime importance, as the absence of war constitutes a prerequisite for non-military CSBMs.

1.4 Arms Control and National Security

Nations may feel free from fears of aggression – defined by the UNGA as the use of armed force by a state against the sovereignty, territorial integrity or political independence of another state, or in any other manner inconsistent with the UN Charter – only in conditions of international security.

There are two ways of achieving security without reliance on arms build-ups: through arms control agreements and through collective international security arrangements. The two are closely intertwined.

Functions of Arms Control

Arms control may do the following: (a) reduce the risk of war started by accident or by design; (b) slow down global and regional arms races; (c) increase predictability in relations between hostile states and reduce fears of the intentions of a potential adversary; (d) pre-empt the development of new types of weapon and means of warfare; (e) minimize the disparities between heavily and lightly armed states, thereby removing an important source of instability; (f) encourage states to resort to peaceful means in solving their disputes; (g) save resources needed for economic and social development; (h) mitigate the destruction and suffering in armed conflicts which may break out despite negotiated arms limitations; (i) diminish the dangers to the environment; and (j) promote better understanding among nations.

Arms control negotiations are an important component of international diplomacy. Obviously, certain conflicts – such as those provoked by revolutionary or national liberation movements – cannot be directly affected by interstate arms control agreements. On the other hand, international controls on the spread of weapons, weapon technologies and weapon-usable materials may circumscribe the scope and effects of such conflicts as well.

Arms control is normally not a matter for negotiation among friendly nations. It is needed, above all, where relations among states are characterized by enmity. However, a modicum of sanguine expectation from negotiations is indispensable; it is hard to imagine parties engaged in an armed conflict with each other discussing how to destroy the weapons they are using in that conflict. Apart from such extreme cases, and short of a complete breakdown of communication between states on a collision course, there are few situations that could justify abandoning efforts to control armaments by states that claim not to harbour aggressive intentions.

To the extent that arms control is meant to serve the security and other interests of all parties participating in negotiations, it cannot be seen as a favour rendered by one state to another, or as a reward for international 'good behaviour'. It is also risky to link arms control with the domestic policies of the negotiating partners: this may impede progress in arms control without necessarily promoting the solution of other issues. Even when the negotiating climate is not conducive to early results, a continuous intergovernmental communication channel to deal with matters of armament may be important to ensure the preservation of peace.

Incentives and Disincentives

In entering into arms control agreements, states demonstrate their dedication to the cause of peace; agreements may also reinforce their international political standing. First and foremost, however, what guides states are security and economic interests.

Arms control agreements provide for mutual rights and obligations, but these rights and obligations are not necessarily equal for all. For example, agreements freezing the deployment and/or qualitative or quantitative levels of armaments favour those parties which enjoy military superiority over others. Similarly, agreements which proscribe transfers of certain militarily important items may be qualified as discriminatory by states which do not or cannot produce the items in question. In such cases, there is a need for positive incentives to induce the militarily disadvantaged states to enter into an arms control agreement.

Universal adherence to multilateral arms control agreements is desirable but not indispensable. Nonetheless, to be meaningful, the agreements must attract most, if not all, of the militarily and economically significant states. It is to these states that positive incentives are often addressed. Such incentives specifying the advantages of the parties may be either endogenous (included in the text of the treaty) or exogenous (included in a separate document or documents not forming part of the treaty). For example, to compensate for the self-imposed nuclear arms denial of the non-nuclear-weapon states under the NPT, the nuclear-weapon states assumed an obligation to contribute to the development of peaceful uses of nuclear energy in non-nuclear-weapon states, with due regard for the needs of the developing areas of the world. In addition, in statements not directly linked to the NPT, the nuclear-weapon powers pledged not to use nuclear weapons against parties not possessing such weapons (except in some special circumstances) and to provide assistance to any state that has become a victim of nuclear aggression or been threatened with such aggression. Similarly, the CW Convention prohibits restrictions that would impede trade in chemicals for peaceful purposes. In addition, the industrially under-developed countries, which are less well prepared to protect themselves against the consequences of chemical warfare, are granted the right to participate in the exchange of protective equipment and material as well as of relevant scientific and technological information, and to obtain assistance if chemical weapons are used against them.

When positive incentives prove insufficient, recourse may be had to negative incentives specifying the disadvantages of *not* joining a given arms control agree-ment. Thus, in 1992 the Nuclear Suppliers Group agreed that transfer to a non-nuclear-weapon state of nuclear facilities, equipment, components, material and technology should not be authorized unless that state behaved like a party to the NPT. This agreement among the nuclear suppliers may have influenced the decision of certain countries engaged in constructing nuclear power stations to renounce their nuclear-weapon option and accede to the NPT. Similarly, the CW Convention stipu-lates that three years after its entry into force the transfer of certain chemicals which possess properties enabling them to be used as chemical weapons, but which are of substantial economic interest to many countries, may take place only among parties. This stipulation may have played a major role in speeding up the ratifications of the Convention.

Not only accession to, but also compliance with, an arms control agreement may sometimes be 'bought' with economic assistance. This was the case of North Korea, a party to the NPT, which was promised two modern nuclear power reactors in exchange for ceasing activities suspected to be part of a nuclear-weapon pro-gramme. All the same, it should be borne in mind that even countries suffering from an economic crisis are not likely to join agreements which they believe might affect their security – as exemplified by the refusal of India and Pakistan to join the NPT.

1.5   The Negotiating Machinery

Policy decisions in the field of arms control are a function of the interaction of vari-ous sectors of government. Consequently, in arms control negotiations each side tries first to enlist the support of its own political and military establishments. The support of allied governments is often sought as well.

Negotiations may be conducted through exchanges of concessions from the diver-gent starting positions of the opposing sides, with a view to reaching convergence of views and eventually a treaty. Negotiations may also involve a joint search for a broadly worded agreement in principle, to be developed in detail in the course of treaty drafting. Both methods are in use, although – depending on the nature of the negotiation – one of the two will usually predominate. Procedures are not of decisive importance for the outcome of negotiations, but the existence of adequate institu-tional mechanisms may help to further the cause.

Multilateral negotiations, the results of which are of global interest and may apply to all states, have always been conducted in specially established forums. However, there is no fixed pattern for the conduct of regional negotiations, which aim at agreements to be observed only (or mainly) by countries in a particular geographical area. The set-up of bilateral or trilateral arms control talks is, as a rule, decided ad hoc.

The Conference on Disarmament

The multilateral arms control negotiating mechanism is provided by the Conference on Disarmament (CD), based in Geneva. The CD is the successor to the Ten-Nation Committee on Disarmament, established by the foreign ministers of France, the United Kingdom, the United States and the Soviet Union (1959–60), the Eighteen-Nation Committee on Disarmament (1962–69), the Conference of the Committee on Disarmament (1969–78) and the Committee on Disarmament (1979–83).

Structure. In 1978 the membership of the Geneva negotiating body was increased from 31 to 40 states. It included all five acknowledged nuclear-weapon powers plus 35 other states representing all geographical regions and political groupings. In 1990, as a result of the unification of Germany, the CD membership was reduced to 39, and when Czechoslovakia, after its breakup, ceased to be a CD member, it fell to 38. In 1996 the CD decided to admit 23 more states. Since one of them was to be Iraq, a country subject to UN sanctions for its aggression against Kuwait, the United States insisted that the newly admitted states commit themselves not to obstruct any action of the Conference by resorting to the rule of consensus provided for in the CD Rules of Procedure. A 'solemn' commitment to this effect was included in a joint letter of the 23 countries to the President of the CD. This commitment was to cease to apply if a consensus decision were reached in the CD that the 'circum-stance' which had given rise to the situation requiring it no longer existed. For any of the new members not subject to comprehensive enforcement measures under Chapter VII of the UN Charter, the above commitment was to cease to apply two years after the decision to enlarge the CD had been adopted. A few delegations questioned the appropriateness of creating a class of CD members whose rights of participation would be restricted, but they did not formally oppose it. In 1999 five more states (out of over 20 requesting membership) were allowed to join the CD, this time without conditions attached to the admission. Thus, the CD membership was brought up to 66. The Socialist Federal Republic of Yugoslavia, although an original member, ceased to participate when no agreement could be reached on successor arrangements; the understanding was that representatives of the Federal Republic of Yugoslavia (Serbia–Montenegro) would not attempt to occupy Yugoslavia's seat. Although the dissolution of the Warsaw Treaty Organization (WTO) and the expansion of the North Atlantic Treaty Organization (NATO) had seriously affected the design of the initial composition of the CD, some appearances of solidarity were maintained among the members of each of the three regional–political groupings inherited from the Cold War alignments – the Western Group, the Eastern European Group and the 'Group of 21' (non-aligned states).

The CD holds annual sessions, each session being divided into three parts. The presidency rotates among all members, each president exercising his functions during a period of four working weeks. Representatives of non-member-states may attend plenary meetings and, if the Conference so decides, other meetings as well. They may submit written proposals or working documents on the subjects of negotiation. The Conference may invite non-member-states, upon their request, to express their views in both formal and informal meetings. Non-governmental organizations (NGOs) do not have such rights. Their communications are retained by the Secretariat of the CD and made available to delegations only upon request. As a consequence, the CD is more immune to the pressure of public opinion than multilateral forums dealing with human rights or protection of the human environment.

The CD is not a UN organ but has close links with the world organization. Although formally autonomous in its activities, the CD often (although not always) takes into account resolutions of the UNGA and regularly submits reports to it. It is taken for granted that the texts of agreements worked out in the CD should be transmitted to the UNGA with the request to have them recommended for signature and ratification by member-states. The budget of the CD is included in the budget of the United Nations. The Conference holds its meetings on UN premises and is serviced by UN personnel. The Secretary-General of the CD is appointed by the UN Secretary-General and acts as his personal representative.

As deemed necessary for the performance of its functions, especially when a draft treaty is to be elaborated, the CD establishes subsidiary bodies: ad hoc committees, working groups, technical groups or groups of governmental experts. The Conference defines the mandate for each subsidiary body, a mandate valid only for a given session of the Conference. Meetings of subsidiary bodies are closed, whereas plenary meetings of the Conference are normally held in public. The CD conducts its work and adopts its decisions by consensus.

Agenda. When it was established, the CD was mandated to deal with arms control and disarmament in the following areas: (a) nuclear weapons in all aspects; (b) chemical weapons; (c) other weapons of mass destruction; (d) conventional weapons; (e) reduction of military budgets; (f) reduction of armed forces; (g) disarmament and development; (h) disarmament and international security; (i) collateral measures, confidence-building measures, effective verification methods in relation to appropriate disarmament measures; and (j) a comprehensive programme of disarmament leading to general and complete disarmament under effective international control.

Within the above terms of reference, the CD adopts an agenda for each session. In 2002 this agenda included: (a) cessation of the nuclear arms race and nuclear disarmament; (b) prevention of nuclear war, including all related matters; (c) prevention of an arms race in outer space; (d) effective international arrangements to assure non-nuclear-weapon states against the use or threat of use of nuclear weapons; (e) new types of weapons of mass destruction and new systems of such weapons, including radiological weapons; (f) a comprehensive programme of disarmament; (g) transparency in armaments; and (h) consideration and adoption of the annual report and any other report, as appropriate, to the UN General Assembly.

Not all items figuring on the agenda are dealt with at the Conference. Only those that are specified in the programme of work adopted for each session are subject to in-depth consideration and negotiation. Other items are occasionally referred to in the delegates' statements. Several annual sessions have ended without agreement on the programme of work because of linkages made between different, often unrelated, measures of arms control.

Shortcomings. Upon the termination of the Cold War, the CD succeeded in working out two important treaties – the CW Convention and the CTBT. However, after the latter had been elaborated, the CD proved unable to agree on what other measures should be taken up and, in fact, interrupted its negotiating activities. The reasons for this critical situation were multiple. They could be found in the outdated membership set-up based on the geopolitical and military realities of the 1970s, in the inability of the CD to negotiate more than just one arms control measure at any given session and in its inflexible rules of procedure.

The requirement of consensus, understood as unanimity, enables any participant to block decisions on any matter, whether substantive or procedural, thereby paralysing all CD work. This virtual right of veto has frequently been resorted to in order to prevent the CD from dealing with issues of paramount importance to a number of states. It has been used to thwart the appointment or extension of the mandate of special coordinators that elicit the views of delegations on issues under discussion and assist the president in conducting informal consultations. It has also been used to hinder the establishment of working committees for items included in the CD agenda or the appointment of chairpersons of these committees. It was grossly abused when a delegation prevented the CD from informing the United Nations that consensus on the text of a treaty had not been reached. As a result of these various problems, the CD began losing its credibility, and its enlargement did not improve the situation.

Prospects. In its report, issued in July 1999, the Tokyo Forum for Nuclear Nonproliferation and Disarmament – an independent international panel of experts – recommended that the CD should update its work programme and revise its procedures or else suspend its operations. In fact, more is needed to revitalize the arms control negotiating machinery. There is no reason why global arms control problems should be dealt with in only one international forum, while global economic or environmental problems are taken up in a number of forums. Nor is there any reason why only certain countries, those selected by the CD itself, should be 'privileged' to negotiate global arms control agreements. The present single negotiating body could be replaced by specialized open-ended negotiating conferences, to be convened by countries interested in or directly affected by certain specific arms control measures. The 'Ottawa Process', set in motion by Canada and a group of like-minded states to deal with the ban on anti-personnel landmines, has demonstrated that such an approach can bear fruit. To be effective, such conferences would have to be autonomous, not accountable to other international bodies. The UNGA may recommend signature and ratification of treaties, but it should not be given authority to invalidate agreements reached by groups of states.

One of the major weaknesses of the CD could be avoided if the arms control conferences adopted flexible rules of work. The consensus rule should not apply to procedural or organizational matters. It is even arguable whether it should apply to substantive matters. There is no risk in adopting veto-free procedures, because no conference or organization can impose treaty obligations upon sovereign states through voting. Treaty texts negotiated internationally are not automatically binding on the negotiating states; they remain to be signed by individual governments and subsequently approved by legislative bodies. In other words, if there is to be rapid and meaningful progress in the field of multilateral disarmament, the entire negotiating machinery must be completely revamped.

The Organization for Security and Co-operation in Europe

The most elaborate regional negotiating mechanism for politico-military affairs, including arms control, is the mechanism of the OSCE.

Composition. From the 1970s, 33 European states plus Canada and the United States were involved in negotiating measures to strengthen confidence, stability and security in Europe. In the late 1980s, members of the two military alliances – 16 from NATO and seven from the WTO – embarked on a parallel negotiation, under the auspices of the CSCE, on conventional armed forces in Europe. When the CFE Treaty was concluded, the CSCE (subsequently the OSCE) established a Forum for Security Co-operation (FSC) open to all participating states. Inaugurated in 1992 in Vienna, the Forum originally consisted of a Special Committee and the Consultative Committee of the Conflict Prevention Centre. In 1993 the Consultative Committee was dissolved, and two years later the Special Committee was renamed the Forum for Security Co-operation. With the unification of Germany, the accession of Albania and Andorra, and the dissolution of the Soviet Union as well as of the Socialist Federal Republic of Yugoslavia and the Czech and Slovak Federal Republic (Czechoslovakia), the number of the OSCE participating states rose from the initial 34 to 55, including the former Soviet Asian republics.

Agenda of the FSC. The 1992 OSCE Helsinki Document outlined a 'Programme for Immediate Action' for the FSC. It mandated the Forum to conduct consultations and negotiations on, among other issues, harmonization of the obligations contracted under various agreements in the fields of arms control and confidence and security building, considering that not all OSCE states were parties to these agreements; exchange of military information; cooperation on non-proliferation of armaments; cooperation in defence conversion; development of military contacts; and transparency in force planning (size, structure and equipment of the armed forces as well as defence policy, doctrines and budgets). In carrying out this programme, in 1993 the FSC adopted documents on stabilizing measures for localized crisis situations; on principles governing conventional arms transfers; on military contacts and cooperation; and on defence planning. Two additional documents were adopted in 1994: on global exchange of military information; and on principles governing non-proliferation. In 1996 the FSC agreed on a framework for arms control, which set guidelines for arms control negotiations, and on the development of the agenda of the Forum to address the implementation of agreed arms control measures and the development of new ones.

Procedures. The FSC meets weekly under a rotating chairmanship, each chairman exercising his functions during a one-month period. Like the OSCE itself, the FSC takes its decisions by consensus, but these decisions are only politically, not legally, binding on the participating states. Given the heterogeneous composition of the OSCE, it may be difficult for its participants to agree on all measures, especially those which are not of equal interest to all. However, a limited number of OSCE states may form a working group to consider and negotiate among themselves certain regional agreements. There are thus, within the framework of the OSCE, opportunities not only for tri-continental (Europe, Asia and North America) but also for regional and even sub-regional arms control negotiations to be conducted and treaties concluded. This was the case of the arms control agreements concerning Yugoslavia.

Review Conferences

Most arms control agreements provide for review conferences to be convened at regular intervals and/or whenever so requested by the parties. The function of these conferences is to review the operation of the agreement with a view to ensuring that its purposes and provisions are being realized.

A review may reveal shortcomings, gaps or even loopholes facilitating circumvention of the obligations and conclude that the text of the treaty ought to be modified. As a rule, a review conference is not authorized formally to adopt the necessary modifications; treaties contain special clauses detailing the amendment procedure. Since amending a treaty may be a difficult undertaking, parties sometimes resort to a simpler and safer practice – that of strengthening the treaty provisions and removing the ambiguities through agreed understandings, without tampering with the text of the treaty itself. Such understandings can be negotiated in the course of the review conference and included in its final declaration. Often, however, states seek to use the review process to impose their own interpretations or to raise issues not directly related to the treaty under review. A number of conferences proved useless when they only reiterated the existing obligations or recorded an agreement to disagree on certain basic issues.

The rules of procedure of review conferences envisage the possibility of voting on a final declaration. However, in the practice followed so far, whenever consensus cannot be reached the participants prefer to include the dissenting views in a separate document or documents annexed to the common declaration or to admit failure. Unlike procedure at the CD, NGOs are allowed to present their views and proposals to the governmental delegations participating in review conferences. They usually do so at conference meetings specifically devoted to this purpose.

2

Historical Overview

The practice of negotiating arms control among sovereign nations in an international forum and in time of peace, with a view to making the measures agreed upon applicable to several or all nations, is relatively recent. Among the earliest efforts in this field were the two International Peace Conferences held at The Hague at the turn of the past century.

2.1　The Hague Peace Conferences

The Hague Conferences of 1899 and 1907 were convened at the initiative of the Emperor of Russia, which was lagging in the European arms race and could not afford to catch up with its rivals because of its economic weakness. Russia's declared aim was to ensure universal peace and bring about a reduction of 'excessive' armaments. The diplomatic note circulated by the Russian Foreign Minister prior to the First Hague Conference stated that the armed peace had become a burden for the peoples of Europe because intellectual and physical forces, as well as labour and capital, were to a large extent diverted from their natural applications to unproductive ends. One hundred and eight delegates from 26 countries participated in the First Hague Conference, whereas as many as 256 delegates from 44 countries participated in the Second Conference.

The disarmament goals of the Hague Conferences were not achieved. Proposals for limiting the calibre of naval guns, the thickness of armour plate and the velocity of projectiles were rejected. Very few politicians were at that time interested in halting the competition in arms. A resolution was adopted declaring that a restriction on military expenditure was highly desirable, and governments were asked to examine the possibility of an agreement on the limitation of armed forces and war budgets. However, military expenditures in practically all the participating states continued to grow, and the arms race went on.

Nevertheless, the Hague Conferences contributed to the evolution of international law by codifying the rules of war, including those which prohibit or restrict the use of certain insidious types of weapon – asphyxiating gases, expanding bullets or submarine contact mines. The territory of neutral countries was declared inviolable. Another achievement was the establishment of the Permanent Court of Arbitration, the forerunner of today's International Court of Justice. The need for collective action to settle disputes between states which could not be solved by diplomatic means and to control the effects of warfare was thus internationally recognized. These achievements were – to a considerable extent – due to pressure exerted by non-governmental peace advocates, such as Baroness Berta von Suttner, the 1905 Nobel Peace Prize Laureate. Plans for a third peace conference had to be abandoned in view of the intensified interstate antagonisms that preceded World War I.

In 1994 the Russian Foreign Minister proposed to celebrate the 100th anniversary of the First Peace Conference by convening another such conference. The declared aim was to improve the system of peaceful settlement of disputes and further develop the international humanitarian law of warfare. This proposal did not find sufficient international support to materialize. However, in 1999, non-governmental organizations, meeting in The Hague, adopted an 'Agenda for Peace and Justice' that dealt with the root causes of war; international humanitarian and human rights law and institutions; prevention, resolution and transformation of violent conflict; and disarmament and human security.

2.2   The Post-World War I Peace Treaties

The 1919 Treaty of Versailles

After World War I, which ended with an armistice, the victorious Allies, led by the Prime Ministers of Great Britain, France and Italy, as well as the President of the United States, drafted a peace settlement that called for a substantial disarmament of the defeated Germany. The 1919 Treaty of Versailles stipulated that the German Army was to be limited to 100,000 men and was not to be allowed to have tanks or heavy artillery. The German Navy was to be reduced to six battleships, six light cruisers, 12 destroyers and 12 torpedo boats, and was to be deprived of submarines. No military or naval air forces were permitted. Permissible arms munitions and other war material were specifically enumerated and could be produced only in Allied-approved factories; their import was prohibited. Strictly forbidden were both the manufacture and imports of asphyxiating, poisonous or other gases and all analogous liquids, materials or devices. The same applied to materials especially intended for the manufacture, storage and use of the said products or devices.

Germany's General Staff was to be dissolved, universal compulsory military service abolished, and any measures of mobilization excluded. Restrictions were imposed on German military schools; educational establishments or associations were not allowed to occupy themselves with military matters. In the Baltic and North Seas, German fortifications were to be demolished. The left bank of the Rhine River as well as a 50-kilometre-wide zone east of the Rhine were to be demilitarized.

The Treaty of Versailles was largely circumvented or openly violated. The General Staff continued to exist, although in a different form; military personnel were retained in excess of the set limits, while new personnel were illegally trained; paramilitary groups were created for reserve duty; arms were maintained in secret depots; and weapons prohibited by the Treaty were developed and manufactured in Germany or imported. The supervision of the Treaty entrusted to the Inter-Allied Commissions of Control was never fully effective and gradually ceased to be exercised. However, verification of compliance was not the major problem: the British and French governments were quite aware that the Treaty was being violated. It was rather the inability or unwillingness of these governments to enforce compliance that made it possible for Germany to rearm. By 1936 the arms control clauses of the Treaty of Versailles ceased to be operative.

Other Peace Treaties

Post-World War I peace agreements imposed by the Allied Powers on Germany's allies paralleled the disarmament clauses of the Treaty of Versailles. They limited the size of armies and armaments, reduced the navies and prohibited air forces. Thus, the 1919 Peace Treaty, signed at St Germain-en-Laye, limited the Austrian Army to 30,000 men; the 1919 Peace Treaty, signed at Neuilly, limited the Bulgarian Army to 20,000 men and required the surrender of most of its arms and war material; and the 1920 Peace Treaty, signed at Trianon, reduced the Hungarian Army to 35,000 men. The 1920 Peace Treaty, signed at Sèvres, imposed limitations on Turkey but was never implemented owing to Turkey's internal upheaval and Turkish–Greek fighting. It was replaced by the Treaty of Lausanne, signed in 1923.

In introducing restrictions on the armaments of the vanquished nations, the victorious powers committed themselves to limit their own armaments. This was to take place in accordance with the principles set out by the newly founded League of Nations.

2.3   The League of Nations

The Covenant

The Covenant of the League of Nations, which formed Part I of the Treaty of Versailles, required the reduction of armaments of all nations 'to the lowest point consistent with national safety and the enforcement by common action of international obligations'. Members of the League undertook to exchange information regarding the scale of their armaments, their military, naval and air force programmes, and the condition of those of their industries that were adaptable to warlike purposes. The Council of the League was to formulate plans for armaments reduction for the consideration of and action by governments, taking account of the geographical situation and the circumstances of each state. A Permanent Court of International Justice was to be created. Arms build-up ceased to be a matter of purely national concern.

To advise the Council of the League on implementation of the disarmament provisions of the Covenant, a Permanent Advisory Commission was set up, composed of military, naval and air force representatives appointed by each state member of the Council. Moreover, a Temporary Mixed Commission was formed to examine the relevant political, social and economic questions. In 1925, a Preparatory Commission consisting of representatives of both members and non-members of the League started its deliberations regarding the envisaged Disarmament Conference. This Commission held six sessions and was dissolved in 1930 after it had submitted a draft Convention on the Reduction and Limitation of Armaments.

The League of Nations Yearbooks

The League of Nations spent more time and effort on disarmament than on any other subject. In 1924 its Secretariat began to publish the *Armaments Year-book* on the strength and equipment of the states' armed forces. The yearbook was based on public sources; certain editions included data on the production and exchange of goods related to national defence, as well as information on paramilitary formations and police forces. An indication of the size and trends of military spending was also given. Yet another publication, the *Statistical Year-book* of the League of Nations, contained data on the international transfers of arms and ammunition and showed the values of imports and exports according to official national statistics. The figures were approximate, incomplete and generally non-comparable, while trade in certain important categories of arms was not covered at all. Nonetheless, both yearbooks made it possible to bring the problem of armaments within the reach of the general public for the first time. They also provided a tool for the League activities aimed at controlling the international trade in arms and the manufacture of arms.

Attempts to Regulate Arms Trade and Production

Earlier attempts to regulate the arms trade had been limited to one continent, or a part of it, as under the 1890 Brussels Act prohibiting the introduction of firearms and ammunition to Africa between latitudes 20° North and 22° South (except under effective guarantees), or to one country, as in the case of the 1906 Act of Algeciras repressing the contraband of arms to Morocco. The League of Nations was the first international body entrusted (by its Covenant) with general supervision of the trade in arms and ammunition and with prevention of the 'evil effects' attendant upon the private manufacture of munitions and implements of war.

The 1919 St Germain Convention. Under the 1919 St Germain Convention for the Control of the Trade in Arms and Ammunition, which was worked out in conformity with the relevant provisions of the League of Nations Covenant, there was to be no arms export, save for exceptions to be permitted by means of export licences granted by governments. A comprehensive list of armaments to which different regulations were applicable was drawn up, and transparency, or what was then called 'publicity', for the arms trade was required. However, the Convention never came into force, mainly because of the refusal of the United States to ratify it. This meant that the original aim of the Convention – that of preventing an uninhibited spread throughout the world of those weapons which the belligerent powers had accumulated during World War I and for which they had no further use – could not be achieved.

The 1925 Geneva Convention on the Arms Trade. Subsequent efforts in this field led to the signing in Geneva, in 1925, of the Convention for the Supervision of the International Trade in Arms and Ammunition and in Implements of War. This Convention distinguished five categories of arms: (a) arms exclusively designed and intended for land, sea and air warfare; (b) arms capable of use for both military and other purposes; (c) war vessels and their normal armament; (d) aircraft (assembled or dismantled) and aircraft engines; and (e) gunpowder, explosives and arms not covered by the first two categories.

An export licence or declaration was required for export of any item in the first category; authorization by the government of the importing country was also necessary if these items were exported to private persons. Similarly, items in the second category could be exported only when accompanied by an export document, but prior authorization of the government of the importing country was not necessary. In the case of the third category, detailed information was to be published regarding vessels transferred and those constructed on behalf of the government of another state, including armaments installed on board. As regards the fourth category, a return was to be made public giving the quantities of aircraft and aircraft engines exported as well as the country of destination. Trade in items belonging to the fifth category was not to be subject to any restriction, unless the commodities were destined for certain territorial and maritime zones in Africa or the Middle East, referred to as 'special zones' and specified in the Convention.

The purpose of the 1925 Convention was not to reduce the international trade in arms, which was seen as a legitimate activity, but to prevent illicit traffic. This was to be accomplished through universal export licensing by governments and through publicity in the form of statistical returns. However, no supervision of arms production was provided for. This inequity was the main reason why many countries, especially non-producing, arms-importing countries, refused to ratify the Convention, which consequently never entered into force.

Of the documents signed simultaneously with the 1925 Convention, only the Protocol for the Prohibition of the Use in War of Asphyxiating, Poisonous or Other Gases, and of Bacteriological Methods of Warfare became effective and remains in force.

The 1929 Proposal for Supervision of Arms Production. In 1929, a special committee set up by the Council of the League of Nations submitted a draft convention for the 'supervision of the private manufacture and publicity of the manufacture of arms and ammunition and of implements of war'. According to the draft, no private manufacture of arms belonging to the first four categories established by the 1925 Geneva Convention, referred to above, would be permitted, unless licensed by governments. Moreover, data were to be published showing the value, quantity and weight of arms of the first, second and fourth categories which had been manufactured in private enterprises (under licence) or in state-owned establishments.

Objections were raised with regard to different provisions of the draft, mainly those related to restrictions on private manufacture of arms and disclosure of data on arms industry. Demands were put forward by some governments to abolish private manufacture of arms or to internationalize all arms production. In this situation, it became impossible to reach agreement.

Organizing the Peace

The 1924 Geneva Protocol. In 1924, the Assembly of the League of Nations adopted a plan for the organization of peace, known as the 1924 Geneva Protocol. The Protocol prohibited recourse to war under any circumstances; it determined that a state which refused to resort to arbitration, or to comply with the provisional measures prescribed by the Council, should be presumed to be the aggressor; it made compulsory the application of sanctions; and it stipulated that all disputes should be terminated by a binding decision pronounced by the Permanent Court of International Justice, the Council of the League or a board of arbitration.

The Protocol proved unacceptable to many states, which objected to the requirement of compulsory arbitration for all disputes. (Under the Covenant, only grave interstate disputes were to be submitted to arbitration or judicial settlement, or to enquiry by the Council.) Opponents of the Protocol were also reluctant to assume the burdens inherent in the application of sanctions.

The 1928 Kellogg–Briand Pact. The most remarkable agreement reached in the inter-war period to abolish the use of violence in relations among nations was the Pact for the Renunciation of War as an Instrument of National Policy signed in Paris in 1928 and in force since 1929. The parties to this treaty, also known as the Kellogg–Briand Pact (after the US Secretary of State and the French Foreign Minister, who had negotiated it), gave up recourse to aggressive war. Without renouncing the right to self-defence, they agreed that the settlement of all disputes or conflicts which might arise among them would always be sought by peaceful means. With the participation of Germany, Japan and the United States, the Kellogg–Briand Pact managed to achieve a higher degree of universality than the Covenant of the League of Nations. Unlike the Covenant, however, it did not establish a permanent supervisory organization; nor did it envisage sanctions in the case of breaches.

2.4   The First World Disarmament Conference

In depriving war of legitimacy, the 1928 Kellogg–Briand Pact provided an impetus for the 1932 Disarmament Conference, the only conference held prior to World War II to discuss a universal reduction and limitation of all types of armament. Convened in Geneva under the auspices of the League of Nations, it was attended by representatives of over 60 states. Without prejudging the decisions of the Conference, the participating governments were asked to refrain, for a period of one year, from any measure involving an increase in their armaments. This so-called armaments truce was later extended for a few months.

Public opinion was very active throughout the Disarmament Conference. Prior to the opening of the Conference, several international organizations adopted resolutions in which they set out their views as to the way in which various problems of disarmament should be approached. Their representatives were allowed access to the Conference, and a special plenary meeting was held at which petitions were presented. Thousands of letters and messages were addressed to the President of the Conference from all over the world.

The following questions were examined in detail by specialized commissions, sub-commissions and committees of the Disarmament Conference: (a) establishment of a system of collective security; (b) limitation of the strength of the armed forces; (c) limitation of land, naval and air armaments; (d) limitation of national defence expenditures; (e) prohibition of chemical, incendiary and bacteriological warfare; (f) control of arms manufacture and trade; (g) supervision and guarantees of implementation of the obligations contracted by the parties; and (h) 'moral disarmament' intended to create an atmosphere favourable to the peaceful solution of international problems.

A draft convention, drawn up by the Preparatory Commission, was first to be submitted to the Conference for consideration. Subsequently, a British draft was accepted as the basis for the future convention, and a provisional text taking account of the modifications to this draft was published in September 1933, along with amendments proposed and statements made by various delegations. Summaries of the points of agreement and disagreement revealed at the Conference follow below.

Renunciation of War

The participating states were willing to enter into immediate consultation in the event of a breach, or threat of breach, of the 1928 Kellogg–Briand Pact, with the purpose of preserving peace and averting conflict. Such consultation could be set in motion by the Council or the Assembly of the League of Nations or by a state not member of the League. A draft undertaking not to resort to force, to be signed by all the European states, was adopted, and various delegations expressed the hope that the undertaking would subsequently assume a universal character. In a message to the Conference, the President of the United States proposed that all nations should conclude a 'solemn and definite' pact of non-aggression.

Positions were less clear, however, as regards the definition of an aggressor, the procedure for establishing facts constituting aggression and the problem of mutual assistance.

Armed Forces, Armaments and Defence Expenditures

The negotiators agreed, in principle, that a quantitative limitation and subsequent reduction of armed forces should be brought about. Nevertheless, no common decision could be reached on how to assign definite figures of effectives to individual states. It was also generally understood that qualitative disarmament should apply, in the first place, to those weapons which were most specifically offensive, most efficacious against national defence or most threatening to civilians, but controversies arose regarding the applicability of these criteria to individual categories of arms.

Among the proposals for the limitation of land armaments, the most remarkable was that submitted by the United States, requiring that tanks and heavy mobile land-guns should be abolished. However, the draft convention went no further than to suggest maximum limits for the weight of a tank and for the calibre of mobile land-guns; only tanks and guns exceeding the fixed limits would be abolished. Various suggestions were made concerning the numerical ceilings to be prescribed, as well as the time limits for destruction of excess material. The French delegation moved that weapons exceeding the prescribed limits should be internationalized. It also made its acceptance of the provisions relating to land war material conditional upon the organization of an effective system of supervision, particularly with regard to the manufacture of arms.

The discussions of naval armaments were determined largely by the 1922 Washington and 1930 London Naval Treaties, which had limited the sizes of the major powers' navies and were subject to revision at an international conference scheduled for 1935. Pending this conference, Great Britain proposed that the stipulations of both treaties should be retained; states not bound by these treaties would pledge to observe the status quo, meaning that any new warship construction undertaken before 1935 could only replace 'over-age' tonnage. Measures proposed by other delegations went considerably further. It was, for example, suggested that submarines and aircraft carriers should be abolished by all states. Strong objections were raised against attempts to incorporate the provisions of the two above-mentioned treaties, which had been concluded by a few naval powers, into what was intended to be a universal disarmament convention.

The draft submitted by the Preparatory Commission of the Disarmament Conference provided for limitations on the number and the total horsepower of military
aircraft. In the course of the Conference, several proposals were put forward with a
view to strengthening these provisions. Certain delegations suggested that military
aviation should be abolished altogether, while others only recommended a ban on
bombing from the air. It was assumed that internationalization or other regulatory
measures would be needed to prevent states from using civil aviation for military
purposes. According to the British draft, the adoption of concrete undertakings in
this field was to be left to the next disarmament conference, whereas limits on the
numbers of aeroplanes capable of use in war would be accepted without delay.

As regards limitations on national defence expenditures, a technical committee of
the Conference recognized that it was possible for states to draw up, for all practical
purposes, a complete statement of such expenditures, and for an international supervisory body to verify, with a high degree of accuracy, how these amounts had been
calculated. However, certain members of this committee pointed out the difficulties
arising from currency fluctuations and from the different methods of accountancy
used by governments. The need for periodic publicity to be given to the parties'
defence expenditures – irrespective of the nature and origin of the resources from
which these expenditures were met – was thoroughly discussed. The instruments
necessary for the application of the system of publicity were specified.

Chemical, Incendiary and Bacteriological Warfare

The draft convention prohibited the use of chemical weapons, including lachrymatory, irritant or vesicant substances as well as incendiary or bacteriological weapons,
against any state and in any war, whatever its character. Lachrymatory substances
intended for use in police operations, as well as appliances for the use of these substances, would have to be declared by the parties. All preparations for chemical,
incendiary or bacteriological warfare would be prohibited in time of peace as in time
of war. Accordingly, the manufacture, import, export or possession of appliances or
substances suitable exclusively for chemical or incendiary warfare or suitable for
both peaceful and military purposes but intended for use in violation of the convention would be banned. Similarly, instruction and training of armed forces in the use
of chemical, incendiary or bacteriological weapons and means of warfare would be
forbidden. A procedure of enquiry and on-the-spot investigation of the alleged uses
of the prohibited weapons was provided for. The right of reprisal, however, was recognized, as was the right to possess material and installations necessary to ensure
individual or collective protection against the effects of chemical, incendiary or bacteriological weapons, and to conduct training with a view to such protection.

Arms Trade and Manufacture

In taking up the subject of the trade in and manufacture of arms, the Conference had
before it the 1925 Convention for the Supervision of the International Trade in Arms
and Ammunition and in Implements of War (not in force) providing for control and
publicity in respect of exports of certain categories of arms, as well as the 1929 draft
convention subjecting private manufacture of arms to a system of licensing and publicity. Many delegations argued that since these two documents had been formulated
new facts and ideas had emerged and that there was therefore a need for more com-

plete regulations. Others were unwilling to accept stricter controls. The main questions concerned the principle of state responsibility for, and the kind of publicity to be given to, the trade in and manufacture of arms as well the principle of qualitative and quantitative limitations on manufacture.

A report published in 1935 included texts reflecting unanimity on the need for an effective system of control and regulation of the arms trade and manufacture. However, considerable differences remained with regard to the character of the measures necessary to bring such a system into being. The requirement of equality between countries producing arms and those not producing them was acknowledged, but opinions differed as to how to achieve such equality. Certain delegations made their position on arms trade and manufacture conditional upon the nature and extent of the obligations which the parties would undertake under a general disarmament convention.

Verification and Sanctions

The need for effective international control of compliance with the obligations assumed by the parties was strongly emphasized throughout the debates of the Conference. It was agreed that a Permanent Disarmament Commission to be set up at the seat of the League of Nations and composed of representatives of the parties should be ready to assume its duties as soon as the convention entered into force. These duties were to include investigations of alleged infractions of the convention. Moreover, there were to be regular inspections of armaments of each state, at least one per year, on the basis of equality between the parties.

As regards guarantees of implementation, it was assumed that in case of an established breach of the provisions of the convention, the Council of the League would exercise its rights under the Covenant. However, the French delegation insisted on defining more precisely the action to be taken in such an event. It proposed that the Permanent Disarmament Commission should demand that the party at fault fulfil its undertakings within a fixed period. The Commission should also appoint an inspection committee to check whether this demand had been met. If the violation continued, the parties could jointly use the necessary means of pressure against the defaulting state to ensure implementation of the convention. If war should ensue, the defaulting party was to be subject to sanctions in accordance with the provisions of the Covenant. These sanctions could include mandatory economic measures, such as severance of trade and financial relations or interruption of postal and railway communications, as well as non-mandatory military measures to be recommended by the League's Council.

Moral Disarmament

Under the heading of moral disarmament the Conference discussed questions relating to education, cooperation among intellectuals, the press, broadcasting, theatre and cinema. The committee dealing with moral disarmament adopted a text stating that parties should undertake to ensure that education at every stage should be so conceived as to inspire mutual respect between peoples and emphasize their interdependence. The parties would further undertake to ensure that persons entrusted with education and preparing textbooks were inspired by these principles, to encourage the use of cinema and broadcasting for increasing the spirit of goodwill among nations and to use their influence to avoid the showing of films, broadcasting of programmes or organization of performances obviously calculated to offend the legitimate sentiments of other countries.

A proposal to adapt municipal laws to the development of international relations was also submitted. It provided for legislation to be introduced by the parties, enabling them to inflict punishment for certain acts detrimental to good relations among states. Such acts would include preparation and execution of measures directed against the security of a foreign power, attempts to induce a state to commit a violation of its international obligations, aiding or abetting armed bands formed in the territory of one state and invading the territory of another state, dissemination of false information likely to disturb international relations and false attribution to a foreign state of actions likely to bring it into public contempt or hatred. It was also suggested that the parties should pledge themselves to consider introducing into their state constitutions an article prohibiting resort to force as an instrument of national policy, embodying thereby the principles of the 1928 Kellogg–Briand Pact.

Suspension of the Disarmament Conference

After several years of work, agreement seemed to have been achieved on the following points: certain methods of warfare should be prohibited; armaments should be limited both qualitatively, through the abolition of some particularly powerful types of weapon, and quantitatively, through a reduction in the numbers of the weapons retained; manufacture of and trade in arms should be placed under supervision; publicity should be given to national defence expenditures; inspections should make it possible to establish violations; and implementation of the disarmament obligations should be guaranteed. However, the withdrawal of Germany from both the Disarmament Conference and the League of Nations, as well as German rearmament in violation of the Treaty of Versailles, brought about a breakdown of attempts to transform these agreed points into a generally acceptable disarmament convention. In early 1936, the Council of the League decided to suspend the Disarmament Conference.

The Conference never reconvened. However, much can be learned from the record of its deliberations, which includes a thorough examination of the political, technical, economic, legal and moral aspects of disarmament. Many ideas put forward at the League of Nations, both before and during the Disarmament Conference, have been revived in recent years, and a number of points made at that time remain topical.

2.5   The Post-World War II Peace Treaties

Treaties with Bulgaria, Hungary, Finland, Italy and Romania

In the early post-World War II years, a major international problem was the demilitarization of the vanquished states. The Peace Treaties concluded by the Allied Powers in 1947 with Bulgaria, Hungary, Finland, Italy and Romania imposed the following arms restrictions.

Each of these five states was prohibited from possessing, constructing or testing any atomic weapon, any self-propelled or guided missiles or apparatus connected with their discharge (other than torpedoes and torpedo-launching gear comprising the normal armament of naval vessels permitted by the treaty), sea mines of non-contact types, torpedoes capable of being manned, submarines or other submersible craft, motor torpedo boats or specialized types of assault craft.

As regards limitations on land forces, including frontier troops, Italy was not allowed to exceed 185,000 combat, service and overhead personnel and 65,000 carabinieri; Bulgaria – 55,000 personnel plus 1,800 for anti-aircraft artillery; Hungary – 65,000 personnel, including anti-aircraft and river flotilla personnel; Romania – 120,000 personnel plus 5,000 for anti-aircraft artillery; and Finland – 34,400 personnel, including anti-aircraft personnel.

As regards limitations on naval forces, Italy was not authorized to have more than 25,000 personnel and 67,500 tons of the total displacement of war vessels; Bulgaria, 3,500 personnel and a total of 7,250 tons; Romania, 5,000 personnel and a total of 15,000 tons; and Finland, 4,500 personnel and a total of 10,000 tons.

As regards limitations on air forces, Italy was forbidden to possess more than 200 fighter and reconnaissance aircraft and 150 transport, air–sea rescue, training and liaison aircraft, with a total personnel strength of 25,000; Bulgaria – 90 aircraft, of which not more than 70 could be combat types of aircraft, with a total of 5,200 personnel; Hungary – 90 aircraft, of which not more than 70 could be combat types of aircraft, with a total of 5,000 personnel; Romania – 150 aircraft, of which not more than 100 could be combat types of aircraft, with a total of 8,000 personnel; and Finland – 60 aircraft, with a total of 3,000 personnel. All five countries were barred from possessing or acquiring any aircraft designed primarily as bombers with internal bomb-carrying facilities.

Because of the division of Europe into two antagonistic military blocs, full implementation of the military clauses of the 1947 Peace Treaties proved impossible. In April 1949 Italy became a founding member of the North Atlantic Treaty Organization (NATO) and considered itself released from the obligations under the military clauses of its Peace Treaty, which it denounced in 1952. In September 1990 Finland stated that the Peace Treaty stipulations restricting Finnish military capabilities had become null and void, with the exception of the ban on the acquisition of nuclear weapons. Bulgaria, Hungary and Romania practically abrogated the military provisions of their Peace Treaties when they signed treaties of mutual assistance with the Soviet Union in 1948, and when they joined the Warsaw Treaty Organization (WTO) in 1955. However, none of the latter three countries has formally denounced its Peace Treaty with the Allied Powers.

The Austrian State Treaty

The 1955 State Treaty for the Re-establishment of an Independent and Democratic Austria stipulated that Austria should not possess, construct or experiment with any atomic weapon; any other major weapon adaptable to mass destruction and defined as such by the appropriate organ of the United Nations; any self-propelled or guided missiles or torpedoes, or apparatus connected with their discharge or control; sea mines; torpedoes capable of being manned; submarines or other submersible craft; motor torpedo boats; specialized types of assault craft; guns with a range of over 30 kilometres; asphyxiating, vesicant or poisonous materials or biological substances in quantities greater than, or of types other than, those required for legitimate civil purposes, or any apparatus designed to produce, project or spread such materials or substances for war purposes. The Allied Powers reserved the right to add to this list prohibitions of any new weapons that might result from scientific development.

In November 1990, in a formal communication sent to the signatories of the State Treaty, Austria declared that the military clauses of the Treaty had become obsolete, with the exception of those concerning atomic, biological or chemical weapons.

Restrictions on Germany's Armament

Concluding a peace treaty with Germany, the country responsible for the outbreak of World War II, proved impossible in the atmosphere of the Cold War, which began between the major victorious powers soon after the termination of hostilities. The imposition of communist regimes in Eastern Europe, subversive activities in Greece and, in particular, the blockade of Berlin in 1948–49 had generated Western fears of aggressive intentions on the part of the Soviet Union. The response of the Western Allies was to seek closer unity among themselves as well as cooperation in defence matters with their former German enemy. The first moves in this direction had been the proposals of the early 1950s for a unified Western European Army. These failed when France refused to ratify the European Defence Community Treaty of 1952. The idea was then put forward to allow the Federal Republic of Germany (FRG) to join the 1948 Treaty of Economic, Social and Cultural Collaboration and Collective Self-Defence among Western European States (the Brussels Treaty), in return for controls over German armaments and force levels. This new, less federalist formula was conceived with a view to making West German rearmament acceptable to those in Western Europe who feared a resurgence of German military power, thereby removing the political obstacles to West German membership of NATO.

At conferences in London and Paris, held in 1954, the Brussels Treaty was modified, in particular through the creation of the Council of Western European Union (WEU) and the requirement that the parties and the organs established by them work in close cooperation with NATO. Several protocols to the Treaty were agreed as part of the so-called Paris Agreements. By May 1955 these protocols had been ratified by all the countries concerned – Belgium, France, the FRG, Italy, Luxembourg, the Netherlands and the United Kingdom – which thus formed the WEU. The armaments and force levels of its members were to be submitted to control – albeit to varying degrees – by the Agency for the Control of Armaments (ACA).

In Annex I of Protocol No. III to the Treaty, the Federal Chancellor declared that the FRG undertook not to manufacture in its territory any atomic, chemical or biological weapons. Another undertaking of the Federal Republic was not to manufacture in its territory long-range missiles, guided missiles and influence mines (defined as 'naval mines which can be exploded automatically by influences which emanate solely from external sources'); large warships, including submarines; and bomber aircraft for strategic purposes – all specified in Annex III. Modification or cancellation of the latter undertaking could, upon request of the Federal Republic, be carried out by a resolution adopted by a two-thirds majority of the WEU Council if, in accordance with the needs of the armed forces, an appropriate recommendation were made by the Supreme Commander of NATO. As regards other categories of armament, the FRG was subject to the same type of control as other WEU members:

stocks of specified weapons maintained on the mainland of Europe were not to exceed NATO requirements nor levels approved by the WEU Council. The ACA thus had to exercise two different types of control: non-production control with regard to the Federal Republic, and level-of-stock control with regard to all WEU members.

The restrictions on West German conventional armament were subject to continuous revisions and cancellations. The last items to be removed from the list of prohibited weapons – following the decision adopted in June 1984 – were guided weapons with ranges exceeding 70 kilometres and bomber aircraft for strategic purposes. As regards atomic, chemical and biological weapons, both the Federal Republic of Germany and the German Democratic Republic had for years been internationally bound by the 1968 Non-Proliferation Treaty, the 1925 Geneva Protocol and the 1972 Biological Weapons Convention. In 1990, on the eve of German unification, the governments of both German states reaffirmed their contractual and unilateral undertakings not to manufacture, possess or have control over nuclear, biological and chemical weapons. The united Germany became a party to the 1993 Chemical Weapons Convention.

Restrictions on Japan's Armament

In June 1947, the representatives of nations that had been engaged in the war against Japan met as the Far Eastern Commission and adopted a decision on basic post-surrender policy for Japan. Japan was not to have any army, navy, air force, secret police organization, civil aviation or gendarmerie; it was, however, allowed to have adequate civilian police forces. Japan's ground, air and naval forces were to be disarmed and disbanded, and the Japanese General Staff was to be dissolved. Military and naval *matériel*, military and naval vessels, and military and naval installations as well as military, naval and civilian aircraft had to be surrendered to the Allied commanders in the zones of capitulation of the Japanese troops and disposed of in accordance with decisions of the Allied Powers. Inventories were to be made and inspections authorized to ensure complete execution of these provisions.

In a more specific policy decision on the prohibition of military activity in Japan and the disposition of Japanese military equipment, adopted in February 1948, the Far Eastern Commission imposed bans on: the possession of arms, ammunition and implements of war by any Japanese citizen, except for police and hunting purposes; the development, manufacture, import and export of arms, ammunition and implements of war and materials intended for military use, except for the import of arms and ammunition for non-military purposes mentioned above; the manufacture of aircraft of all kinds; the construction of any naval combatant or auxiliary vessel or craft, the conversion of any commercial vessel or craft to military purposes, or the reconstruction or remodelling of commercial vessels or craft so as to render them more suitable for military purposes; and military training of the civilian population and military instruction in schools. The Constitution of Japan provides for the renunciation of war and non-possession of a war potential.

In September 1951, Japan regained its international status when its former enemies – with the exception of China, India and the Soviet Union – signed a peace treaty. The Allied military occupation ended in 1952, after which US armed forces remained in Japan under a special agreement.

Most of the severe restrictions imposed on Japan in the military field were lifted relatively quickly, and Japan established Self-Defence Forces (SDF). As early as 1955, in a joint US–Japanese statement, the Foreign Minister of Japan indicated that Japan's defence strength had reached a considerable level. He agreed with the US Secretary of State that efforts should be made to establish conditions in which Japan could, as rapidly as possible, assume primary responsibility for the defence of its homeland and be able to contribute to the preservation of peace and security in the Western Pacific. Already at the end of the 1980s Japan found itself among the world's leading military spenders, and its SDF, comprising the army, air force and navy, had reached a high degree of technological sophistication. In 1992, the Japanese Parliament passed a controversial bill permitting SDF to participate in UN peacekeeping operations.

3

The United Nations

International endeavours to regulate armaments on a worldwide scale, which had been interrupted by World War II, resumed in 1945 within the framework of the United Nations.

3.1   The Charter

Unlike the Covenant of the League of Nations, which had attached considerable importance to disarmament and to the means needed to achieve it, the UN Charter, signed in June 1945 and in force since October 1945, made few references to disarmament. Principles 'governing disarmament and the regulation of armaments' were included among the general principles of international peace and security to be considered by the UN General Assembly (Article 11). The UN Security Council, consisting (since 1965) of 15 members, of which five occupy permanent seats, is to formulate plans for the establishment of a 'system' for the regulation of armaments, to the extent that there would be the least diversion for armaments of the world's human and economic resources (Article 26). One reason for this difference in emphasis lies in the fact that when the League Covenant was written, many believed that World War I had been caused by the arms race that preceded the war, whereas a few decades later the prevalent belief was that World War II could have been avoided if only the great powers had maintained an adequate military potential as well as a readiness to use it. Unlike the League Covenant, the UN Charter was drafted when war was still in progress and when planning a system of disarmament might have seemed ill timed. Furthermore, the system of enforcement measures envisaged in the Charter is predicated on the continued existence of national armed forces. These are to be made available to the Security Council to maintain or restore international peace and security, but may be used for self-defence in the case of armed attack against a UN member before the Security Council takes the necessary measures. This implies that the term 'disarmament', used in the Charter, was not meant to denote the absence of arms.

Notwithstanding the Charter provisions, the United Nations quickly became involved in arms control. This was prompted chiefly by the use of atomic bombs shortly after the signing of the UN Charter and by the fear that this new weapon of unprecedented destructiveness might be used again. Indeed, the very first UN General Assembly resolution, unanimously adopted in January 1946, established a commission to deal with the problem of atomic energy and atomic weapons. This Atomic Energy Commission was composed of one representative from each of the 11 states then represented on the Security Council and Canada when that state was not a member of the Security Council. The terms of reference of the Commission included making specific proposals for the elimination from national armaments of atomic weapons and of all other major weapons 'adaptable' to mass destruction and for effective safeguards against the hazards of violations. In December of that year, the United Nations recommended a general regulation and reduction of armaments and armed forces. Since then, disarmament has been an item of central importance on the UN agenda.

In 1959, in a remarkable demonstration of the progressive expansion of its mandate (exercised with equal force in such areas as decolonization and human rights), the UN General Assembly went far beyond the original language of the Charter in adopting a resolution calling for 'general and complete disarmament'. In 1961 it approved the principles for negotiation on universal disarmament.

3.2   The Main UN Arms Control Bodies

The General Assembly

The UN General Assembly – the most representative body of the world community – is the principal arena for international policy debates. It is also the chief deliberative organ of the United Nations in the field of disarmament.

Regular Sessions. Arms control issues are debated in one of the main committees (the First Committee) of the regular sessions of the UN General Assembly or, less frequently, directly in the plenary sessions without recourse to a subsidiary body. The Assembly provides opportunities for governments to state their official arms control policies, as well as to establish new intergovernmental contacts and hold informal talks on a wide range of questions. It adopts resolutions which contain proposals and recommendations. Several UN General Assembly resolutions represented landmarks in the arms control deliberative process. In a number of instances they have provided an impetus to arms control negotiations and agreements.

In 1982, in an effort to implicate the UN Secretary-General more directly in the arms control process, the General Assembly empowered him to investigate alleged violations of the ban on the use of chemical and biological weapons. In 1991 it requested him to establish a universal register of conventional arms to include data on international arms transfers as well as information on military holdings and procurement through national production. The intention was to pave the way towards global conventional arms control, largely neglected for many years.

The General Assembly may also decide that conferences should be held under UN auspices to negotiate certain arms control measures. Such a special conference, convened in 1979, discussed conventional weapons that are excessively injurious or have indiscriminate effects; this led to the opening for signature, in 1981, of the 'Inhumane Weapons' Convention. Another conference, directly organized by the United Nations and dealing with arms control-related issues, took place in 1987 to consider the relationship between disarmament and development. During the same year a UN conference discussed ways of promoting international cooperation in the peaceful uses of nuclear energy. A UN conference organized in 2001 adopted a Programme of Action to combat the illicit trade in small arms and light weapons.

However, an overwhelming majority of General Assembly recommendations concerning arms control have had little effect on national policies or on the course of arms control negotiations. The proliferation of resolutions, dealing year after year with the same issues, has considerably reduced their value; in some cases, two or more resolutions adopted on the same issue contained divergent recommendations. General Assembly resolutions do not, therefore, adequately play the role originally assigned to them, that of serving as a sounding board for governmental proposals. Moreover, the number of states voting against or abstaining on crucial questions is sometimes considerable, whereas the affirmative votes often do not include all the militarily significant states, that is, states whose consent is indispensable for reaching an arms control agreement. As a result, the other important role of the General Assembly, that of providing guidance for arms control talks, has been weakened. The situation could improve if steps were taken to streamline the arms control agenda, which lacks a logical structure, and to enable the Assembly to concentrate on priority issues requiring multilateral consideration.

Special Sessions. Special UN General Assembly sessions may be convened to deal exclusively with disarmament matters. The first such session, held in 1978, elaborated principles of disarmament and agreed on a programme of action that formed a broad frame of reference for the arms control negotiators. It improved the machinery for discussing and negotiating disarmament by making it more representative. It postulated *inter alia* that member-states should be informed of all disarmament efforts, including those made outside the auspices of the United Nations. This point was particularly significant because the most vital arms control negotiations had been conducted, and were likely to continue to be conducted, among the great powers, without UN involvement.

The first UN General Assembly special session on disarmament enhanced the role of non-nuclear-weapon states in world affairs. It also helped non-governmental organizations to mobilize public opinion for the cause of disarmament. For the first time, representatives of these organizations as well as of research institutions were allowed to address the UN General Assembly on issues of universal importance. The value of non-governmental scientific research in the field of armaments and disarmament was acknowledged, and the need for educational programmes for disarmament was recognized.

By contrast, the second session, which took place in 1982, failed to meet the expectations of its initiators. It was unable to adopt a comprehensive programme of disarmament or to agree on other substantive items on its agenda. Instead of furthering the processes initiated by the first session, it reopened the discussion on points that had been agreed upon four years earlier. Considerable time and effort were needed simply to reconfirm the validity of the Final Document of the first session. Nevertheless, the second session became the focus of public attention as well as a rallying point for worldwide demonstrations in favour of peace. This may have helped in reaching consensus on a World Disarmament Campaign, one of the very few tangible results of the session. (In 1992, the World Disarmament Campaign was renamed the UN Disarmament Information Programme.)

The third session, held in 1988, proved a complete disappointment, even though it brought together more heads of state or government than any previous disarmament meeting. Among the principal disagreements that blocked consensus on a final document were those related to regional disputes. Even the modest goal of activating multilateral arms control efforts, which were then increasingly substituted by bilateral US–Soviet transactions, was not achieved.

Special disarmament sessions of the General Assembly should not be regarded as substitutes for, or as complements to, regular sessions. They must arise from special circumstances, for example, when it is generally felt that a representative gathering of high-level officials could remove some fundamental obstacles to a multilateral agreement. They would then have to deal with specific issues rather than with generalities. Special sessions could also be held to seek urgent approval of treaties, worked out in negotiating bodies, to accelerate their entry into force. Once convened, they might serve as a clearinghouse for new ideas and approaches and help to set up some improved deliberative and negotiating mechanisms.

The Security Council

As mentioned above, the UN Security Council has a statutory duty to formulate plans for the establishment of a system for the regulation of armaments. It is to be assisted in this work by the Military Staff Committee, consisting of the chiefs of staff of the Security Council permanent members or their representatives. In the early post-war period the Council was actively engaged in arms control negotiations, but since the 1950s its role in this field has been considerably reduced.

Nevertheless, in 1968, the Security Council adopted a resolution providing for immediate assistance to any non-nuclear-weapon state party to the 1968 Non-Proliferation Treaty (NPT) that is a victim of an act, or of a threat, of nuclear aggression. In 1995 the Council formally took note of the assurances given by four nuclear-weapon states not to use nuclear weapons against non-nuclear-weapon states parties to the NPT, except under certain circumstances. Moreover, in several arms control agreements the Security Council has been given a role in dealing with complaints about breaches of obligations. Parties to these agreements have agreed to cooperate in any investigation that the Council may initiate on the basis of an official complaint, and the Council must inform the parties of the results of the investigation. Each party is obliged to provide or support assistance, in accordance with the UN Charter, to any other party which so requests, if the Security Council decides that the latter has been harmed or is likely to be harmed as a result of a violation of the agreement. Those treaties that allow withdrawal in the case of extraordinary events jeopardizing the supreme interests of a party oblige that party to notify the Security Council in advance of the decision to withdraw.

In 1991, following the cessation of hostilities in the Gulf, the Security Council took a series of arms control and disarmament measures in the context of its responsibility for the maintenance of international peace and security. Thus, by Resolution 687 of 3 April 1991 (the so-called ceasefire resolution), the Security Council decided that Iraq should destroy, remove or render harmless: (a) all chemical and biological weapons and all stocks of agents, all related sub-systems and components, and all research, development, support and manufacturing facilities; and (b) all ballistic missiles with a range greater than 150 kilometres and related major parts, as well as repair and production facilities. The task of overseeing the implementation of this decision was entrusted to the UN Special Commission on Iraq (UNSCOM). The Commission – a subsidiary organ of the Security Council – was to be accorded unconditional and unrestricted access to all areas, facilities, equipment, records and means of transportation which it wished to inspect. Subsequently, in Resolution 715 of 11 October 1991, the Security Council approved a plan for monitoring Iraqi compliance with the obligations under the ceasefire regime not to use, develop, construct or acquire the prohibited weapons.

Moreover, Iraq was to undertake unconditionally not to acquire or develop nuclear weapons or nuclear-weapon-usable material or any sub-systems or components or any related research, development, support or manufacturing facilities. All relevant items were to be destroyed, removed or rendered harmless under international supervision. Verification of compliance with these obligations was to be carried out by the International Atomic Energy Agency (IAEA) with the assistance and cooperation of UNSCOM. Export of arms and related *matériel* to Iraq was prohibited until the Security Council decided otherwise. In 1999, after UNSCOM had encountered insurmountable obstacles in fulfilling its tasks, it was replaced – again by a decision of the Security Council – by the UN Monitoring, Verification and Inspection Commission (UNMOVIC) reporting to the Council through the Secretary-General. However, this Commission, too, was denied by the Iraqi authorities the possibility to properly perform its duties.

In their Declaration of January 1992, the members of the Security Council committed themselves to work to prevent the spread of technology related to the research for or production of weapons of mass destruction.

The Disarmament Commission

The UN Disarmament Commission (DC) was established in 1952 as a successor to the Atomic Energy Commission and the Commission for Conventional Armaments. It laid dormant from 1965, until the 1978 First Special Session of the UN General Assembly devoted to disarmament decided to reactivate it. The task of this subsidiary, deliberative, inter-sessional organ of the General Assembly, composed of all UN members, is to consider and make recommendations on various problems in the field of disarmament and to follow up the decisions of the special sessions – a wide and far-reaching but very imprecise mandate. In fact, the DC, meeting annually for a session of a few weeks, has largely replicated the debate held in the General Assembly and elsewhere.

Since 1990, the functioning of the DC has been somewhat improved, owing to its decisions to limit the working agenda to a maximum of four substantive items for in-depth consideration, not to maintain any subject on the agenda for more than three consecutive years, and not to establish more than four subsidiary bodies for the consideration of substantive issues. However, the DC has produced few agreed recommendations. It has not done much that could not be entrusted to the First Committee of the General Assembly which, since 1978, has dealt exclusively with disarmament matters and related international security questions.

During the past decades several other UN bodies have been established to deal with arms control issues. Some ceased to function upon completion of their tasks; others adjourned *sine die* or were simply dissolved.

Studies

The United Nations has made a number of studies dealing with technical, economic and political aspects of arms control. These studies have been initiated by the General Assembly and, since 1978, also by the Secretary-General's Advisory Board on Disarmament Studies (in 1989 renamed the Advisory Board on Disarmament Matters). Their purpose, as defined by the Board, is to assist ongoing negotiations; to assist in the identification of specific topics with a view to initiating new negotiations; to provide a general background to current deliberations and negotiations; and to assess and promote public awareness of the threat posed by nuclear weapons and the arms race.

Studies carried out by qualified experts often contain a thorough analysis of the problems as well as relevant suggestions. Some studies have succeeded in promoting specific measures and in defining the parameters of proposed negotiations. Others have provided useful information normally not available to many nations. However, several studies, especially those prepared by groups with the same composition as the formal UN bodies, contained merely a collection of well-known official government views. In a few cases, the groups so composed failed to produce a report because of their inability to overcome political and ideological differences.

The quality of the UN-initiated studies could improve if expert groups appointed by the Secretary-General included a higher proportion of independent scholars and if they were given more time to prepare their reports. Since the early 1980s certain studies have been entrusted to the UN Institute for Disarmament Research (UNIDIR), which has an autonomous status and cooperates with relevant national and international research institutions.

Within the UN system several specialized agencies and organizations carry out arms control-related activities, including studies. The most important of them are: the IAEA, the United Nations Educational, Scientific and Cultural Organization (UNESCO), the World Health Organization (WHO), the World Meteorological Organization (WMO), the International Labour Organization (ILO) and the United Nations Environment Programme (UNEP).

3.3   UN Involvement in Arms Control Negotiations

Nuclear Disarmament

The danger posed by nuclear weapons has been at the centre of United Nations attention from the very start.

The Baruch Plan. At the inaugural meeting of the Atomic Energy Commission, in 1946, Bernard Baruch, the US delegate, put forward a proposal which came to be known as the Baruch Plan. According to this plan, an International Atomic Development Authority would be entrusted with managerial control or ownership of all atomic energy activities potentially dangerous to world security; with the power to control, inspect and license all other atomic activities; and with the duty to foster the beneficial uses of atomic energy. In particular, the Agency was to conduct continuous surveys of supplies of uranium and thorium and bring these materials under its control. It was to possess the exclusive right both to conduct research in the field of atomic explosives and to produce and own fissionable material. All nations were to grant the freedom of inspection deemed necessary by the Agency.

The Baruch Plan was based on the 1946 Acheson–Lilienthal Report (named after the US Secretary of State and the future first chairman of the US Atomic Energy Commission) but differed from it on the important point of sanctions. The Acheson–Lilienthal Report did not provide for measures to be taken against violators; the goal of the envisaged organization was only to sound a warning signal in the event of danger. In the Baruch Plan, however, the United States stressed the importance of immediate punishment for infringements of the rights of the Agency and maintained that there must be no veto to protect those who violated the prohibition on the development or use of atomic energy for destructive purposes.

The United States later explained that it had in mind the ownership and exclusive operation by the international authority of all facilities for the production of uranium-235 and plutonium. Once a system of control and sanctions was operating effectively, production of atomic weapons would cease, existing stocks would be destroyed, and all technological information would be communicated to the authority. In other words, control would have to come first; atomic disarmament would follow.

The Gromyko Plan. The Soviet Union rejected the US plan on the premises that it would interfere with the national sovereignty and internal affairs of states and that the provision denying a permanent member of the Security Council the right of veto was contrary to the UN Charter. At the second meeting of the Atomic Energy Commission, in 1946, it submitted a draft convention, called the Gromyko Plan (after the Soviet delegate, later Foreign Minister), which reversed the priorities put forward by the United States. The production and use of atomic weapons were to be prohibited and all atomic weapons were to be destroyed within three months, whereupon an international system would be established to supervise the implementation of these commitments. Violations would be considered a serious crime against humanity, and severe penalties would be provided by domestic legislation. The convention would be of indefinite duration and would enter into force after approval by the UN Security Council and ratification by its permanent members.

According to the Gromyko Plan, the composition, rights and obligations of the International Commission for the Control of Atomic Energy, to be established within the framework of the UN Security Council, would be determined by a special international convention. The Commission would periodically inspect facilities for the mining of atomic raw material and for the production of atomic materials and atomic energy. It would carry out special investigations of suspected violations and would have the right to submit recommendations to the Security Council on measures to be taken against violators of the convention on the prohibition of atomic weapons and of the convention on the control of atomic energy.

US–Soviet Differences. The basic differences between the two positions concerned, first, the stage at which atomic weapons were to be prohibited – that is, whether a convention outlawing these weapons and providing for their destruction should precede or follow the establishment of a control system; and, second, the role of the UN Security Council in dealing with possible violations – that is, whether the rule of veto would be applicable. Breaking the deadlock in the negotiations proved impossible, mainly because the Soviet Union was at that time considerably less advanced in the atomic field than the United States and did not want to accept a plan which would lead to a US monopoly of atomic weapons for at least several years, until the envisaged destruction of these weapons could take place. Indeed, the United States would have retained the atomic bomb until the end of the final stage of the Baruch Plan, whereas the Soviet Union would have been barred from even trying to build the bomb. Similarly, given the international climate of mistrust and suspicion, the Soviet proposal for abolishing atomic weapons before an effective international control to ensure compliance with the ban had been established was unacceptable to the United States and, for that matter, to other Western countries as well. Moreover, US insistence that 'immediate punishment' be inflicted for infringements, in circumvention of the UN Security Council, implied readiness to launch an attack against another great power and, thereby, start another world war. It was, therefore, considered unrealistic, even by some high US officials.

In 1948, at the insistence of the United States, the UN General Assembly approved the Baruch Plan by an overwhelming majority. Despite the adoption of what was subsequently called the 'United Nations Plan' for the control of atomic energy, hopes for taking effective measures in this field and for averting a nuclear arms race were dissipated.

Atoms for Peace. Talks on disarmament, in particular the US–Soviet dialogue on nuclear arms control, resumed a few years later when, in 1953, US President Eisenhower, speaking at the UN General Assembly, proposed the so-called 'Atoms for Peace' plan. The idea was to promote disarmament by an indirect approach, that of building up the peaceful uses of atomic energy. The atomic powers were to contribute fissionable material for such uses to an agency which would be set up under the aegis of the United Nations and which would help countries to obtain the benefits of atomic energy. The proposal, which was so formulated as to render it attractive to most countries and make it difficult for the Soviet Union to object, led to the establishment, in 1956, of the IAEA. This Agency went into formal operation in 1957, with the following main functions: to assist research, development and practical application of atomic energy for peaceful purposes; to make provision for relevant materials, services, equipment and facilities, with due consideration for the needs of the underdeveloped areas of the world; to foster the exchange of scientific and technical information and to encourage the exchange and training of experts in the field of peaceful uses of atomic energy; to administer safeguards designed to ensure that relevant materials, equipment and information were not used in such a way as to further any military purpose; and to establish standards of safety for the protection of health and the minimization of danger to life and property.

Since 1970, the IAEA has had a key role in safeguarding compliance with the 1968 NPT and the treaties which have established nuclear-weapon-free zones in various parts of the world, as well as with the Security Council resolutions concerning Iraq (see above).

Limiting Armed Forces and Armaments

Parallel to the consideration of atomic weapons, efforts were made in a separate UN commission to reach agreement on limiting conventional weapons.

Soviet and Western Approaches. In 1948, the Soviet Union proposed, as a first step, that the permanent members of the UN Security Council (China, France, the United Kingdom, the United States and the Soviet Union) should immediately reduce by one-third all land, naval and air forces; that atomic weapons be prohibited; and that an international control body be established, within the framework of the Security Council, to supervise and control the implementation of these measures. At that time, the Soviet Union insisted that atomic weapons and conventional weapons be dealt with together in any plan for disarmament, while the United States and its allies argued that a start should be made on conventional disarmament.

In 1949 the Western powers presented a plan for a census and verification of information on armed forces and conventional armaments, and envisaged a central control authority to be placed directly under the Security Council. The Soviet Union opposed this plan, because it considered it to be an unacceptable preliminary condition for the reduction of armaments and armed forces and because the plan had no provision for collecting information on atomic weapons. The Western plan was approved by the UN General Assembly but was never implemented.

In the Disarmament Commission, set up in 1952, the argument continued as to whether disarmament should begin with atomic or conventional weapons, and as to whether the disclosure of information on armed forces and armaments as well as verification of the accuracy of this information should be carried out before or after the adoption of a programme of disarmament. Neither side was prepared to compromise on priorities; each side accused the other of wishing to retain the weapons in which it was stronger. In any event, the political climate of the early 1950s was hardly propitious for arms control talks, as the main protagonists deeply distrusted each other. The war in Korea threatened to spread into a worldwide conflagration, and recourse to atomic weapons was being considered. An additional irritant was the charge put forward in the United Nations by the Soviet Union that during the Korean War the United States had used bacteriological and chemical weapons.

Only in 1953, with the end of the Korean War and the changes in the government of the Soviet Union following the death of Stalin, did the international atmosphere improve sufficiently to allow reconsideration of the problem of disarmament. Moreover, the new relationship of forces between the two great powers seemed to favour arms control talks. The United States, which before the Korean War had been greatly inferior to the Soviet Union in conventional arms, rearmed considerably in the early 1950s, while the Soviet Union achieved an important atomic capability. A five-power (Canada, France, the Soviet Union, the United Kingdom and the United States) subcommittee of the UN Disarmament Commission was established to seek, in private, agreement on a 'comprehensive and coordinated' disarmament programme with adequate safeguards. There was an explicit understanding that efforts to reach such an agreement were to be made concurrently with progress in the settlement of international disputes.

Western and Soviet Disarmament Programmes in 1954–55. In 1954, France and the United Kingdom jointly put forward a programme based on the following principles: (a) measures of reduction, of prohibition and of disclosure and verification, regarding military manpower, military expenditure, conventional armaments and nuclear weapons had to be linked together in order to increase the security of all parties at all stages; (b) the transition from one stage of the programme to another should be automatic, subject to the competence of the control organ to verify the next stage; and (c) measures prohibiting weapons of mass destruction should be subdivided among use, manufacture and possession, and should take effect at different stages. At the outset, the nuclear powers would regard themselves as prohibited from using nuclear weapons except 'in defence against aggression'. (After the invention of the thermonuclear fusion weapon – the 'hydrogen bomb' – the term 'nuclear weapons' came to be used to include both this and the atomic fission weapon.)

A few months later, the Soviet Union submitted a draft international convention based on the French–British proposal but with certain amendments. In particular, the Soviet plan set specific time limits for reductions and required a total and unconditional ban on the use of nuclear weapons. The main concession to the West consisted in accepting that half of the agreed reductions in armed forces and conventional armaments might take place before any action to prohibit nuclear weapons.

In 1955, Canada and the United States joined France and the United Kingdom in submitting a memorandum which repeated in general terms the French–British programme of 1954. France and the United Kingdom further suggested that the ceilings for the armed forces of China, the Soviet Union and the United States should be between 1 million and 1.5 million men each and that those of France and the United Kingdom should be 650,000 men each. For other countries, the permitted levels were to be considerably lower. France and the United Kingdom also proposed that a total prohibition on the use of nuclear weapons should be effected when 75% of the reduction of conventional armaments and armed forces had been completed (not at the end of the disarmament programme, as proposed by them earlier). An effective system of control was to operate throughout the entire disarmament programme.

At first the Soviet Union opposed the Western plan. Then, on 10 May 1955, it put forward its own plan in which it accepted the specific ceilings for armed forces, as proposed by France and the United Kingdom, as well as the suggested postponement of the total prohibition on the use of nuclear weapons.

The Soviet plan was to be completed in two stages of one year each. In the first stage, the five great powers would reduce their armed forces and armaments by 50% of the difference between the levels at the end of 1954 and the ceilings of 1–1.5 million men for China, the Soviet Union and the United States and 650,000 men for France and the United Kingdom. A world conference would establish ceilings for other countries. In carrying out the agreed reductions of armed forces, states possessing nuclear weapons would undertake to discontinue tests of these weapons. They would also commit themselves not to use nuclear weapons except for purposes of defence against aggression, once a decision to that effect had been taken by the Security Council. Finally, some of the military bases on the territories of other states would have to be eliminated. During the second stage, the second half of the reductions would be carried out. When 75% of the total reduction had been completed, a total prohibition on the use of nuclear weapons would come into force. These weapons would be destroyed simultaneously with the final 25% of the reduction of armed forces.

A separate section of the Soviet plan, dealing with international control, stated that there was no way of assuring that all stocks of nuclear weapons had been eliminated and that there were therefore possibilities whereby some nuclear weapons could be hidden. Hence the Soviet Union proposed setting up an early-warning system to monitor large troop movements, arguing that a surprise nuclear attack was likely to be preceded by a considerable build-up and movement of conventional forces. A control agency would install in the territories of all states concerned, on a basis of reciprocity, control posts at major ports, at railway junctions, on main highways and at airfields, so that the observers could alert the world to possible dangers. The control agency would have the right to request from states information on the implementation of measures of reduction of armaments and armed forces as well as the right of unhindered access to documents pertaining to budgetary appropriations for military purposes. It would also have the power to exercise control, including inspection, on a permanent basis and on a scale necessary to ensure implementation of the disarmament programme.

This Soviet proposal was the most comprehensive and detailed programme of general disarmament thus far submitted to the United Nations. The timing for its presentation seemed opportune, as the world situation began to look hopeful. The year 1955 saw the conclusion of the State Treaty re-establishing an independent Austria and prohibiting the possession, construction or testing by Austria of weapons of mass destruction and of certain other types of weapon, as well as the entry into force of the formal undertaking by the Federal Republic of Germany, under the 1954 Paris Agreements, not to manufacture on its territory atomic, chemical or biological weapons. (See Chapter 2.) In the same year, the first international conference on the peaceful uses of atomic energy took place, and a meeting of the heads of government of France, the Soviet Union, the United Kingdom and the United States was held in Geneva, creating a relaxed international atmosphere known as the 'Geneva spirit'.

The 1955 Geneva Summit. The 1955 Geneva Summit Conference discussed the Soviet programme for the reduction of armaments and the prohibition of nuclear weapons, a British memorandum on joint inspection of forces confronting each other in Europe, a French proposal for reductions in military budgets and using the savings to assist underdeveloped countries, and the US plan for 'open skies' to guard against a large-scale surprise attack.

Under the US plan, the United States and the Soviet Union were to exchange military 'blueprints', that is, information about the strength, command structure and disposition of personnel, units and equipment of all major land, sea and air forces, as well as a complete list of military plants, facilities and installations. Verification of information was to be conducted by ground observation and by mutual, unrestricted aerial reconnaissance. The Soviet Union saw this as 'control without disarmament', which would increase international mistrust and tension. The United States emphasized that an effective method of inspection and control was the first requirement for an agreement.

Shortly thereafter, the United States placed a reservation on all of its 'pre-Geneva substantive positions' pending the outcome of a study of inspection methods. This in fact amounted not only to the withdrawal of the Western disarmament proposal, after a very large and essential portion of it had been accepted by the Soviet Union, but also to the formal abandonment of the Baruch Plan, which had been approved by a majority of UN members. Thus, efforts to achieve agreement on a programme of arms reduction and disarmament involving all armaments in a coordinated manner were brought to a standstill.

Later, attention shifted to partial arms control approaches, such as: halting nuclear-weapon tests; restricting the production of fissionable materials exclusively to non-weapon purposes; establishing a European zone of arms limitation; reducing force levels; reducing military budgets; prohibiting the use of nuclear weapons; ensuring that the launching of objects through outer space was exclusively for

peaceful purposes; safeguarding against the possibility of surprise attack; and eliminating foreign military bases. There were sharp disagreements in each of these
fields. The Sub-Committee of the UN Disarmament Commission, which had been
negotiating measures of arms control for over three years, ended its work on a note
of acrimony in 1957.

General and Complete Disarmament

On 17 September 1959 the United Kingdom submitted to the UN General Assembly
a plan for 'comprehensive' disarmament, based on the principle of balanced stages
towards the abolition of all nuclear weapons and the reduction of all other weapons
to levels which would rule out the possibility of aggressive war. The next day, the
Soviet Union proposed a disarmament programme aimed at eliminating all armed
forces and armaments within four years. (It is worth noting that already in 1928, in
the Preparatory Commission for the World Disarmament Conference, the Soviet
Union proposed, in a draft convention for 'immediate, complete and general disarmament', that all armed forces should be disbanded, existing armaments destroyed,
military training stopped, war ministries and general staffs abolished, military
expenditure discontinued and military propaganda prohibited.) A revised, detailed
version of the 1959 Soviet programme, in the form of a draft treaty on general and
complete disarmament under strict international control, became a basis for discussion in the Committee on Disarmament in Geneva along with the US proposed out-
line of basic provisions of a treaty on general and complete disarmament in a peaceful world. The term 'peaceful world', appearing in the US text, was an important
qualification. It conveyed the US conviction that disarmament might be possible
only in conditions of assured universal peace, in contrast to the Soviet thesis that
disarmament per se would create a peaceful world.

The negotiating parties had before them a set of principles, as agreed between the
Soviet Union and the United States in a joint statement of 1961 (the so-called
McCloy–Zorin Statement), which were to guide them in finding solutions to the
complex problem of general and complete disarmament. The main agreed principles
were those regarding a balanced, staged and verified elimination of all armed forces
and armaments. However, the parties could not agree on how to apply these principles. The plans were amended by each side in the course of the following years,
but the differences remained unresolved. The main divergences are summarized
below.

The Principle of Balance. The Soviet Union placed the main emphasis on the
completion of the disarmament process within a short, fixed period of time: the
more quickly nuclear delivery vehicles were eliminated, the sooner balance would
be achieved. The United States proposed to keep the relative military positions and
the pattern of armaments within each military establishment similar to those at the
beginning of the disarmament process. To this end, disarmament, starting with a
freeze, was to be gradual; as confidence developed, the military establishment
would, by progressive reductions, shrink to zero.

Duration and Stages. Both sides envisaged three stages of the disarmament process and made the transition from one stage to the next dependent on the completion
of previous disarmament measures. The Soviet Union proposed a four-year pro-

gramme, with 15 months for each of the first two stages, but was later prepared to extend the period for implementing the whole programme to five years and the first stage to two years. The US plan provided for two stages of three years each, to be followed by a third stage, the duration of which would be fixed at the time the treaty on general and complete disarmament was signed.

Reduction of Armed Forces and Conventional Armaments. The US plan provided for a reduction of the armed forces of both the Soviet Union and the United States to 2.1 million and 1.05 million men in the first and second stages, respectively, with a 30% reduction of all major armaments by categories and types of weapon in the first stage and a 35% reduction in each of the second and third stages. Subsequently, the United States amended its proposal to prohibit the production of certain major armaments in the first stage except for replacement purposes, in order to ensure that the 30% reduction would in fact reduce both the quantity and quality of all armaments covered by the reduction. A reduction of agreed military bases, without distinction between foreign and domestic bases, would take place in the second stage. The Soviet Union originally provided for the reduction of Soviet and US armed forces to the level of 1.7 million and 1 million men in the first and second stages, respectively, but later proposed a compromise first-stage level of 1.9 million men. The revised draft envisaged reductions of 30%, 35% and 35% of conventional armaments in the respective successive stages, and a reduction in the production of conventional armaments, parallel to the reductions of armed forces, through the elimination of factories engaged in such production. Total elimination of all foreign military bases would take place in the first stage, starting with the liquidation of such bases in Europe.

Nuclear Disarmament. Both plans contained first-stage obligations for the nuclear powers not to transfer control of nuclear weapons or information on their production to non-nuclear-weapon states. In all other respects they differed. The original Soviet draft provided for the complete elimination of nuclear-weapon delivery vehicles and the cessation of the production of such vehicles in the first stage, whereas total elimination of nuclear weapons as well as of fissionable material for weapon purposes and the discontinuance of their production would take place during the second stage. The plan was subsequently amended to permit the Soviet Union and the United States to retain on their own territories a so-called nuclear umbrella, that is, a limited number of intercontinental missiles, anti-missile missiles and anti-aircraft missiles of the ground-to-air variety until the end of the third stage. The US plan envisaged, in the first stage, the ending of the production of fissionable material for weapon purposes and the transfer, for peaceful uses, of certain agreed quantities of such material already produced and stockpiled. The number of nuclear-weapon delivery vehicles would be reduced by 30% in the second stage, while stocks of nuclear weapons would be reduced by an agreed percentage and the production of nuclear weapons would be subject to agreed limitations. Total elimination of such weapons would take place in the third stage.

Verification. Both sides agreed on the need to verify what was being reduced, destroyed or converted to peaceful uses, as well as to control the cessation of production of armaments. In addition, the United States stressed the need to verify the remaining quantities of armaments and forces and to ensure that undisclosed, clandestine forces, weapons or production facilities did not exist. The Soviet Union was opposed to the inspection of remaining stocks but was willing to consider budgetary controls.

Peacekeeping. The United States proposed that in the first stage a UN peace observation corps should be established. At the start of the second stage, a UN peace force would be created; during the remainder of that stage the jurisdiction of the International Court of Justice would become compulsory for legal disputes, and measures would be adopted against indirect aggression and subversion. The question of whether the peace force, to be fully developed in the third stage, should be equipped with nuclear weapons was to be left open for future decision. The Soviet draft provided that in the course of and following the disarmament process, force contingents with non-nuclear weapons would be made available to the Security Council under Article 43 of the UN Charter. The Soviet Union opposed the creation of supra-national institutions and objected to any possibility of providing the UN peace force with nuclear weapons.

Failure of the Concept. The talks on general and complete disarmament failed. They were doomed to fail, among other reasons, because no one could provide a satisfactory answer to such a fundamental question as what would be the political order governing international relations in a completely disarmed world. The same applies to mechanisms and procedures for settling disputes among states and maintaining peace. The more immediate obstacle was that the negotiators were unable to agree on how much disarmament should be undertaken in the first stage of a disarmament process. The Soviet Union claimed that only a very substantial reduction in military power during the first stage could eliminate the danger of nuclear war, whereas the Western powers maintained that they could not accept radical first-stage measures or give up their nuclear deterrent until confidence was established between East and West and until an international peace force was formed to replace national forces.

Realization of the insuperable difficulties in agreeing on a programme for general and complete disarmament had the effect of turning attention once again to specific partial measures of disarmament. In fact, a few first-stage measures proposed in the Soviet and US plans, such as a ban on nuclear-weapon testing and prevention of nuclear-weapon proliferation, now became the subjects of separate negotiations. These negotiations were held either directly among the nuclear-weapon powers or in the multilateral Committee on Disarmament. General and complete disarmament has remained for the United Nations an ultimate goal worth striving for, rather than a practical policy objective.

3.4   UN Involvement in 'Micro-Disarmament'

In the present-day world many armed conflicts are different from those which the United Nations was created to deal with. The drafters of the UN Charter had in mind, in the first place, wars between states, whereas current wars are often of an intra-state nature. The weapons used in the latter are described as 'small' or 'light', but they are nonetheless responsible for millions of dead and wounded, both military and civilians. In several cases, UN peacekeeping forces, active in the areas of conflict, have been given the task of assembling the arms voluntarily surrendered by the warring factions and disposing of them. However, most of these so-called micro-disarmament operations have proved ineffective. The main reasons can be summarized as follows.

The means of self-defence are given up by the citizenry only if the authorities are able to provide a secure environment. This is not the case in the 'failed states', whose governmental law and order functions have collapsed, and where the most devastating civil strife takes place. Moreover, in the absence of a comprehensive and enforceable political settlement of the disputes that caused the armed conflict, there is a powerful incentive for the parties to retain a certain amount of weapons, for each side fears that the other could gain advantage in the post-disarmament period. Finally, the UN peacekeeping forces, which manage the weapons collection programmes, have neither the capacity nor the resources to verify compliance with the disarmament commitments of the warring factions. It has happened that weapons presumed to have been eliminated or safely stored have reappeared in the possession of one faction or another.

3.5 Assessment

By virtue of its universal character, the United Nations is the only forum in which universal consensus on key security issues can be worked out. It therefore bears primary responsibility in the field of arms control. This means that it must set goals for, and assist in the conduct of, both regional and global arms control negotiations, as well as stand ready to facilitate the implementation of the agreements reached. The UN Secretariat helps in fulfilling these tasks by servicing international conferences, working together with experts engaged in disarmament-related studies, following up UN General Assembly resolutions, administering a programme of fellowships on disarmament for government officials, maintaining liaison with non-governmental organizations, publishing the *Disarmament Yearbook* and disseminating relevant information.

In accordance with its responsibility for the progressive development of international law, the United Nations can perform the important function of codifying the principles of the law of arms control, already accepted internationally, as well as of elaborating new principles. The latter could include extending the rule of customary law of armed conflict – that the right of belligerents to choose methods and means of warfare is not unlimited – by providing that the right of states to possess arms is not unlimited either. As a logical corollary to the adoption of such a principle, all 'excess' weapons, those which are not indispensable for the defence of the national territory or for collective self-defence, would have to be banned.

4

Nuclear-Weapon Explosions

The issue of nuclear-weapon test explosions had been on the agenda of multilateral, bilateral (US–Soviet) or trilateral (British–US–Soviet) arms control negotiations since 1954, when India proposed a so-called 'standstill agreement' on testing. The proposal was put forward after a major radiation accident which followed a US nuclear test in the Pacific. Before a comprehensive nuclear test ban was signed in 1996, three limited agreements circumscribed the environment in which testing was allowed and the size of permitted explosions.

4.1   The 1963 Partial Test Ban Treaty

The Treaty Banning Nuclear Weapon Tests in the Atmosphere, in Outer Space and Under Water, usually referred to as the Partial (or Limited) Test Ban Treaty (PTBT or LTBT), was signed on 5 August 1963. It resulted from talks conducted since the late 1950s, chiefly between the Soviet Union on the one side and the United Kingdom and the United States on the other. The resolutions adopted by the UN General Assembly and the discussions held at the Conference of the Eighteen-Nation Committee on Disarmament (ENDC) had stimulated these tripartite exchanges and had given them a semblance of being international multilateral negotiations.

   As confirmed by subsequent events, the conclusion of the PTBT was prompted less by an urge to turn the tide of arms competition than by the need to improve US–Soviet relations, which had been severely strained by the 1962 Cuban Missile Crisis, and to bring about a general relaxation of international tensions. An additional incentive may have been the desire shared by the United States and the Soviet Union to make it more difficult for China and France to build their own nuclear arsenals. The nuclear testing issue was deemed to be well-suited to all these purposes: world opinion was aroused by the risks of radioactive contamination, and public pressure for a test ban was increasing as more evidence on the biological effects of nuclear fallout became available. The fact that both major powers had by then already carried out extensive series of tests in the atmosphere and made certain that testing could be continued underground reduced the cost of their mutual 'sacrifice'.

   The PTBT proved to be a popular move. It was well received by most governments and entered into force very quickly – in October 1963.

Scope of the Obligations

The PTBT bears the mark of a transitional arrangement. In the preamble, the 'original parties' – the United Kingdom, the United States and the Soviet Union – pledged themselves to seek to 'achieve the discontinuance of all test explosions of nuclear weapons for all time', and in one of the five articles they stated their determination to conclude a treaty resulting in the 'permanent banning of all nuclear test explosions, including all explosions underground'.

Environments Covered. The prohibition under the PTBT covers nuclear-weapon test explosions, as well as 'any other' nuclear explosion in three environments – the atmosphere, outer space and under water – at any place under the jurisdiction or control of the parties, without qualification as to the yield. Whereas the ban on nuclear-weapon test explosions appears clear, the ban on other nuclear explosions may appear equivocal. As evidenced by the negotiating history, the term 'other' was inserted in order to prevent explosions for peaceful purposes in the specified environments – whether tests or otherwise – in view of the difficulty of differentiating between military and civilian explosions. However, the Treaty is not interpreted as restricting the use of nuclear weapons in armed conflicts. The phrase 'under its jurisdiction or control' was understood as extending the prohibition to non-self-governing territories administered by states parties as well as territories under military occupation.

Since there exists no commonly accepted definition of 'atmosphere' and 'outer space' and no agreement on where one ends and the other begins, the two environments are considered, for the purpose of the Treaty, as one continuous environment. Hence the language used: 'in the atmosphere; beyond its limits, including outer space'. It may be added that the 1967 Outer Space Treaty contains an explicit ban on the testing of any type of weapon on celestial bodies, a ban that was reiterated and reinforced with regard to the moon in the 1979 Moon Agreement.

The underwater environment is also understood comprehensively. The enumeration in the PTBT of 'territorial waters or high seas' was not meant to be exhaustive but illustrative; all bodies of water are included in the ban, both inland waters, lakes and rivers, and the seas. High seas were singled out to remove the possibility of an argument being put forward that these parts of the seas were not under the 'jurisdiction or control' of any party and thus not covered by the prohibition. In any event, the parties undertook to refrain from conducting nuclear explosions 'anywhere' in the environments described.

Nuclear explosions conducted underground, whatever their purpose, are not covered by the Treaty. However, there is a prohibition on any such explosion causing radioactive debris to be present outside the territorial limits of the state under whose jurisdiction or control the explosion is conducted. This may mean that an underground explosion which broke the surface of the ground would still be considered as 'underground' as long as it did not produce radioactivity detectable outside the boundaries of the country that conducted it. It is not clear whether any amount of radioactive material travelling beyond the borders of a testing state constitutes a violation, or only what might be considered a dangerous amount. In the latter case, a threshold of radiation hazard would have to be defined using some objective criteria, but this has not been done. The matter was rendered even more complicated by the fact that in the Russian-language version of the PTBT the term used for 'debris' is '*osadki*', which means deposit or fallout, whereas not all radioactive debris is necessarily deposited on the ground as fallout. The relevant clause clearly favours large countries as there is a chance that radioactive material that might vent from an underground test to the surface would not travel beyond their borders. In practice, even that could not be prevented.

Assistance in Testing. The parties to the PTBT also undertook to refrain from 'causing, encouraging, or in any way participating in' the carrying out of nuclear explosions by other nations in the prohibited environments. Of the three terms employed, 'encouraging' is the least definite. If it were to include moral support or economic help indirectly used by the recipient to pay for the cost of nuclear explosions, it would be difficult to prove a breach.

Assistance in carrying out underground tests was not prohibited as long as the tests did not produce the radioactive effects described above. Thus, the United Kingdom could conduct its nuclear explosions jointly with the United States at the US Nevada Test Site without breaching its international obligations.

The Right of Withdrawal

The PTBT is of unlimited duration, but each party, 'in exercising its national sovereignty', has the right to withdraw from it if 'extraordinary events, related to the subject matter of this Treaty, have jeopardized the supreme interests of its country'. A party which considered withdrawing would decide for itself whether such events had occurred and would not need to justify its action to any external authority: a simple notice addressed to all other parties to the PTBT three months in advance would suffice. This clause was included (for the first time in an arms control agreement) over initial objections raised by the Soviet Union, which claimed that a provision for withdrawal was not necessary because it was its inherent right as a sovereign nation to abrogate any treaty at any time if its national interests so required.

A material breach of the PTBT would certainly be treated as an 'extraordinary event' in the meaning of the Treaty, but no international mechanism was established to verify whether the ban was being observed. There was a presumption that the parties would check compliance with the Treaty unilaterally, using their own means. It also appeared unlikely that any of the nuclear-weapon parties would break away from the Treaty to restore its freedom to test in all environments. Even China and France, the nuclear-weapon states which had not signed the PTBT, gave up atmospheric testing through unilateral statements of renunciation: France – in 1975, after a suit had been brought against it in the International Court of Justice by the Australian and New Zealand governments, which complained about the pollution of the South Pacific environment with radioactive fallout from French nuclear tests; and China – several years later, after a series of protests against its tests were made by both neighbouring and distant countries.

Assessment

The PTBT complicated the development of large thermonuclear weapons. It also made it impossible for the parties to conduct full-scale operational testing (including the measurement of certain effects) of nuclear weapons, already developed, in the environments in which these weapons were meant to be used. However, the agreed restrictions did not prevent the United States, the United Kingdom and the Soviet Union from satisfying most of their military requirements since they could still test underground and, at the same time, deny to others important intelligence information about the characteristics of the explosions (and thus of the weapons) that can be gathered from atmospheric tests.

The PTBT helped to curb the radioactive pollution of the atmosphere and reduce the health hazards associated with nuclear fallout, thereby making an important contribution to the environmental protection regime. In national policies it marked the first major success of the proponents of arms control, who managed to overcome the resistance of the proponents of an uncontrolled arms race. In international policies it ratified a major improvement in US–Soviet relations, became an obstacle to the wider spread of nuclear weapons and paved the way for the 1968 Non-Proliferation Treaty (NPT). Wide participation in the PTBT, the passage of nearly four decades without established material breaches or withdrawals from the Treaty, and the fact that even non-parties (China and France) stopped testing in the environments specified in the PTBT, may all lead to the conclusion that the ban on nuclear explosions in the atmosphere, outer space and under water has become customary law binding on all states.

4.2 The 1974 Threshold Test Ban Treaty

After the PTBT entered into force, appeals were made, mainly in the UN General Assembly, for further measures of restraint that would suspend nuclear-weapon testing, or limit or reduce the size and number of nuclear-weapon tests pending the conclusion of a comprehensive ban. These appeals, however, were ignored by the testing powers. The United States argued that a partial approach would not remove the obstacles to resolving the problem of adequate verification, while the Soviet Union insisted on dealing with the testing problem as a whole and contended that a quota or a threshold magnitude for tests would not put a stop to the build-up of nuclear arsenals. In the summer of 1974, both countries retreated from their positions. On 3 July they signed the Treaty on the Limitation of Underground Nuclear Weapon Tests, which came to be called the Threshold Test Ban Treaty (TTBT).

Scope of the Obligations

The TTBT established a limit on the amount of energy that may be released by underground nuclear explosions, that is, on their explosive yield. The two parties undertook to 'prohibit, to prevent and not to carry out' any underground nuclear-weapon test having a yield which exceeds 150 kilotons (the equivalent of 150,000 tons of trinitrotoluene, TNT, high explosive) at any place under their jurisdiction or control, beginning on 31 March 1976. The term 'test' applied to one underground nuclear explosion or to two or more underground explosions taking place within one-tenth of a second and separated from each other by no more than two kilometres. The yield attributed to a test made up of more than one explosion is the aggregate of the yields of the individual explosions within that test.

The official justification for setting a distant date for the entry into force of the yield limitation was that considerable time was needed to make all verification arrangements. A more important reason, however, was that some warheads already under development were planned to have a yield exceeding the agreed limit. Their testing, therefore, had to take place before the restrictions became effective. Test explosions with yields exceeding the threshold were in fact conducted by both the

United States and the Soviet Union in the period from July 1974, when the TTBT was signed, to the end of March 1976, when it was to enter into force.

In addition to the limit placed on the size of underground nuclear-weapon tests, each party to the TTBT pledged to restrict the number of its tests to a minimum.

Entry into Force and Duration

In signing the TTBT, the United States expressed confidence that it would be able to recognize violations by using its national means of verification and owing to the data-exchange provision of the Treaty. Later, however, the United States concluded that it could not rely on unchecked information supplied by the other side. It then proposed that the verification clauses contained in the Protocol to the TTBT be strengthened so as to ensure that the 150-kiloton threshold was being observed. Only then, the United States stated, would it be prepared to ratify the TTBT. For these reasons, ratification of the TTBT was postponed for 16 years, but the parties announced that they would observe the agreed limitation throughout the pre-ratification period.

Negotiations aimed at working out new procedures and methods of verification, additional to those included in the TTBT, started in 1987. In 1988, a joint US–Soviet verification experiment was conducted at the Soviet and US test sites. Subsequently, the foreign ministers of the two sides, meeting in 1989 at Jackson Hole, Wyoming, agreed that the parties could use techniques for on-site measurement of explosion yields, in-country seismic monitoring, as well as on-site inspection. This agreement led to the signing of a verification protocol, which replaced the original 1974 Protocol to the TTBT, and the entry into force of the Treaty in December 1990.

Notifications and other information relevant to the TTBT were to be transmitted through the Nuclear Risk Reduction Centers established by the 1987 Agreement between the United States and the Soviet Union. A Bilateral Consultative Commission (BCC) was set up to discuss questions relating to the implementation of, or compliance with, the TTBT or its Protocol, as well as possible amendments to either of these documents. A coordinating group of the BCC was to coordinate the activities of the verifying party with those of the testing party with regard to each test.

The TTBT was to remain in force for a period of five years, unless replaced earlier by an agreement on the cessation of all underground nuclear-weapon tests. If such an agreement was not achieved, the Treaty could be extended for successive five-year periods, unless either party notified the other of its termination no later than six months prior to the expiration of the Treaty. A possibility was, nevertheless, provided for withdrawing from the Treaty at any time on six months' notice, if 'extraordinary events' had jeopardized the supreme interests of either of the parties; such notice would have to include a statement of the relevant events.

Assessment

The TTBT further constrained the development of high-yield nuclear warheads by the United States and the Soviet Union. The United Kingdom also committed itself to abide by the provisions of the TTBT, even though it was not a signatory. Cessation of explosions in the megaton range also had a positive environmental effect by reducing the danger of geological disturbances and, more importantly, by minimizing the risks of radioactive venting. Furthermore, the TTBT requirement for an exchange of detailed information concerning sites and yields of nuclear explosions was a step towards greater international openness.

However, the TTBT did not contribute to the cessation of the nuclear arms race. The 150-kiloton yield threshold was too high to be meaningful; the parties did not experience onerous restraints in continuing their nuclear-weapon programmes. In any event, for many years the trend had been to improve the effectiveness of nuclear-weapon systems by increasing the accuracy of missiles rather than by increasing the yield of warheads. Nor did the agreed threshold reflect the verification capabilities: it was generally recognized, even at that time, that nuclear explosions of much lower size than 150 kilotons could be detected and identified.

One cannot avoid the impression that the idea of a threshold treaty was hastily conceived for purposes only loosely related to arms control considerations. The TTBT seems to have served chiefly the public relations needs of the parties by giving the appearance of progress in arms control, when it was politically expedient to do so, and to cover up the inability of the leaders of the two great powers to reach, at their meeting in June 1974, a more important agreement on strategic offensive arms limitations. The conclusion of the TTBT was certainly also motivated by a desire to pre-empt the charge expected to be voiced at the approaching first Review Conference of the parties to the Non-Proliferation Treaty that the nuclear-weapon powers were not fulfilling their disarmament pledges under that Treaty.

The TTBT was criticized at both the Conference on Disarmament and the United Nations as inadequate. Unlike other nuclear arms control agreements, it was not formally welcomed by the UN General Assembly.

4.3   The 1976 Peaceful Nuclear Explosions Treaty

The provisions of the TTBT did not extend to underground nuclear explosions for peaceful purposes. Since such explosions cannot be distinguished from explosions serving military ends, the possibility remained of circumventing the threshold limitation on weapon tests. To remove the loophole, the United States and the Soviet Union decided to work out a separate agreement to become effective simultaneously with the TTBT.

The Treaty on Underground Nuclear Explosions for Peaceful Purposes, which came to be called the Peaceful Nuclear Explosions Treaty (PNET), was signed on 28 May 1976. It regulated explosions carried out by the United States and the Soviet Union at locations outside their nuclear-weapon test sites (and therefore presumed to be for peaceful ends) as from 31 March 1976, the date valid also for the TTBT.

Scope of the Obligations

To ensure that the underground explosions declared to be for peaceful purposes do not provide weapon-related benefits not obtainable from limited weapon testing, the parties had no other choice than to establish the same yield threshold for peaceful applications as had been imposed on weapon tests under the TTBT, namely, 150 kilotons. A higher threshold would have allowed circumvention of the TTBT, while a lower one would have made it difficult or impossible to plan most of the applications then envisaged.

The yield restriction was to apply to individual explosions, as distinct from group explosions. The possibility of carrying out individual explosions with a yield greater than 150 kilotons was left open for future consideration 'at an appropriate time to be agreed'. However, a threshold for peaceful explosions could not be raised without affecting the threshold for weapon tests. Indeed, the US interpretation of the provision in question was that any change in the yield threshold for peaceful nuclear explosions would require an amendment of the PNET and that such amendment would have to be ratified.

Different PNET rules were to govern a 'group explosion' – defined as two or more individual explosions for which the time interval between successive individual explosions does not exceed five seconds and for which the emplacement points of all explosives can be interconnected by straight line segments, each of which joins two emplacement points and each of which does not exceed 40 kilometres. A group explosion was permitted to exceed the 150-kiloton limit and reach an aggregate yield as high as 1,500 kilotons (1.5 megatons) if carried out in such a way that individual explosions in the group could be identified and their individual yields determined to be no more than 150 kilotons. Certain envisaged peaceful applications of nuclear energy, such as large-scale excavation projects, might indeed require many nuclear blasts of varying size, but the PNET required that they be consistent with the PTBT, which prohibits any explosion causing radioactive debris to be present outside the territorial limits of the state conducting the explosion.

As in the case of the TTBT, all notifications and other relevant information were to be transmitted through the US–Soviet Nuclear Risk Reduction Centers. The Joint Consultative Commission (JCC) established by the PNET could be used by the parties to facilitate implementation of the verification provisions. In addition, for each explosion for which verification activities were to be carried out, a coordinating group was to be established under the auspices of the JCC.

Entry into Force and Duration

The Protocol to the PNET, signed on 1 June 1990 by the United States and the Soviet Union simultaneously with the Protocol to the TTBT, replaced the 1976 Protocol to the PNET. The new document expanded and strengthened the procedures and methods of verification originally agreed upon. The provisions of the two new protocols were in many respects identical.

The exchange of instruments of ratification of the PNET and the TTBT took place simultaneously, and the duration of the two treaties was to be the same. Their close interrelationship, or rather subordination of the PNET to the TTBT, was emphasized by the clause excluding the possibility to terminate the PNET while the TTBT remained in force, but allowing withdrawal from the former at any time upon the termination of the latter.

Assessment

Peaceful nuclear explosions with the same yields as those set in the TTBT could not produce militarily significant information which was not obtainable through weapon tests permitted by the TTBT. Consequently, the nuclear-weapon powers had no incentive to seek such information through explosions regulated by the PNET.

The PNET did not increase the very limited arms control value of the TTBT. It may even have had a negative impact on the policy of preventing nuclear-weapon proliferation by providing respectability to the arguments of those states that sought to develop a nuclear-weapon capability under the guise of an interest in peaceful explosions. The PNET envisaged US–Soviet cooperation, on the basis of reciprocity, in areas related to underground nuclear explosions for peaceful purposes, but this clause was a dead letter already at the time of signing.

In the mid-1970s, after 27 tests, the United States terminated its programme of nuclear explosions for civilian purposes, because it had found it impossible to establish applications which would be technically feasible, economically viable and publicly acceptable. This so-called Plowshare Program then disappeared from the US federal budget. The Soviet Union, however, pursued its programme of peaceful nuclear explosions. By the end of the 1980s it had conducted well over 100 explosions, outside its known weapon test sites, for cavity construction (mainly for storage of gas condensates), seismic sounding (to map the geological structure at great depth), oil and gas extraction, extinguishing burning oil wells, canal building, ore fragmentation, waste burial, coal mining, and some other purposes. The Soviet Union (and later Russia) has not conducted any explosions for non-military purposes since 1989.

4.4   Negotiations for a Comprehensive Test Ban

The nuclear test limitation treaties, analysed above, did not significantly reduce the freedom of the great powers to develop improved nuclear-weapon designs and did not, therefore, affect the weapon programmes of the parties. Nor did they render it considerably more difficult for non-nuclear-weapon states to develop a nuclear-weapon capability. Especially flawed were the bilateral US–Soviet TTBT and PNET. Efforts to negotiate an end to all nuclear-weapon tests, begun in the 1950s, continued amidst active public interest.

Negotiations in 1958–63

In 1958, an East–West conference of seismic experts produced a report on the feasibility of detecting nuclear explosions. The report called for a large network of specially constructed, land-based and sea-based international control posts manned by thousands of experts. This very elaborate and costly scheme would have, supposedly, been able to detect nuclear tests in the atmosphere and underwater down to small yields, but would not have been able to detect underground events below a seismic magnitude corresponding to a 20-kiloton explosion, or even a higher-yield explosion, if the seismic signals were deliberately muffled by the testing state. The proposed verification was not only technically unwieldy but also politically unacceptable to many.

Later in 1958 the United Kingdom, the United States and the Soviet Union engaged in tripartite negotiations at the Conference on the Discontinuance of Nuclear Weapon Tests. The conference centred its debate almost exclusively on verification of compliance, but the divergent ideas proved irreconcilable. Consequently, the moratorium on testing, then in force, could not be converted into a formal treaty.

When the conference adjourned in 1962, the newly established Eighteen-Nation Committee on Disarmament became the principal forum for test ban negotiations, where the dispute about verification continued. Apart from the controversy over the number of unmanned seismic stations (the so-called 'black boxes') to be located in each country, the main bone of contention was the number of annual mandatory on-site inspections, the United States insisting on seven and the Soviet Union accepting no more than two or three inspections. It was not clear whether all parties to what was intended to be a multilateral treaty would have the right to ask for an agreed number of inspections, and whether each party would be obliged to accept them. However, irrespective of verification, the great powers were far from reaching agreement on a comprehensive ban because of the conflicting strategic interests related to the development of new types of nuclear weapons. There is good reason to believe that, even if either of the superpowers had accepted the other's figure for on-site inspections, they would still not have stopped all testing. The modalities for carrying out such inspections – far more controversial than the numbers – were not even seriously considered. At that time, only partial solutions to the problem of nuclear testing appeared realistic. Eighteen months after the adjournment of the Conference on the Discontinuance of Nuclear Weapon Tests, the negotiators signed the PTBT.

Negotiations in 1977–80

In 1977, the United Kingdom, the United States and the Soviet Union again engaged in trilateral talks for a comprehensive nuclear test ban treaty. Despite serious headway made on several controversial issues, the negotiators were still far from reaching the declared goal. Since the duration of the projected multilateral comprehensive treaty was to be limited to three years, the adherence of non-nuclear-weapon states, particularly those parties to the 1968 Non-Proliferation Treaty, would have been impossible, as the latter had already renounced the possession and thereby also testing of nuclear explosive devices. The policies of the negotiating parties towards China and France – the two nuclear-weapon powers not participating in the talks – were not determined either. As regards verification, the negotiating parties failed to resolve the complex questions relating to the instrumentation of the so-called national seismic stations (NSS), which were to be automatic and tamper-proof, as well as the number of such stations to be installed in each of the three states. Also unresolved were problems regarding procedures for the emplacement of the NSS and their maintenance, as well as for the transmission of data.

In 1980, with the change of US Administration upon the election of President Reagan, the trilateral talks were adjourned *sine die*. The United States made public its view that nuclear testing was important for the security of the Western alliance and that, consequently, a comprehensive test ban could be only a 'long-term objective' – to be sought in the context of radical nuclear arms reductions, maintenance of a credible nuclear deterrent, expanded confidence-building measures and improved verification capabilities. The question of nuclear testing returned to the multilateral Conference on Disarmament, where several years were spent on arguing whether a special working committee of the Conference should be set up and, if so, what its mandate should be. Only the Ad Hoc Group of Scientific Experts, created in 1976 to work out international cooperative measures to detect and identify seismic events, held substantive discussions and submitted periodic reports. In 1986 and 1987, the UN General Assembly requested the nuclear-weapon powers to provide notification of their nuclear explosions and asked states not conducting nuclear explosions but possessing data on such events to make the data available to the United Nations.

The 1991 PTBT Amendment Conference

According to the provisions of the PTBT, any party may propose an amendment to the Treaty. Upon request from one-third or more of the parties, a conference must be convened by the depositary governments (the Soviet Union, the United Kingdom and the United States) to consider the amendment. In the late 1980s, in view of the continuous deadlock in the consideration of a comprehensive test ban, the UN General Assembly recommended in several resolutions that advantage be taken of the relevant provision of the PTBT in order to convert the partial ban into a total ban.

The PTBT Amendment Conference was held in January 1991. The amendment proposed by a group of non-aligned countries consisted of an additional article and two protocols. The new article would state that the protocols constituted an integral part of the Treaty. Under Protocol I, the parties would undertake – in addition to their obligations under the PTBT – to prohibit, to prevent and not to carry out any nuclear-weapon test explosion or any other nuclear explosion under ground or in any other environment. In addition, each party would undertake to refrain from causing, encouraging or in any way participating in carrying out any nuclear explosion anywhere in any of the environments described in Protocol I. Protocol II would deal with the verification of compliance with a comprehensive ban, including monitoring techniques, international cooperation for seismic and atmospheric data acquisition and analysis, on-site inspection and procedures to consider ambiguous situations. The setting up of an organization to assist in the verification of compliance was also envisaged.

The proposed amendments were not submitted to a vote. Instead, the conference mandated its president to conduct consultations with a view to achieving progress towards a comprehensive ban and resuming the work of the conference at an 'appropriate time'.

To be binding, an amendment to the PTBT must be accepted and ratified by a majority of the parties, including all three depositaries. However, long before the Amendment Conference had convened, the United States announced that it was opposed to modifying the Treaty; the United Kingdom held the same view. Moreover, China and France – the other testing states – could not be involved in the amendment process because they were not parties to the PTBT. The conference had therefore no chance to succeed.

A Breakthrough

The situation changed radically in 1992, when the US Congress, following the example of Russia and France, declared a nine-month suspension of nuclear testing. It also resolved that the US testing programme should be terminated by 30 September 1996, after a limited number of explosions designed primarily to improve the safety of nuclear weapons had been carried out. Resumption of testing

Table 4.1   *Nuclear Explosions 1945–98*

I. 16 July 1945 (the first nuclear explosion) to 5 August 1963
(the signing of the Partial Test Ban Treaty)

USA	USSR	UK	France	China	India	Pakistan	Total
331	221	23	8	0	0	0	583

II. 6 August 1963 to 30 May 1998

USA	USSR/Russia	UK	France	China	India	Pakistan	Total
701	494	22	202	45	3	2	1,469

III. 16 July 1945 to 30 May 1998

USA	USSR/Russia	UK	France	China	India	Pakistan	Total
1,032	715	45	210	45	3	2	2,052

Notes:

1. The number of the nuclear test explosions listed here takes into account the definition of an underground test, given in the 1974 TTBT (see section 4.2 above).

2. All British tests from 1962 on were conducted jointly with the USA at the Nevada Test Site; the number of US tests is, therefore, actually higher than that indicated here.

Source: *SIPRI Yearbooks: Armaments, Disarmament and International Security.*

would be allowed only if another country conducted a test after that date. In the meantime, the US Administration was to prepare and submit to Congress a schedule for the resumption of talks on tests with Russia and a plan for achieving a multilateral comprehensive ban on testing nuclear weapons by September 1996.

Some high officials of the US Administration regretted the decision of the Congress – which was signed into law by the President in October 1992 – and called it unwise. They reiterated the view that testing was important for improving the safety and reliability of nuclear weapons. Consequently, the US government opposed the 1992 UN General Assembly resolution urging a comprehensive test ban. The British government shared the US view. China expressed the opinion that a nuclear test ban may be achieved only in the framework of complete nuclear disarmament.

Nevertheless, in 1993 the new US Administration decided that the United States would use other means than test explosions to ensure the safety and reliability of its nuclear arsenal. It then extended the moratorium on testing. France, Russia and the United Kingdom followed suit. Thus, after decades of fruitless efforts, the way was opened for the termination of all nuclear tests

4.5   The Comprehensive Nuclear Test-Ban Treaty

The talks for a comprehensive test ban resumed in January 1994 at the Conference on Disarmament (CD), which set up for this purpose an ad hoc committee. The 50th UN General Assembly called upon the CD to complete the text of the treaty as soon as possible in 1996, so as to enable its signature by the outset of the

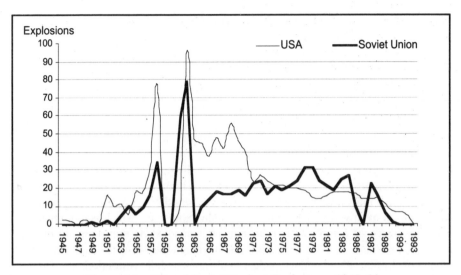

Figure 4.1 *US and Soviet Nuclear Explosions, 1945–92*

51st session of the Assembly. In addition to the 'rolling text', subject to continuous changes, the CD had before it drafts submitted by Australia and Iran. The main contentious points were the scope of the obligations of the parties, entry into force of the treaty and its duration, and verification of compliance. After two and a half years of work, the negotiators succeeded in removing most obstacles to the agreement. In June 1996 the chairman of the ad hoc committee proposed a draft which included compromise formulations for the remaining controversial issues. In August 1996 an overwhelming majority of the CD participants arrived at the conclusion that the chairman's draft represented the maximum common ground among the negotiators. They were, therefore, prepared to accept the text without further changes. Amendments, which had been suggested by a few delegations, were discussed but were not subject to negotiations, with one exception: it was agreed to meet the Chinese delegation's request to modify the voting procedure concerning the initiation of on-site inspection. This agreement, reached between China and the United States, was included in the chairman's concluding statement and, subsequently, in the revised draft treaty.

According to the CD rules of procedure, all decisions of the Conference must be taken by consensus. Since, for a variety of reasons, India was opposed to the draft treaty and decided to use its right of 'veto', the text of the Treaty – although supported by most CD participants – could not be recognized as a product of the CD. Even the transmittal to the United Nations of a special CD report on the test ban negotiations, which would have noted the failure to reach consensus, was vetoed by India (with the support of Iran). In this situation, a group of states supporting the treaty decided to turn directly to the UN General Assembly and ask it to endorse the treaty text – circulated, on behalf of Belgium, as a document of the CD – and recommend it for signature and ratification by all states. On 10 September 1996 the UN General Assembly resolved with an overwhelming majority to adopt the treaty text.

The Comprehensive Nuclear Test-Ban Treaty (CTBT) was opened for signature on 24 September 1996.

Scope of the Obligations

From the start of the negotiations, the CD participants were agreed that nuclear-weapon test explosions should be banned at any place in the atmosphere, in outer space, under water and under ground. However, a controversy arose – and lasted for quite a long time – as to whether the CTBT should or should not cover so-called hydronuclear experiments (HNEs), which release small amounts of nuclear energy. (The prefix 'hydro' means here that the core of the nuclear device behaves like a fluid under compression by the chemical explosive.) For some, 'small' meant a yield equivalent to a few kilograms of high explosive; for others, it meant a yield equivalent to tens or even several hundred tons of high explosive – which is more than the explosive force of certain battlefield nuclear weapons. At one point during the negotiations, France and the United Kingdom wanted to reserve the right 'in exceptional circumstances' to conduct nuclear explosions without restriction on yield, but they later withdrew this demand. The higher the yield, the greater the value of the explosion for nuclear-weapon development.

It was argued by some negotiators that certain kinds of test explosion were absolutely necessary to keep nuclear weapons safe and reliable, and that such explosions should be treated as 'activities not prohibited' by the CTBT. In fact, HNEs may be useful to check the so-called 'one-point' safety of a nuclear weapon, that is, whether an accidental explosive disassembly of the weapon, caused by penetration of a bullet or the shock wave of a sudden impact, will occur without producing a significant nuclear yield. HNEs may also serve to assess the significance of unforeseen physical changes in the warhead. However, their value is relatively small. The tests so far conducted by the nuclear-weapon powers must have already ensured a high degree of safety of nuclear weapons; no accidental nuclear explosion has occurred since the beginning of the nuclear age. A few detonations of the non-nuclear explosive component of nuclear weapons did take place, causing the dispersal of radioactive materials, but the risks of such occurrences were considerably lowered when the conventional explosive initiating the fission or fission-fusion reaction was replaced in most weapons with an 'insensitive' high explosive, less prone to accidental detonation. Safety concerns can also be addressed by prohibiting, in peacetime, all flights of aircraft carrying nuclear weapons, as such flights present the greatest danger of mishap. Many nuclear weapons are equipped with so-called Permissive Action Links (PALs) permitting the use of weapons only by authorized personnel, as well as with use-denial mechanisms that disable the weapons when unauthorized persons attempt their use. Improvement of these devices does not require explosive testing and can be made without affecting the weapon design. Nor is nuclear explosive testing indispensable to ensure the reliability of ageing stockpiles. This can be achieved through visual and electronic examination of warheads disassembled in the course of routine maintenance operations and possible correction or replacement of faulty components.

Indeed, in the so-called JASON Report, prepared in 1995 for the US Department of Energy, a group of senior American non-governmental scientists advising the government on technical security issues concluded that the United States can have confidence in the safety, reliability and performance of the nuclear weapons that are designated to remain in its stockpile. Age-affected warheads could be remanufactured, using the original materials. The present weapon designs are deemed to be sufficiently robust to tolerate the inevitable changes that would occur in remanufacture.

What can hardly be done without test explosions is the development of entirely new or significantly modified designs of nuclear weapons. However, if development testing were permitted, the test ban would be deprived of the arms control value. Yet another yield limitation agreement, following the 1974 TTBT, would be incompatible with the states' commitment under the 1963 PTBT – reiterated in the 1968 NPT – to achieve the discontinuance of *all* nuclear test explosions. Moreover, non-nuclear-weapon parties to the NPT are prohibited from engaging in any nuclear testing activity; a multilateral threshold treaty would undermine this prohibition. For all these reasons, HNEs had to be completely prohibited. Legitimizing them would have promoted proliferation rather than inhibit it.

To ensure that the CTBT would contain no loopholes allowing further development of nuclear weapons, India proposed that it prohibit not only nuclear test explosions but also 'any release of nuclear energy caused by the assembly or compression of fissile or fusion material by chemical explosive or other means'. It favoured, together with Indonesia, Egypt, Iran and a few other countries, the outlawing of all tests of nuclear devices, whether explosive or not. Such proposals were unacceptable to the nuclear-weapon powers, which argued that activities intended to maintain the safety and reliability of nuclear weapons, not involving nuclear explosions, should continue. They also proved unacceptable to certain non-nuclear-weapon nations in so far as they would amount to prohibiting activities serving civilian purposes, in particular, laboratory-scale experiments to develop means of producing commercial energy by creating nuclear fusion. These so-called inertial confinement fusion (ICF) experiments do have some weapon applications, because they involve the same physical processes as those occurring in thermonuclear weapons, but – according to a study published in 1995 by the US Department of Energy's Office of Arms Control and Nonproliferation – they cannot be used as a substitute for a nuclear explosive testing programme.

On 10 August 1995, France – thus far the main proponent of HNEs – declared that it would accept a prohibition on 'any nuclear-weapon test explosion or any other nuclear explosion'. This declaration, which reproduced the language formally proposed by Australia, was understood as a renunciation of the postulate to exempt small-yield explosions from a future global nuclear test ban. Also the United Kingdom announced its acceptance of the Australian text on the scope of the CTBT. On 11 August 1995 the United States made a similar announcement, setting the goal of achieving a 'true zero yield' CTBT. The Director of the US Arms Control and Disarmament Agency made it clear that the United States would rule out all HNEs, even those of a few pounds' nuclear energy release. In October 1995, after a meeting between Presidents Clinton and Yeltsin, it was announced that Russia supported a treaty banning all nuclear explosions, whatever their yield.

There remained the problem of nuclear explosions meant for peaceful, that is, civilian, purposes, which China (with practically no support from other delegations) proposed to exclude from the scope of the CTBT. However, such explosions –

although permitted under the NPT for nuclear-weapon states – cannot be tolerated under a comprehensive ban, because there are no means to distinguish them from explosions conducted for military purposes. To meet the Chinese postulates, at least partially, it was agreed that the review conference, to be held ten years after entry into force of the CTBT, should, upon request by any party, consider the possibility of permitting the conduct of underground nuclear explosions for peaceful purposes. If the conference decides by consensus that such explosions may be permitted, it must commence work with a view to recommending to the parties an appropriate amendment to the Treaty. The probability of reaching such a decision is low, as is the probability of amending the Treaty; amendments may be adopted only by a positive vote of a majority of parties with no party casting a negative vote. Nonetheless, the compromise regarding peaceful nuclear explosions, which China considers to be merely a 'temporary' solution, is unfortunate. It may reopen the debate on an issue which has a potential of subverting not only the test ban but also the NPT, for it could be understood as justifying research and development in the field of nuclear explosives (short of conducting actual explosions) not only by the nuclear-weapon powers, but by other states as well.

The agreed language on peaceful nuclear explosions led to the acceptance of the undertaking not to carry out 'any nuclear weapon test explosion or any other nuclear explosion', and to prohibit and prevent any such explosion at any place under the jurisdiction or control of the parties. Each party must refrain from causing, encouraging, or in any way participating in the carrying out of such explosions.

To deter potential violators, several representatives demanded that a ban be imposed not only on the conduct of nuclear explosions but also on preparations for such explosions. This demand, related to the proposal for closing down all nuclear test sites, was categorically rejected by the United States, Russia and China. Only France decided to close its testing site.

Entry into Force

Setting a mere number – without qualification – of ratifications needed for the CTBT to enter into force was considered inappropriate by many, because those actually testing or capable of testing could remain unconstrained. To avoid such a situation, it was necessary to obtain ratifications from all the nuclear-weapon powers and all nuclear-threshold states. Nuclear-weapon states have been defined in the NPT as those that have manufactured and exploded a nuclear weapon or other nuclear explosive device prior to 1 January 1967. However, there is no generally accepted definition of nuclear-threshold states. It would have been, therefore, politically awkward to single them out by name. This impediment could be circumvented if one required, as an indispensable minimum, ratification by the declared nuclear-weapon states plus all those states that conduct nuclear activities not subject to comprehensive IAEA nuclear safeguards. All other nuclear-capable non-nuclear-weapon states are already prohibited by the NPT or the nuclear-weapon-free-zone treaties from acquiring and, *ipso facto*, from exploding nuclear weapons, and are subject to comprehensive IAEA nuclear safeguards. Their participation in the CTBT is, of course, desirable but not indispensable. With this approach, the number of states whose ratifications would be necessary for a CTBT to become effective could have been reduced to eight: China, France, Russia, the United Kingdom, the United

States, India, Israel and Pakistan. Alternatively, when it became clear that not all nuclear-threshold countries would subscribe to the CTBT, one could have required ratification only by the five nuclear-weapon powers. No other country would then be in a position, by withholding its ratification, to prevent the Treaty from entering into force. Verification arrangements could be introduced gradually, *pari passu* with the increase in the number of adherents; they did not need to be, and actually could not be, global from the very beginning.

Straightforward solutions, such as those indicated above, were not seriously considered. Those that were considered provided for a number of ratifications much higher than eight, which number would, in most cases, include the eight countries specified above. According to one formula, all participants in the CD would have to ratify the CTBT. According to other formulae, all of the several dozen states possessing or building nuclear power or research reactors, or all countries providing facilities for monitoring the test ban, would have to ratify the Treaty. According to yet another formula, countries which had ratified the Treaty could subsequently decide to waive, individually or collectively, any requirement for its entry into force that might be stipulated in the text (including the requirement of ratification by all eight countries in question), thus making the Treaty effective immediately only for them.

In an effort to reconcile the divergent positions, the chairman of the ad hoc committee proposed that the CTBT should enter into force 180 days after the date of deposit of the instruments of ratification by all states listed in an annex to the Treaty, but not earlier than two years after its opening for signature – the time estimated for the establishment of the verification machinery. The annex enumerated 44 states, those which were members of the CD as at 18 June 1996 (date of the effective expansion of the CD membership from 38 to 61) and formally participated in the work of the 1996 session of the CD (that is, excluding Yugoslavia, which was formally a member of the CD but had not for several years participated in its work), and which, according to the IAEA publications of December 1995 and April 1996, possessed nuclear power or research reactors. The nuclear-threshold states – India, Pakistan and Israel – were included in this number as meeting the above criteria. Most negotiators found the above formula suitable for the Treaty. However, India stated categorically that it would not subscribe to the draft under consideration and dissociated itself from the envisaged monitoring system. The treaty was not – in its opinion – conceived as a measure towards universal nuclear disarmament and was, therefore, 'flawed'. (Also several other, mainly non-aligned, countries complained that the goal of nuclear disarmament was not adequately stated in the Treaty preamble; they did not however, find this omission serious enough to reject the draft.) India also said that the treaty language affected its 'sovereign right to decide', in the light of its national interest, whether or not it should accede to the Treaty. This was understood as a warning that entry into force of the CTBT could be blocked by India for an indefinite period of time. In fact, any country, out of the 44 mentioned above, could do so, even for reasons not related to the subject matter of the Treaty.

Recognizing the above predicament, the CTBT stipulates that, in case the Treaty does not become effective three years after the date of the anniversary of its opening for signature, the UN Secretary-General, the depositary of the Treaty, will, upon request of the majority of states that have deposited their instruments of ratification, convene a conference of those states. The conference shall consider and decide by consensus which measures 'consistent with international law' might be taken to accelerate the ratification process. The nature of such measures is not specified. It is understood, however, that the conference may not amend the entry into force provision. India interpreted this clause as a threat of UN sanctions against non-parties, but the Chairman pointed out that the clause in question did not refer to the UN Security Council action under Chapter VII of the UN Charter. The envisaged procedure could be repeated at subsequent anniversaries to persuade the recalcitrant countries to accede to the Treaty.

Duration

From the beginning of the negotiations, a view was widely shared that the CTBT should be of unlimited duration. This corresponded to the pledges made by the parties to the PTBT and the NPT to stop nuclear-weapon tests for all time. As in other arms control agreements, the possibility to withdraw from the CTBT is provided for when the country's supreme interests are in jeopardy. The withdrawing party would then have to give prior notice with an explanation and justification for its action.

The US suggestion to make unilaterally decided withdrawals a simple formality (without citing reasons of supreme national interests) already at the time of the first review conference was strongly criticized and had to be retracted. Indeed, a provision for an 'easy exit' from the Treaty would have jeopardized its survivability. However, this US retreat might be of no real consequence, should some future problems with the US nuclear-weapon stockpile – such as the uncertainty about the safety or reliability of weapons – justify withdrawal from the CTBT, as envisaged in the 1995 US Comprehensive Test-Ban Treaty Safeguards. Russia said that it would withdraw from the CTBT to conduct tests if there were no other means to confirm confidence in the safety or reliability of the key types of its nuclear weapons. Such interpretations of 'supreme interests' could facilitate arbitrary decisions not subject to international scrutiny.

Verification of Compliance

One of the controversies which arose in connection with the verification provisions of the CTBT was the composition of the Executive Council, a body destined to play an important political role in decision making within the framework of the Comprehensive Nuclear Test-Ban Treaty Organization (CTBTO). To ensure an equitable geographical distribution of seats on the Council, it was decided that it should consist of 51 members elected by the Conference of the States Parties: ten from Africa; seven from Eastern Europe; nine from Latin America and the Caribbean; seven from the Middle East and South Asia (Iran and some Arab countries objected to the inclusion of Israel in this regional grouping); ten from North America and Western Europe; and eight from South-East Asia, the Pacific and the Far East. It was made clear that the composition of these six geographical regions was CTBT-specific; it was not to set a precedent for other multilateral agreements or negotiating forums. To ensure, furthermore, that no party was a priori excluded from membership of the Council, at least one-third of the seats allocated to each geographical region would have to be filled by states designated on the basis of their nuclear capabilities relevant to the Treaty, as determined by international data as well as all or any of the following criteria: number of monitoring facilities of the International Monitoring System (IMS); expertise and experience in monitoring technology; and contribution to the annual budget of the Organization. One of the seats allocated to each geographical region must be filled on a rotational basis by the party that is first in the English alphabetical order among the parties that have not served as members of the Executive Council for the longest period of time. The remaining seats are to be filled by states designated from among all parties in a given region by rotation or elections.

The IMS is to comprise facilities – listed in an annex to the CTBT – for seismological monitoring, radionuclide monitoring including certified laboratories, hydroacoustic monitoring, infrasound monitoring, and respective means of communication. Doubts were expressed about the usefulness of infrasound monitoring and about the cost-effectiveness of radionuclide monitoring in detecting underground explosions. It was, nevertheless, agreed that the synergy of different monitoring technologies should enable verification of events well below one-kiloton yield – the adopted seismic threshold of detectability.

Suspicious events that cannot be clarified through consultations may be subject to international on-site inspection – an admittedly exceptional occurrence. A request for such inspection must be based on information collected by the IMS, on any relevant technical information obtained by national technical means of verification in a manner 'consistent with generally recognized principles of international law', or on a combination thereof.

China, supported by Pakistan, Iran and a few other countries, insisted that 'human intelligence' and espionage must be specifically excluded from the purview of national technical means of verification used to trigger on-site inspections, and that the information gathered by such means should be verifiable. However, most delegations appeared satisfied with the assurances, written into the Treaty, that 'verification activities shall be based on objective information' and 'shall be carried out on the basis of full respect for the sovereignty' of states parties. However, as a concession to China, it was agreed that the decision to approve an on-site inspection would be made by at least 30 affirmative votes of members of the Executive Council, instead of a simple majority.

The likelihood that the CTBT – which in practical terms affects only a handful of nations – would be violated, is not high. A single small explosion, difficult to detect, may not be sufficiently important from the military point of view to justify the risk of exposure (it would certainly not be useful for thermonuclear weapon development), whereas a larger explosion or a series of small ones could probably be detected with the technical and other means which are already in the possession of several countries. The envisaged international verification machinery seems, therefore, to be excessively complex.

Assessment

The degree of importance attached to a test ban by different countries or groups of countries has varied over the years, reflecting major changes in the world political situation as well as the evolving strategic interests of the nuclear-weapon states. However, the concerns of the world community about nuclear testing remain unchanged. They relate to the proliferation of nuclear weapons, the race for qualitative improvement of nuclear arsenals and the contamination of the human environment.

For the cause of inhibiting the proliferation of nuclear weapons, the CTBT does not carry the same significance now as it would have carried in the early years of the nuclear age. Today any state having an indigenous modern technological base or the financial resources to buy the necessary technology can manufacture, without testing, a fission atomic device of a relatively simple design (although of uncertain yield) with a high degree of confidence that the device will work. Thermonuclear devices are more complicated. Developing them without testing would be very difficult although not impossible, but there can be no certainty that such non-tested devices will function as envisaged. In any event, by imposing the same obligation not to test on all parties, the CTBT is bound to strengthen the non-proliferation regime, as it eliminates an important asymmetry between the rights and obligations of the nuclear 'haves' and 'have-nots' under the NPT.

If an emerging nuclear-weapon state decides to test a newly developed nuclear device, it will do so chiefly for political reasons, namely, to demonstrate to the world that it has acquired a workable nuclear weapon and thereby claim some special international status. However, when a recognized nuclear-weapon power conducts test explosions, it does so primarily to validate modifications in the existing designs of nuclear warheads. The main purposes of these – often sophisticated – modifications are to achieve greater efficiency in the use of fissionable and fusionable materials and, at the same time, make the weapon assembly compatible with the means of delivery, as required by current military needs. Simulation with supercomputers cannot meet all these objectives. Warheads of designs not tested through explosions are not deemed sufficiently reliable to be deployed. At least one explosion of a new or significantly re-designed warhead at or near full yield is considered indispensable by technical experts and military establishments of the nuclear-weapon powers. Testing is needed not only to modernize the first two generations of nuclear weapons – the fission and fusion explosive devices – but also to develop so-called 'third generation' nuclear weapons. These constitute a refinement of the techniques involved in fission/fusion processes for the purpose of achieving special weapon effects, such as earth penetration, enhanced electromagnetic pulse (EMP) or enhanced radiation. It is thus evident that the complete cessation of nuclear testing will bring arms control benefits by putting a stop to substantial qualitative improvements of nuclear weapons. The test ban should also make it unlikely that something completely new, unpredictable and exotic would suddenly emerge in the nuclear field.

Whereas one of the central purposes of the PTBT was to reduce the radiation hazards from nuclear tests in the atmosphere, venting of radioactivity from underground nuclear tests could not be avoided. In some cases, radioactive fallout was detected beyond the national borders of the testing states. An almost permanent legacy of underground tests is the inventory of radioactive substances deposited underground, and there is grave concern about their possible long-term effects. The CTBT will at least put an end to further such contamination of the environment. It will also free human and material resources which are spent on the development of nuclear weaponry.

Meaningful progress in nuclear disarmament – the goal set by the United Nations in numerous resolutions – is improbable as long as nuclear test explosions are not definitively and universally banned. Entry into force of the CTBT, as quickly as possible, is, therefore, imperative.

Implementation

In May 1998, less than two years after the signing of the CTBT, two non-signatories – India and Pakistan – carried out a few underground nuclear test explosions. India, which had exploded a nuclear device 24 years earlier, stated that the new tests had proven its capability for a 'weaponized' nuclear programme and had, thereby, provided reassurance to the people of India that their national security interests would be protected. Pakistan, which had been engaged in a nuclear-weapon programme since the 1970s, stated that India's tests had destabilized the security balance in South Asia, and that it was obliged to establish the balance of mutual deterrence by its own tests.

The Indian and Pakistani tests were criticized by many states as a serious challenge to the nascent globally binding norm against nuclear testing. Some states terminated their assistance to, and imposed economic sanctions on, both countries. Moreover, in a unanimous resolution, adopted on 6 June 1998, the UN Security Council condemned the nuclear tests conducted by India and Pakistan, demanded that both countries refrain from further testing, and urged them to become parties to the CTBT. The resolution encouraged all states to prevent the export of equipment, materials or technology that could in any way assist programmes in India and Pakistan for nuclear weapons and for ballistic missiles capable of delivering such weapons.

After their nuclear test explosions, India and Pakistan promised to observe unilateral voluntary moratoria on testing. India went even further in saying that it might convert its moratorium into a legal obligation. However, the widely publicized aspirations of the Indian military strategists to build a 'triad' of ground-, sea- and air-based nuclear forces may lead to a new series of test explosions necessary to develop new designs of nuclear warheads. If India conducts further tests, Pakistan will certainly follow suit.

Since three years after its signing the CTBT was not yet in force, a conference of states that had deposited their instruments of ratification was convened in accordance with Article XIV of the Treaty. The purpose of this conference, held in Vienna in October 1999, was to consider measures to bring the Treaty rapidly into effect. (See the subsection *Entry into Force* above.) The conference ended with the adoption of a Final Declaration calling on states which had signed but not ratified the CTBT, in particular those whose ratification was needed for the Treaty's entry into force, to accelerate their ratification processes with a view to their early successful conclusion. The declaration also appealed to non-signatories to sign and ratify the CTBT as soon as possible and to 'refrain from acts which would defeat the Treaty's object and purpose in the meanwhile'. A week later, in spite of the above appeals, the US Senate refused to ratify the CTBT. (This was the first rejection by the US Senate of a security-related international treaty since the 1919 Treaty of Versailles.) Consequently, the United States decided not to attend, in any capacity, the

second Article XIV Conference, held in New York in 2001, and opposed the retention of the CTBT issue on the UN General Assembly's agenda.

The US opponents of the CTBT see a need for explosive tests to improve the
nuclear capabilities of the United States, to maintain the reliability and safety of its
nuclear warheads and perhaps even to manufacture small, low-yield tactical nuclear
weapons, which are more usable than strategic nuclear weapons and, in particular,
more suitable for the destruction of deep, hardened underground facilities. They
argue that a complete cessation of tests would not be verifiable. They thus ignore the
Stockpile Stewardship and Management Program of the US Department of Energy,
intended to preserve the US nuclear deterrent without further nuclear testing, as well
as the letter transmitting the CTBT to the US Senate, in which President Clinton
said that, in his judgement, the Treaty is 'effectively verifiable'. They also disregard
the Russian proposal for the development of additional verification measures upon
entry into force of the CTBT. By mid-2002 only three nuclear-weapon powers,
namely, France, Russia and the United Kingdom, had ratified the Treaty.

A sharp controversy arose over the admissibility, under the CTBT, of the so-
called subcritical experiments. In these experiments, chemical high explosives generate high pressures that are applied to nuclear-weapon material. As a consequence,
some atoms of this material undergo fission, but no self-sustaining fission chain
reaction occurs. Among states possessing nuclear weapons, at least the United States
and Russia are engaged in such activities. Subcritical experiments do not produce
nuclear explosions and are, therefore, not specifically prohibited. However, according to the widespread opinion among non-nuclear-weapon states, their conduct may
contribute to the qualitative improvement of nuclear-weapon designs, which would
contradict one of the chief purposes of the CTBT, as defined in its preamble. Moreover, the pursuit of subcritical experiments without international control could
undermine confidence in the CTBT, as it may be difficult to distinguish them from
the unambiguously prohibited hydronuclear experiments.

5

Nuclear Arms Limitation

5.1 Nuclear Doctrines

Military doctrines describe the conditions under which force may be used and provide general guidelines for the structuring of armed forces. Nuclear doctrines define the role of nuclear weapons in both deterring and waging nuclear war. Evolving in accordance with changes in the technological, political and military environment, these doctrines largely determine the decisions of policy makers with regard to the acquisition, deployment, targeting and use of nuclear weapons.

Massive Retaliation and MAD

In 1954, the United States announced that it had adopted the doctrine of massive retaliation. The doctrine implied a threat that Soviet aggression would be met with a major US nuclear attack. It reflected the asymmetry in the military balance in Europe, where the Eastern bloc had a significant superiority over the West in conventional forces.

By the late 1950s, when the Soviet Union had begun acquiring an intercontinental nuclear-weapon capability, it appeared unlikely that the United States would respond with massive use of nuclear weapons to an armed attack carried out solely with conventional weapons. The demise of the US nuclear monopoly made the doctrine of massive retaliation obsolete. The nuclear stalemate in US–Soviet relations in the 1960s led to the adoption of the doctrine of mutual assured destruction (MAD). According to this doctrine, no country would attack another if it knew that the attacked side had the capability to inflict unacceptable damage on the attacker. For the Soviet Union, 'unacceptable damage', as defined by the United States, would have meant the destruction of 20–25% of the Soviet population and 50–70% of its industrial capacity. The credibility of the MAD doctrine rested on the survivability and vulnerability of nuclear forces. If the nuclear forces of one country were not survivable, that country might be tempted in time of crisis to launch a nuclear strike to pre-empt its nuclear adversary. On the other hand, if a country believed that it was invulnerable to a retaliatory strike, it might not be deterred by the nuclear forces of the other side.

In the 1970s, US strategists stressed the need for flexibility in strategic doctrine and in the command-and-control system. They argued that, if deterrence failed, the United States should be able to conduct nuclear war without engaging in wholesale devastation and with as little destruction of its own population and industry as possible. This reasoning led to the NATO-formulated strategy of flexible response, also called 'graduated deterrence', which was based on the military capability to react effectively in a conflict situation by using first conventional weapons and then, only if necessary, nuclear weapons.

Unlike the MAD doctrine, which envisaged countering military aggression at a very high level of destructive power, the new approach required precise targeting against the opponent's military assets and high-accuracy strategic capability. Since at that time not only the United States but also the Soviet Union was acquiring such a capability, the new strategy fuelled the nuclear arms competition, exacerbating tensions between the superpowers.

Nuclear War-Fighting

New more precise and more discriminating weapons gave rise to the counter-force doctrine, which implied the ability of one country to annihilate the war-fighting capability of another (its nuclear weapons, military units and military facilities), as distinct from the counter-value doctrine, which implied the ability to annihilate the cities and civilian industries of the enemy. This distinction was purely hypothetical, because a counter-force attack would also cause enormous civilian casualties, but the new doctrine led to the renewal of interest in nuclear war-fighting.

Indeed, in the early 1980s the US Administration gave some consideration to whether the capacity to wage a limited nuclear war and to control its escalation, coupled with extensive civil defence arrangements, could permit a country to prevail and win a nuclear exchange. The Soviet Union argued that nuclear war could not be considered a practical policy option and that it would be impossible to limit or control it. However, as was obvious from Soviet military deployment patterns, the Soviet leaders believed that it was essential to possess a nuclear war-fighting capability as well as defensive measures against nuclear attack.

Approaches to Nuclear Deterrence

For over five decades, the existence of nuclear weapons was a constraining factor in the behaviour of the great powers. However, there is no way of determining the extent to which nuclear deterrence actually deterred war between them.

With the end of the Cold War, the United States and the Soviet Union formally espoused the thesis that nuclear war cannot be won and should not be fought. Nonetheless, in its 1999 Strategic Concept NATO still envisaged the first use of nuclear weapons, although it described such use as 'extremely remote'. The United States, the United Kingdom and France – the nuclear-weapon states members of NATO – retained the option of introducing nuclear weapons into *any* armed conflicts, that is, not only conflicts with other nuclear-weapon powers but also those with non-nuclear-weapon states. 'Adequate' sub-strategic forces were maintained by NATO in Europe to provide a link with US strategic nuclear forces. The National Security Concept of Russia, which was made public in early 2000, reaffirmed the country's adherence to a doctrine of nuclear deterrence similar to that of NATO. Russia reserved the 'right' to use nuclear weapons in reaction to all attacks carried out with weapons of mass destruction. It also envisaged the use of nuclear weapons in response to a large-scale aggression with conventional arms in situations critical to its national security. As formulated in the US–Russian joint statement on the Strategic Stability Cooperation Initiative of 4 June 2000, the two powers considered that their capability for nuclear deterrence was necessary to maintain strategic stability and to ensure predictability in the international security environment.

After its nuclear explosions, in 1998, Pakistan admitted its reliance on nuclear weapons in view of the military superiority of India in conventional arms. It declared that it was prepared to resort to nuclear weapons in response to an attack carried out with conventional means of warfare.

Only China and India – among the states possessing nuclear weapons – maintained the policy of not using nuclear weapons first and of not using these weapons against a non-nuclear-weapon state under any circumstance.

5.2   The 1972 ABM Treaty

In the early 1960s, the military doctrines notwithstanding, the stockpiles of nuclear weapons and the means of their delivery had already reached levels high enough for the Soviet and US political leaders to independently conclude that mutual arms constraints might serve their national interests. Both powers started looking for a less risky and less costly way to preserve the balance of nuclear terror than the unlimited accumulation of weapons.

In 1969, the United States and the Soviet Union initiated bilateral negotiations on possible restrictions on their strategic nuclear arsenals. One agreement concluded in the first phase of these Strategic Arms Limitation Talks (SALT I) was the US–Soviet Treaty on the Limitation of Anti-Ballistic Missile Systems, which came to be called the ABM Treaty.

Main Limitations

Signed on 26 May 1972, and in force since October of the same year, the ABM Treaty prohibits the deployment of anti-ballistic missile (ABM) systems for the defence of the territory of the United States and the Soviet Union, as well as the provision of a base for such a defence. The deployment of ABM systems for defence of an individual region is also prohibited, except when expressly permitted by the Treaty. The permitted deployments were originally limited to two sites in each country, one for the protection of the national capital and the other for the protection of an intercontinental ballistic missile (ICBM) complex, and the centres of these two ABM deployment areas for each party were to be at least 1,300 kilometres apart. No more than 100 ABM fixed launchers and 100 ABM single-warhead interceptor missiles may be deployed in an ABM deployment area. ABM radars are not to exceed specified numbers and are subject to qualitative restrictions. The Treaty permits early-warning radars but limits future deployments of such radars to locations along the periphery of the national territory, where they must be oriented outward.

The ABM Treaty prohibits the development, testing or deployment of ABM systems or components which are sea-based, air-based, space-based or mobile land-based. This ban is particularly important, because ABM systems based on mobile components would be expandable beyond the permitted sites, creating a danger of sudden breakout towards the prohibited nationwide defence. In addition, the Treaty prohibits the development, testing and deployment of multiple-launch or rapidly reloadable ABM launchers. It also prohibits giving non-ABM systems or their components the capabilities to counter strategic ballistic missiles or their elements in flight trajectory as well as testing them in an ABM mode. The parties may not transfer to other states, nor deploy outside their national territories, the ABM systems or components thereof which are limited by the Treaty. An agreed statement by the parties extended this no-transfer provision to include technical descriptions or blueprints specially worked out for the construction of ABM systems and their components.

The term 'ABM system' has been defined in the Treaty as any system designed to counter strategic ballistic missiles or their elements in flight trajectory. The components of such a system are listed as 'currently' consisting of ABM interceptor missiles, ABM launchers and ABM radars. This listing is clearly illustrative.

The ABM Treaty, accompanied by agreed and unilateral statements as well as common understandings, is of unlimited duration. However, either side may withdraw from it on six months' notice.

In 1974, in a protocol to the ABM Treaty, the United States and the Soviet Union introduced further restrictions on ballistic missile defence. They agreed to limit themselves to a single area for deployment of ABM systems instead of two areas as allowed by the Treaty. Each party may dismantle or destroy its ABM system and the components thereof in the area where they were deployed at the time of the signing of the protocol and deploy an ABM system or its components in the alternative area permitted by the ABM Treaty, provided that proper advance notification is given. This right may be exercised only once. The deployment of an ABM system within the area selected must remain limited by the levels and other requirements established by the Treaty.

Assessment of the Treaty

Anti-missile systems, in the form in which they existed when the ABM Treaty and its protocol were signed, were deemed unreliable and costly, as well as vulnerable to countermeasures. They were thus patently inadequate for preventing nuclear warheads from reaching the target. However, their modernization was allowed; each side maintained the right to test fixed land-based ABMs at some specified test ranges. Nor was there a ban on the development of ABM systems based on other physical principles than the systems limited by the ABM Treaty and including components capable of substituting for ABM interceptor missiles, ABM launchers or ABM radars. In a statement attached to the Treaty, the parties agreed that, in the event of such new means of anti-ballistic missile protection being created in the future, their specific limitations would be subject to discussion and agreement. In other words, the ABM Treaty did not provide for a complete and unconditional renunciation of defence against ballistic missiles.

In spite of these shortcomings, the ABM Treaty became the cornerstone of strategic arms control. Without it, large-scale deployment of ABMs by one side would certainly have forced the other side to do the same or to increase and improve its strategic offensive forces in order to overcome the defences of the opponent. Moreover, a combination of missile defences with accurate offensive forces could have heightened the risk of war by creating incentives to strike first. In preventing a destabilizing competition between offence and defence, the ABM Treaty provided a sound base for negotiated limitation of offensive arms.

Challenges to the Treaty

Although the ABM Treaty constrained ABM deployment, the parties continued their missile defence technology programmes. The Soviet Union even tried to get around the constraints by constructing in central Siberia, north of the city of Krasnoyarsk, a radar for ballistic missile detection and tracking – which is prohibited – under the guise of a space-tracking radar – which is permitted. The United States modernized its early-warning radars at Thule in Greenland and at Fylingdales in the United Kingdom. This was considered by many as illicit.

SDI. On 23 March 1983, US President Reagan launched an ABM programme, called the Strategic Defense Initiative (SDI), to provide a shield that could effectively protect the United States against a massive Soviet missile attack and render nuclear weapons 'impotent and obsolete'. The programme became known as 'Star Wars' because it aimed at creating space-based systems for directly attacking and destroying re-entry vehicles. A special organization was set up to develop the required technologies.

SDI was the subject of sharp controversies. US domestic critics argued that nothing had altered the strategic reality codified in the ABM Treaty and that, in any event, effective defence against a missile attack was not attainable. (It was later revealed that those responsible for the SDI programme had manipulated the results of some important tests so as to make the programme appear more successful than it actually was.) Soviet leaders described SDI as an effort to acquire the ability to attack the Soviet Union without risk of retaliation. NATO countries expressed concern over a possible US–Soviet ABM race; France and the United Kingdom feared that such a race would decrease the deterrence value of their limited nuclear forces or would compel them to spend more resources on maintaining the effectiveness of these forces.

Most disturbing were the arms control implications of SDI. In the opinion of many authoritative experts, including the negotiators of the ABM Treaty, the ABM deployment planned by SDI (and even the pursuit of certain proposed technologies) would entail abrogation of the ABM Treaty. It would also violate the 1963 Partial Test Ban Treaty and the 1967 Outer Space Treaty, if – as envisaged – X-ray lasers, powered by a nuclear explosion in space, were to be used. To allay these apprehensions, the US Administration announced that, according to its interpretation of the ABM Treaty, the ban on space-based and other mobile ABM systems and components did not apply to lasers or other exotic ABM technologies under development in the SDI programme. This broad interpretation, or reinterpretation, contrasted with the view thus far accepted by the Reagan Administration itself (as reflected in its annual Arms Control Impact Statements prior to 1985) that the ABM Treaty prohibitions were to apply to all mobile ABMs, irrespective of the technology used.

GPALS. With the passage of time, as the once-promising exotic technologies proved disappointing, the SDI programme came to focus on conventional ground-based and space-based interceptors. In his 1991 State of the Union Address, President Bush said that the mission of SDI would be changed from defence against large-scale Soviet attack to protection against limited ballistic missile strikes, regardless of their source. The new concept, called Global Protection Against Limited Strikes (GPALS), would involve the use of space-based rocket interceptors (the so-called Brilliant Pebbles), several ground-based interceptor systems, associated sensors and transportable anti-tactical ballistic missiles (ATBMs). Since GPALS envisaged nationwide missile defences, as well as the development, testing and deployment of space-based anti-ballistic missile components, more than one ABM site for ground-based ABM launchers and more than 100 interceptors per site, it could not be brought into effect without a change in the legal regime established by the ABM Treaty.

US supporters of GPALS referred to the ostensibly successful performance of the US Patriot missiles (originally intended for air defence) in intercepting the Russian-built Iraqi Scud missiles during the 1991 Gulf War. (In fact, the Patriot missiles failed in most or all attempts to destroy the Iraqi missiles in flight.) They argued that a limited defence against ballistic missiles was justified by new threats to the United States emanating from the underdeveloped world and the newly independent states of the former Soviet Union. Opponents of GPALS saw no evidence that the likelihood of undeterrable threats of limited ballistic missile strikes against the United States had increased or that such threats would develop in the foreseeable future.

In June 1992, in a joint statement with US President Bush, Russian President Yeltsin agreed that the concept of a global protection system against ballistic missiles should be developed. The two presidents decided that a high-level group would explore, among other things, possible modifications of the existing agreements that may be necessary to implement the projected system. Since these agreements include the ABM Treaty, the joint statement may have signified a shift from the Russian insistence on maintaining a link between the ABM Treaty's limits on missile defences and significant reductions in strategic offensive forces. A few months later, however, the Russian Foreign Minister openly advocated the preservation of the ABM Treaty and the non-deployment in outer space of any weapons.

In May 1993, in a move that could be understood as the end of the 'Star Wars' era, the US Administration further downgraded the ballistic missile defence programme by restricting it to ground-based components. The US Secretary of Defense announced that, in the future, the programme would focus on creating a system to defend US forces in a theatre of war against battlefield missiles as well as a system to defend the continental United States from limited missile attack, particularly from a nuclear-armed 'terrorist state'. The preceding administrations' 'broad interpretation' of the ABM Treaty was judged incorrect.

THAAD. Another challenge to the ABM Treaty was the US Army's Theater High-Altitude Area Defense (THAAD) system. The THAAD system was designed to intercept, outside or near the upper reaches of the Earth's atmosphere, ballistic missiles with ranges of up to 3,500 kilometres, travelling at speeds of a maximum of 5 kilometres per second, the interceptor's velocity being 3 kilometres per second. The ABM Treaty stipulates that non-ABM systems should not be given capabilities to counter strategic ballistic missiles, but the 'demarcation line' between theatre missile defences and strategic missile defences is not unambiguously clear. Reportedly, THAAD would be capable of intercepting strategic missiles, those with ranges exceeding 5,500 kilometres, which travel at speeds of 6.5–7 kilometres per second. (Missile defence systems that engage targets only at short ranges are not contentious.) This circumstance and the fact that THAAD was intended to be mobile and deployed outside the territory of the United States raised problems of compliance with the ABM Treaty. Nevertheless, tests of THAAD were conducted. They were viewed by some observers as a violation of the ABM Treaty.

Navy Theater-Wide system. This US theatre missile defence (TMD) system, formerly referred to as the 'Upper Tier' system, to be deployed on ships, was yet another challenge to the ABM Treaty. It was to have an interceptor missile whose velocity was likely to exceed 3.3 kilometres per second.

5.3   Further ABM-Related Agreements

In a joint statement issued on 21 March 1997 by a summit meeting held at Helsinki, the United States and Russia reaffirmed their commitment to the ABM Treaty. They also assured each other that theatre missile defence systems, which may be deployed by each side, would not pose a threat to the strategic nuclear force of the other side and would not be tested to give such systems that capability.

Following the Helsinki meeting, the representatives of the United States, Russia, Belarus, Kazakhstan and Ukraine signed in New York, on 26 September 1997, two statements defining the demarcation line between the permitted theatre missile defences and the strategic defences prohibited by the ABM Treaty. They also signed a number of other agreements related to the implementation of the Treaty.

First Agreed Statement

The First Agreed Statement, also referred to as the Agreed Statement Relating to Lower-Velocity TMD Systems, deals with systems with interceptor missiles whose maximum demonstrated velocities do not exceed 3 kilometres per second. Land-based, sea-based and air-based components of lower-velocity TMD systems (that is, interceptor missiles, launchers and radars) are to be deemed compliant with the ABM Treaty if, during the testing of such TMD components or systems, the ballistic target-missile during the flight-test does not exceed a velocity of 5 kilometres per second or a range of 3,500 kilometres. It is understood that the velocity of space-based interceptor missiles shall be considered to exceed 3 kilometres per second.

Attached to the First Agreed Statement are common understandings of the terms used. The statement is to enter into force simultaneously with the Memorandum of Understanding on Succession (see below).

Second Agreed Statement

The Second Agreed Statement, also referred to as the Agreed Statement Relating to Higher-Velocity TMD Systems, deals with systems having interceptor missiles whose velocities exceed 3 kilometres per second. During tests of higher-velocity TMD systems, the velocity of the ballistic target-missile should not exceed 5 kilometres per second, and the range of the ballistic target-missile should not exceed 3,500 kilometres. The higher-velocity TMD agreement does not establish velocity limitations on TMD interceptor missiles and does not impose other restrictions on testing or deployment of such systems.

The parties also agreed not to develop, test or deploy space-based TMD interceptor missiles or space-based components based on other physical principles (OPP), such as lasers, that are capable of substituting for space-based TMD interceptor missiles. However, the development, testing and deployment of air-based, sea-based and land-based TMD or other non-ABM systems based on OPP are not constrained.

Like the First Agreed Statement, the Second Agreed Statement was to enter into force simultaneously with the Memorandum of Understanding on Succession (see below).

Additional agreed documents, attached to or associated with the Second Agreed Statement, include common understandings of the terms used; a joint statement on the annual exchange of information on the status of plans and programmes with respect to systems to counter ballistic missiles other than strategic ballistic missiles; and a non-legally binding unilateral statement declaring that each party has no plans to: flight-test a higher-velocity TMD interceptor missile against a ballistic target-missile before April 1999; develop TMD systems with interceptor missiles exceeding a velocity of 5.5 kilometres per second for land-based and air-based TMD systems or with interceptor missiles exceeding a velocity of 4.5 kilometres per second for sea-based TMD systems; test TMD systems against ballistic target-missiles with multiple independently targetable re-entry vehicles (MIRVs); or test TMD systems against re-entry vehicles deployed or planned to be deployed on strategic ballistic missiles.

Confidence Building

The United States, Russia, Belarus, Kazakhstan and Ukraine reached, also on 26 September 1997, the Agreement on Confidence-Building Measures related to Systems to Counter Ballistic Missiles other than Strategic Ballistic Missiles. They agreed that the TMD systems subject to the provisions of the Agreement were – for the United States – the THAAD system and the Navy Theater-Wide TMD system, and – for Russia, Belarus and Ukraine – the S-300V system, also known as the SA-12 system.

Ninety days after the Agreement's entry into force, the parties must carry out an initial exchange of data – to be subsequently updated annually – on their TMD systems and components subject to the Agreement. They are also obligated to provide notifications regarding the test ranges from which their TMD systems subject to the Agreement will be tested, as well as notifications of any test launches of interceptor missiles of TMD systems subject to the Agreement in which ballistic target-missiles are used. The parties undertook not to release to the public the information provided pursuant to this agreement, except with the express consent of the party that provided such information.

The Agreement was to enter into force simultaneously with the Agreed Statements relating to lower- and higher-velocity TMD systems and to remain in force as long as those statements remained in force.

Succession

When, at the end of 1991, the Soviet Union dissolved and its constituent republics became independent states, the only deployed ABM system was around Moscow, while a number of early-warning radars and an ABM test range were located outside Russia. It became necessary to determine which new states would assume the rights and obligations of the Soviet Union. The 26 September 1997 Memorandum of Understanding on Succession (MOUS) established that not only the United States and Russia but also four other former Soviet republics were to be considered bound by the ABM Treaty. Russia, Belarus, Kazakhstan and Ukraine were thus recognized – for the purposes of the ABM Treaty – as successor states of the Soviet Union and had to collectively assume the rights and obligations of the Soviet Union under the Treaty. This meant that only a single ABM deployment area was to be permitted among the four successor states. It also meant that Russia had the right to continue operating early-warning radars as well as the ABM test range, which were located within other successor states' territories. Regulations of the bilateral Standing Consultative Commission of the ABM Treaty had to be revised to govern the multilateral (five-nation) operation of the Commission. The memorandum was subject to ratification or approval by the signatories. It never entered into force.

5.4   Controversies over National Missile Defence

The obligations under the ABM Treaty notwithstanding, a number of US politicians continued to perceive the need to deploy a national missile defence (NMD) system.

US Plans

The envisaged NMD system was meant to protect the entire territory of the United States, that is, all of its 50 states, against limited nuclear attacks carried out with long-range ballistic missiles – whether accidental, unauthorized or deliberate – as distinct from the ballistic missile defence (BMD) system, combining national and theatre systems. The national system would initially include 100 ground-based interceptors based in Alaska to engage and destroy ballistic missile warheads above the earth's atmosphere by force of impact. The estimated cost of the system would amount to about US$60 billion.

US concerns over ballistic missile attacks were ostensibly centred on 'rogue states' (also called 'states of concern') – deemed irrational and, therefore, impervious to deterrence – in particular, North Korea, Iran and Iraq. However, none of these industrially underdeveloped countries was known to be in possession of a nuclear weapon or of a missile capable of delivering any weapon to an intercontinental target. The probability that they would acquire nuclear capabilities in the foreseeable future was judged by many as low. Moreover, if a 'rogue state' decided to attack the United States with a weapon of mass destruction, it would probably choose delivery means that are more reliable and less expensive than intercontinental ballistic missiles, for instance, ships or aircraft. The dangers of accidental or unauthorized ballistic missile launches by states possessing nuclear weapons as well as ICBMs could be reduced if all strategic forces were taken off alert.

In 1999, the United States adopted legislation making it its policy to deploy an effective NMD system as soon as technologically possible. At the same time it declared its willingness to preserve and strengthen the ABM Treaty. There was, however, no way to reconcile the acquisition of a nationwide system of defence against missiles with the purpose of the Treaty. The Treaty could be amended, but the fundamental modifications of its provisions that would be required to allow NMD appeared impossible to negotiate because of the opposition of Russia, supported by other members of the Commonwealth of Independent States (CIS). China and Russia suspected the United States of trying to gain a decisive strategic advantage and, in joint statements of 10 December 1999 and 18 July 2000, warned that a unilateral US decision to create a national anti-ballistic missile defence system would have a destructive effect on the key international disarmament and non-proliferation agreements.

International Reactions

Russia's main concern in opposing the US plan was not the deployment by the United States of 100 interceptors; the number of interceptors permitted in the original version of the ABM Treaty was much higher. What seemed to be worrying to Russia were the US projects to upgrade the early-warning radars (including those deployed outside the United States) and to orbit space-based sensor satellites, because these would be critical steps towards establishing a base for a nationwide defence: a limited NMD system, initially unable to diminish Russia's nuclear deterrence capability, could be rapidly expanded to deal with large-scale missile attacks. In order to overwhelm the US nuclear 'protective shield', Russia could halt the strategic arms reductions, withdraw from the nuclear arms control agreements (as it threatened to do), multiply the number of nuclear warheads on its ballistic missiles, and keep as many nuclear warheads as possible on launch-on-warning alert to be able to carry out a rapid and massive counterattack. Its proposal to set up, together with NATO, a European theatre missile defence – even if it materialized (which is doubtful) – would not affect the US NMD programme.

In addition, the 'secondary' nuclear-weapon powers – China, France, the United Kingdom (and perhaps also India and Pakistan) – apprehending that their nuclear deterrent would be undercut, might be motivated to increase their nuclear offensive potential. China, in particular, saw the planned US NMD as a weapon system aimed at neutralizing its relatively small nuclear force. It was opposed to the joint development by the United States and Japan of a TMD system, which – it believed – would have the capability to intercept also strategic ballistic missiles. For a variety of reasons, China was also opposed to TMD deployment in Taiwan, *inter alia* because this would encourage pro-independence sentiments in Taiwan. The United Kingdom and Denmark (Greenland) – the basing countries for the future upgraded US sensor systems – appeared reluctant to contribute to the subversion of the ABM Treaty. Other NATO allies, too, had reservations. Some of them were apprehensive that the US NMD system might produce divisive effects by decoupling the security of Europe from the security of the United States and foster nuclear-weapon proliferation. The pressure within the European Union for a common security policy was not unrelated to the unilateralist trend in US foreign policy.

A fear was widespread that nuclear missile defence systems would block the way to nuclear disarmament and set off a new arms race, especially in the field of offensive weapons. To prevent such a development, the UN General Assembly adopted, in 1999, a resolution calling upon the parties to the ABM Treaty to preserve its 'integrity and validity' by refraining from the deployment of anti-ballistic missile systems for the defence of the territory of their country and by not providing a base for such a defence. Only the United States and three other countries opposed the resolution.

Unresolved Questions

The key question, whether there was a threat of nuclear attack with long-range ballistic missiles on any of the parties to the ABM Treaty and, consequently, whether an NMD was needed, was not convincingly answered. The United States insisted that such a threat was real; it even named the potential aggressors (see above). Russia, which first refused to share the US concerns, later admitted that a threat of nuclear attack existed and must be addressed, but it did not specify which country might be likely to carry out such an attack. It argued that the spread of missiles and missile technologies could be countered with political and diplomatic means.

Another important question was whether and, if so, to what extent NMD would be technically and operationally effective. On this score, there was a great deal of scepticism, especially after several US flight-tests had failed. Moreover, the ability of the interceptor missiles to distinguish between enemy warheads and decoys was questionable. Missile defences could also be defeated by other measures. These include the use of chaff, manoeuvrable warheads and low trajectories. The practicality of interception during the missile boost phase (that is, in the early stage of the missile's flight) near the launch points of the identified potential aggressors is very uncertain.

Irrespective of whether the political leaders actually believe that anti-missile defences would work, the possession of such defences could be used in international crisis situations for intimidation or coercion.

5.5   Abrogation of the ABM Treaty

On 13 December 2001, President Bush announced that the United States was withdrawing from the ABM Treaty. Formal notification to this effect was given to the governments of Russia, Ukraine, Kazakhstan and Belarus. The reason for the withdrawal – as specified in the notification – was that a number of states were developing ballistic missiles, including long-range ballistic missiles, as a means of delivering weapons of mass destruction and that this was posing a threat to the territory and security of the United States, jeopardizing its supreme interests. By leaving the Treaty, which – as stated by the president – hindered the US government's ability to develop ways 'to protect our people from future terrorist or rogue-state missile attacks', the United States felt free to conduct the type of research, testing and development that it regarded as necessary to determine whether a workable anti-ballistic missile defence system could be fielded. (A group of members of the US Congress questioned the authority of the president to withdraw from the ABM Treaty without the Senate's consent.)

Russia's reaction to the US withdrawal was surprisingly moderate. President Putin qualified it as a mistake but did not consider it a threat to the national security of Russia, which – unlike the other nuclear-weapon powers – had a system capable of overcoming anti-missile defences. (A few months later, in the Joint Declaration of 24 May 2002 on their new strategic relationship, the United States and Russia agreed to study possible areas for missile defence cooperation, including the expansion of joint exercises related to missile defence and the exploration of potential programmes for the 'joint research and development of missile defence technologies'.)

The US move undermined what was generally acknowledged as the foundation of the global nuclear arms control regime. The only other country that has ever given notice of withdrawal from an arms control treaty is North Korea, but its withdrawal (from the 1968 Non-Proliferation Treaty) was suspended before it became effective. If other states decide to follow the US example, this may have adverse effects on the future of arms control agreements, both bilateral and multilateral, especially the Non-Proliferation Treaty (see Chapter 6).

5.6   The 1972 SALT I Interim Agreement

The first agreement limiting strategic offensive arms, the US–Soviet SALT Interim Agreement, was signed and entered into force in 1972, simultaneously with the ABM Treaty.

Main Limitations

The Interim Agreement provided for a freeze for a period of five years on the aggregate number of fixed land-based ICBM launchers and ballistic missile launchers on modern submarines. The parties were free to choose the mix, except that conversion of land-based launchers for light ICBMs, or for ICBMs of older types, into land-based launchers for modern heavy ICBMs was prohibited. Strategic bombers were not covered by the limitations.

A Protocol, which was an integral part of the Interim Agreement, specified that the United States was to have not more than 710 ballistic missile launchers on submarines and 44 modern ballistic missile submarines, while the Soviet Union was to have not more than 950 ballistic missile launchers on submarines and 62 modern ballistic missile submarines. Up to those levels, additional ballistic missile launchers – in the United States over 656 launchers on nuclear-powered submarines and in the Soviet Union over 740 launchers on nuclear-powered submarines, operational and under construction – could become operational as replacements for equal numbers of ballistic missile launchers of older types deployed before 1964, or of ballistic missile launchers on older submarines. The specified land-based launchers were those capable of firing ballistic missiles at a range in excess of 5,500 kilometres, so as to reach the territory of the other power. Like the ABM Treaty, the Interim Agreement was accompanied by agreed and unilateral statements as well as common understandings.

In September 1977, the United States and the Soviet Union made formal statements that, although the Interim Agreement was to expire on 3 October 1977, they intended to refrain from any actions incompatible with its provisions or with the goals of the then current talks on a new agreement.

Assessment

The Interim Agreement did not cover Soviet intermediate-range rockets aimed at European NATO allies or other countries but unable to reach the United States. Nor did it cover US forward-based aircraft in Europe and bombers aboard US aircraft carriers capable of delivering nuclear strikes against the Soviet Union or its allies. While the number of ballistic missile launchers in the possession of the two sides was not to increase beyond a fixed limit, there were no restrictions in the agreement on the improvement of the quality of these weapons (except for the freeze on the size of ICBM launchers) – on their survivability, accuracy or range. The agreed replacement procedures made it possible for the two parties to substitute modern models for obsolete types of weapon; moreover, the number of nuclear warheads each missile could carry was not circumscribed at all. The absence of qualitative limitations on offensive missiles considerably reduced the value of quantitative limitations on launchers, and the competition in arms continued to be fuelled by technological advances.

Because the Interim Agreement failed to put the US–Soviet strategic relationship on a more stable basis, neither side was fully satisfied with it. Both powers, however, recognized the Interim Agreement as a possible transition to more meaningful measures. Indeed, in the 1973 Agreement on Basic Principles of Negotiations on the Further Limitations of Strategic Offensive Arms, the United States and the Soviet Union undertook to work out a permanent arrangement on more complete measures to limit and subsequently reduce these arms.

The Vladivostok Accord

The essential elements of a new SALT treaty were agreed in 1974. In a joint statement made at the summit meeting held in Vladivostok, the United States and the Soviet Union established the principle of equal ceilings on strategic nuclear delivery vehicles. The agreed aggregate limit for each side was 2,400 intercontinental ballistic missile (ICBM) launchers, submarine-launched ballistic missile (SLBM) launchers and heavy bombers. Of the 2,400 delivery vehicles, only 1,320 launchers of ICBMs and SLBMs equipped with MIRVs would be allowed. Under these ceilings, each side would be free to compose its forces as it wished. Further progress in negotiations was delayed, among other reasons, by disagreement on whether or how cruise missiles (small pilotless aircraft capable of flying at very low altitudes) and the Soviet Backfire bombers (modern supersonic aircraft which could be employed for strategic missions) should be limited.

In March 1977, the US government tried to go beyond the Vladivostok formula by offering the Soviet Union a so-called comprehensive proposal which would have significantly reduced the nuclear arsenals and imposed strict limits on the deployment of new systems and on the modernization of existing ones. In particular, the overall ceiling on strategic nuclear delivery vehicles would have been lowered from the Vladivostok level of 2,400 to 1,800–2,000; the ceiling on launchers of MIRVed strategic ballistic missiles would have been fixed at 1,100–1,200, as compared to 1,320 agreed at Vladivostok; and limitations on the permitted number of MIRVed ICBMs and 'heavy' ICBMs would have been set at 550 and 150, respectively. This approach, concentrating on ICBMs – the most important component of the Soviet nuclear forces – would have had a greater limiting impact on Soviet strategic nuclear-weapon programmes than on US programmes. The Soviet Union therefore immediately rejected it. Another US proposal, which incorporated the Vladivostok terms while deferring consideration of the Backfire bomber and cruise missile issue, was also rejected.

5.7   The 1979 SALT II Agreements

In the negotiations which the United States and the Soviet Union resumed in May 1977, the parties adopted a new framework that permitted a long-term agreement on limits below the overall Vladivostok ceiling, a short-term arrangement for the most contentious issues and a statement of more far-reaching goals to be achieved in the next phase of SALT. This 'three-tier' arrangement was to become the structure of the SALT agreements reached two years later.

Main Limitations

The 1979 US–Soviet Treaty on the Limitation of Strategic Offensive Arms, the so-called SALT II Treaty, set for both parties an initial ceiling of 2,400 on ICBM launchers, SLBM launchers, heavy bombers and air-to-surface ballistic missiles (ASBMs) capable of a range in excess of 600 kilometres. This ceiling was to be lowered to 2,250 and the reduction was to begin on 1 January 1981, while the dismantling or destruction of systems exceeding that number was to be completed by 31 December 1981. A sub-limit of 1,320 was imposed upon each party for the combined number of launchers of ICBMs and SLBMs equipped with MIRVs, ASBMs equipped with MIRVs and aeroplanes equipped to carry long-range (over 600 kilometres) cruise missiles. Moreover, each party was to be limited to a total of 1,200 launchers of MIRVed ICBMs, SLBMs and ASBMs; of this number, no more than 820 could be launchers of MIRVed ICBMs.

A freeze was introduced on the number of re-entry vehicles on current types of ICBMs, with a limit of ten re-entry vehicles on the one new type of ICBM allowed each side, a limit of 14 re-entry vehicles on SLBMs and a limit of ten re-entry vehicles on ASBMs. An average of 28 long-range air-launched cruise missiles (ALCMs) per heavy bomber was allowed, while current heavy bombers might carry no more than 20 ALCMs each. Ceilings were established on the launch-weight and throw-weight of light and heavy ICBMs.

In addition, the following bans were agreed: on testing and deployment of new types of ICBMs, with one exception for each side; building additional fixed ICBM launchers; converting fixed, light ICBM launchers into heavy ICBM launchers; heavy mobile ICBMs, heavy SLBMs and heavy ASBMs; surface-ship ballistic missile launchers; systems to launch missiles from the seabed or the beds of internal waters; as well as on systems for the delivery of nuclear weapons from earth orbit, including fractional orbital bombardment systems (FOBS) capable of launching nuclear weapons into orbital trajectory and bringing them back to earth before the weapons complete one full revolution. The Treaty was to remain in force until 31 December 1985.

The parties also signed a series of agreed statements and common understandings clarifying their obligations under particular articles of the Treaty. Before signing all these documents, the Soviet Union officially informed the United States that its Backfire aircraft was a medium-range bomber and that the Soviet Union did not intend to give this bomber intercontinental capability nor increase its radius of action so as to enable it to strike targets on US territory. The Soviet Union also pledged to limit Backfire production to the 1979 rate of a maximum of 30 per year.

The Protocol to the SALT II Treaty banned, until 31 December 1981, the deployment of mobile ICBM launchers or the flight-testing of ICBMs from such launchers; the deployment (but not the flight-testing) of long-range cruise missiles on sea-based or land-based launchers; the flight-testing of long-range cruise missiles with multiple warheads from sea-based or land-based launchers; and the flight-testing or deployment of ASBMs. At the same time, a Memorandum of Understanding between the United States and the Soviet Union established a database on the numbers of strategic offensive arms.

Finally, in a Joint Statement of Principles and Basic Guidelines for Subsequent Negotiations on the Limitation of Strategic Arms, the parties undertook to pursue the objectives of significant and substantial reductions in the numbers of strategic offensive arms, qualitative limitations on these arms and resolution of the issues included in the Protocol to the SALT II Treaty.

Assessment

Owing to the differences in geography, technology, strategy and defence arrangements with their allies, the United States and the Soviet Union placed different emphasis on various components of their forces. The Soviet Union had more land-based ballistic missiles with larger megatonnage and better air defences, while the United States had more warheads and greater missile accuracy as well as other advantages in submarine and bomber forces. Nevertheless, the fact that the SALT II Treaty established a quantitative parity may have helped the two sides to reach agreement on reductions of force levels by creating an equal basis for such reductions. This was therefore a step forward as compared to the 1972 SALT I Interim Agreement, which did not provide for quantitative parity. The SALT II Treaty required the dismantling, without replacement, of a certain number of nuclear-weapon delivery vehicles: the Soviet Union would have to dismantle some 250 operational missile launchers or bombers, while the United States would have to dismantle 33 strategic nuclear delivery vehicles to comply with the Treaty's overall aggregate limit.

The SALT II agreements also had serious shortcomings. The numerical limits on strategic nuclear forces were set very high. There was a remarkable compatibility between the treaty limitations and the projected strategic nuclear-weapon programmes of both sides. Such destabilizing elements of the strategic nuclear forces as MIRVed ICBMs were allowed to increase, as were the numbers of warheads permitted on ballistic missiles and the numbers of cruise missiles permitted per bomber. The strategic nuclear firepower of both sides was allowed to grow. Nevertheless, mutually regulated arms competition could diminish the stimulus for 'worst-case' military planning, but the significance of the 1979 SALT agreements lay mainly in the promise of more meaningful nuclear arms limitation measures.

The SALT II Treaty never entered into force. The tense international situation at the end of the 1970s, created by the occupation of the US Embassy in Tehran by Iranian extremists, the US 'discovery' of a Soviet troop brigade in Cuba and, in particular, the Soviet armed intervention in Afghanistan – which appeared to validate US distrust of Soviet motives – was not propitious for treaty ratification. US opponents to the SALT II Treaty argued that it was militarily inequitable, and therefore flawed, because it left unaffected the heavy ICBMs deployed by the Soviet Union but not possessed by the United States, because it did not include in its numerical ceilings the Soviet Backfire bomber presumed to have intercontinental strategic capabilities, and because it could not be satisfactorily verified.

Although the SALT II Treaty remained unratified, both sides observed its main limitations for several years. In 1986, when the United States decided to put into service a new heavy bomber equipped with long-range cruise missiles and thereby exceed the limits permitted by SALT II, the Treaty was finally proclaimed invalid.

5.8 The 1987 INF Treaty

The SALT agreements limited only intercontinental (ground- and sea-based) ballistic missiles. Both superpowers took advantage of this incomplete coverage of nuclear delivery vehicles. The United States and especially the Soviet Union continued developing and deploying missiles of intermediate and shorter range.

Origins of the INF Issue

In the late 1970s, the Soviet Union began replacing older intermediate-range SS-4 and SS-5 missiles, not covered by the SALT agreements, with a new intermediate-range missile, the SS-20. This new missile was both mobile and accurate. As distinguished from its predecessors, it carried three independently targetable warheads, instead of one warhead, and used solid fuel instead of liquid fuel – an improvement that contributed to its quick launch capacity. Moreover, the extended range of the SS-20 – up to 5,000 kilometres – enabled it to cover targets in Western Europe, North Africa, the Middle East and, from bases in the eastern Soviet Union, a good part of Asia.

The United States was at that time engaged in modifying its tactical Pershing I missile, deployed in the Federal Republic of Germany (FRG) since the 1960s, in order to give it increased range and accuracy. The United States was also developing ground-launched cruise missiles (GLCMs) for deployment in other European NATO countries. Nevertheless, it perceived new Soviet deployments as an attempt to achieve regional nuclear superiority. To prevent an upset of the military balance in Europe, which – it was feared – could endanger Western security, NATO ministers adopted in 1979 the 'dual-track strategy'. One track called for negotiations between the United States and the Soviet Union to limit deployments of intermediate-range nuclear forces; the other called for deployment in Western Europe, beginning in December 1983, of 464 single-warhead US GLCMs and 108 Pershing II ballistic missiles, in order to redress the imbalance should the proposed negotiations fail. At first, the Soviet Union refused to engage even in preliminary talks unless NATO cancelled its deployment decision. However, in September 1981 it agreed to begin formal negotiations.

INF Negotiations

The main issues raised in the course of the US–Soviet negotiations on intermediate-range nuclear forces (INF) concerned the types of delivery vehicles to be covered by the limitations, the geographic coverage of such limitations, the involvement of third-country forces and the stringency of verification measures.

The US position, approved by the NATO allies, was to limit the scope of these negotiations to the land-based INF missile systems of both countries, to cover such missiles throughout the Soviet Union and to apply full verification. The Soviet position was to include in the negotiations sea-based missiles and aircraft, to deal only with armaments deployed in Europe west of the Urals, to take account of the British and French nuclear forces and to rely mainly on national technical means of verification.

In November 1981, US President Reagan announced the 'zero option' as the Western negotiating position. According to this position, NATO would forgo deployment of INF missiles if the Soviet Union undertook to eliminate all its INF missiles – SS-4s, SS-5s and SS-20s – in both the European and Asian parts of the country. In addition, in a draft treaty presented later, the United States proposed a freeze on the shorter-range Soviet missiles. After President Reagan's announcement, Soviet General Secretary Brezhnev proposed a bilateral freeze on INF missiles in Europe. The Soviet draft treaty, submitted in February 1982, proposed a staged reduction of INF, including some aircraft of both countries, but no new US INF would be allowed to be deployed.

In June–July 1982, as a result of informal consultations in Geneva which came to be known as the 'walk in the woods' talks, US negotiator Nitze and Soviet negotiator Kvitsinsky worked out a compromise. The compromise would have permitted the United States to deploy 75 cruise missile launchers, each with four single-warhead missiles, whereas the Soviet Union would have reduced its intermediate-range forces capable of reaching Europe to 75 SS-20s with three warheads each. The United States would not deploy any Pershing IIs, and the number of Soviet intermediate-range missiles in Asia would be frozen. British and French forces would not be taken into account in this arrangement. The proposal was rejected by both governments: the US Administration was not willing to renounce the deployment of its Pershing IIs, and the Soviet Union was not willing to accept any US INF missiles in Europe. Under another US proposal, made in 1983, deployments of Pershings and GLCMs in Europe would be limited to a specific number of warheads, between 50 and 450, provided that the Soviet Union reduced its total INF forces to the same level. In response, the Soviet Union made several concessions, including a freeze on Soviet SS-20 deployments in Asia, but seemed determined to block any US deployment of INF missiles.

On 23 November 1983, when the first Pershing IIs reached a US unit in the FRG, Soviet negotiators walked out of the INF negotiations. The Soviet Union announced that it would deploy SS-12 missiles forward from Soviet territory into the German Democratic Republic (GDR) and Czechoslovakia. Thus far, the INF negotiations had only brought about increased nuclear deployments on both sides.

In January 1985, agreement was achieved to resume the INF negotiations, along with negotiations on strategic weapons and on weapons in space, in a new bilateral forum called the Nuclear and Space Talks. The Soviet Union insisted that the three issues be dealt with in a single package. However, a few months later, upon the designation of Gorbachev to the post of General Secretary of the Soviet Communist Party, the Soviet Union consented to discuss an INF treaty separately, and it gradually, but relatively quickly, accepted practically all the US postulates. The decisive Soviet concessions were made at the 1986 US–Soviet Reykjavik summit meeting. It was agreed that the Treaty should be confined to US and Soviet armaments only, leaving out the British and French armaments, and that all INF, both those deployed in Europe and those deployed in Asia, should be reduced to zero. Also, missiles of shorter range were to be eliminated.

The INF Treaty between the United States and the Soviet Union on the Elimination of their Intermediate-Range and Shorter-Range Missiles was signed on 8 December 1987. It entered into force in 1988.

Main Obligations

The INF Treaty provided for the elimination by the United States and the Soviet Union of intermediate-range missiles (IRMs) and shorter-range missiles (SRMs). It banned flight-testing and production of all these missiles as well as production of their launchers. IRMs included ground-launched missiles with ranges of 1,000–5,500 kilometres, whereas SRMs included missiles with ranges of 500–1,000 kilometres. The agreed reductions were asymmetrical: the Soviet Union undertook to destroy a greater number of missiles and remove a greater number of warheads from operational status than did the United States. In concrete terms, the Treaty required the destruction of a total of 2,695 IRMs and SRMs, both deployed and non-deployed: 1,836 missiles capable of delivering 3,136 warheads, on the Soviet side; and 859 missiles capable of delivering as many warheads, on the US side. Destruction of missile-operating bases which could be used for systems not controlled by the Treaty was not required.

The elimination of SRMs and their launchers, support equipment and facilities was to take place during the first 18 months. IRMs were to be eliminated in two phases over three years. In the first phase, lasting 29 months, the parties were to reduce their asymmetric IRM forces to an equal level of 200 warheads. During the second phase, lasting seven months, the remaining IRMs and their launchers, support structures and equipment were to be dismantled and destroyed. Strict verification provisions allowed for monitoring compliance.

The INF Treaty was accompanied by two protocols: the Protocol on Procedures Governing the Elimination of the Missile Systems and the Protocol Regarding Inspections. The Memorandum of Understanding established a database.

To permit inspection by the Soviet Union of US missile sites located on the territory of Belgium, the FRG, Italy, the Netherlands and the United Kingdom, a special agreement (the so-called Western Basing Agreement) was concluded between these NATO states and the United States. Agreement was also reached between the Soviet Union and the GDR and Czechoslovakia (the so-called Eastern Basing Agreement) to permit US inspections of Soviet missile sites located on the territory of these Warsaw Treaty Organization states. An exchange of notes took place between the United States and the GDR and Czechoslovakia, confirming inspection procedures for Soviet missile sites in the two states.

Assessment

The INF Treaty eliminated only a small fraction of the nuclear delivery vehicles possessed by the United States and the Soviet Union. Moreover, the warheads and guidance systems removed from the deployed missiles were not eliminated but returned to stockpiles for possible reuse. Nevertheless, the Treaty was highly significant.

The destruction of INF missiles removed an entire category of nuclear weapons which might have been used early and pre-emptively in an East–West armed conflict because of their precision, penetrability and range – shorter than that of strategic nuclear delivery vehicles – as well as vulnerability. Also excluded was the possibility that such missiles might be equipped with conventional or chemical weapons, instead of nuclear weapons, as the INF Treaty provided for the elimination of all ground-based missiles of a specified range.

Motivations for the deployment of 'euromissiles' were never quite clear. Militarily, Soviet missiles appeared redundant, as most targets in Europe were certainly covered by Soviet strategic forces. Similarly, the planned NATO response to the Soviet build-up could only marginally augment the US nuclear potential. Politically, however, by demonstrating its ability to hit Western Europe not only with intercontinental strategic missiles but also with modern, sophisticated non-strategic weapons, specially designed for that purpose, the Soviet Union may have hoped to split the European NATO allies from the United States and force West European subordination or at least accommodation to Soviet interests. NATO reaction may have been also essentially politically motivated: to neutralize the political dividends that the Soviet Union could have derived from its threatening weapons. The INF Treaty put an end to this dangerous 'tug of war' between the superpowers. Although European security was at the centre of the INF problem, West European governments, including those which hosted Pershing II and ground-launched cruise missiles, did not play a major role in bringing about the agreement. It was, to a great extent, the European public opinion that helped to achieve it.

A few complaints were made by both sides in the course of the INF Treaty implementation. They concerned some imprecise notification, storage of missiles at undeclared locations, the improper way in which shipments of certain missiles to the destruction sites had been carried out, as well as certain intrusive methods of inspection. The problems that arose were satisfactorily solved.

5.9   The 1991 START I Treaty

On 31 July 1991, as the result of nine years of negotiations, the United States and the Soviet Union signed the Treaty on the Reduction and Limitation of Strategic Offensive Arms, subsequently called the START I Treaty. This new agreement provided for deep cuts in their nuclear arsenals but, unlike the INF Treaty, it did not require the elimination of an entire category of armaments. The negotiations that led to its conclusion centred on counting rules within agreed limits and sub-limits for both nuclear delivery vehicles and warheads.

The START I Treaty comprises the treaty itself, two Annexes, six Protocols and a Memorandum of Understanding. There are also several associated documents: joint statements, unilateral statements, declarations and an exchange of letters.

Main Provisions

The parties undertook to reduce their strategic offensive arms to equal levels in three phases over a seven-year period from the date on which the Treaty entered into force. After the envisaged reductions, the arsenal of each side was to be limited to 1,600 strategic nuclear delivery vehicles and 6,000 'accountable' warheads, including no more than 4,900 ballistic missile warheads and, in the Soviet (subsequently Russian) case, no more than 1,540 warheads on 154 'heavy' ICBMs. In addition, each side agreed to have no more than 1,100 warheads on deployed mobile ICBMs. The aggregate ballistic missile throw-weight for deployed ICBMs and SLBMs for both sides was not to exceed 3,600 metric tons.

The ceiling of 1,600 strategic nuclear delivery vehicles included deployed ICBMs and their associated launchers, deployed SLBMs and their associated launchers, and deployed heavy bombers. The warhead ceiling of 6,000 included the number of warheads on deployed ICBMs, SLBMs and heavy bombers.

Ballistic Missile Warheads. No missile may be flight-tested with re-entry vehicles (RV) in excess of the attributed number, the RV being that part of the ballistic missile which carries a nuclear warhead. Each side has the right to verify that deployed ballistic missiles contain no more RVs than the number of warheads attributed to them. There is a ban on developing new types of ICBMs and SLBMs that can carry more than ten warheads.

Downloading. The number of warheads on up to three existing types of ballistic missiles may be reduced ('downloaded') up to a total of 1,250 RVs. Any ICBM downloaded by more than two RVs must be equipped with a new front-section platform, and old platforms must be destroyed. If the United States downloads a missile other than its Minuteman III, and if the Soviet Union downloads a missile other than its SS-N-18, they may not, for the duration of the Treaty, build a missile of the same type (ICBM or SLBM) with more warheads than were left on the downloaded missile.

Several reasons were given for downloading – a contentious issue throughout the negotiations. Militarily, the ability to spread over more missiles the total number of warheads set by the START I Treaty allows for a more flexible configuration of the nuclear forces. Economically, a costly enterprise of building new missiles designed to carry fewer warheads may be replaced by a reduction of warheads carried by the existing missiles. Strategically, downloading diminishes the value of each MIRVed missile as a target, reducing the incentive to strike it first.

Heavy ICBMs. The numbers of deployed heavy ICBMs and their warheads were to be cut by half. For such missiles there was to be no downloading, no increase in launch-weight or throw-weight, no new types and no mobile launchers. New heavy ICBM silo construction was allowed only in exceptional cases, but the number of silos was not allowed to exceed 154. Modernization and testing of existing heavy ICBMs could continue, however.

Other ICBMs or SLBMs were to be considered new types if they exceeded certain variances in size, launch-weight and throw-weight. The throw-weight of existing types of ICBMs and SLBMs may not be increased by more than 21%.

Heavy Bombers. Each heavy bomber equipped only for nuclear weapons other than long-range nuclear ALCMs, that is, only for gravity bombs and short-range attack missiles (SRAMs), counts as one warhead. An agreed number of heavy bombers could be removed from accountability if they were converted to non-nuclear capability. Heavy bombers equipped for long-range nuclear ALCMs are to be made distinguishable from other heavy bombers.

The Soviet Backfire (Tupolev 22-M) bomber was not included in START I, but the Soviet Union made a politically binding declaration that it would not deploy more than 300 air force and 200 naval Backfires, and that these bombers would not be given the capability of operating at intercontinental distances.

ALCMs. Conventionally armed cruise missiles that are distinguishable from nuclear-armed ALCMs are not limited under the START I Treaty and may be deployed on any aircraft. Nuclear-armed long-range ALCMs, that is, those with a range of over 600 kilometres, are covered. Each current and future US heavy bomber equipped for long-range nuclear ALCMs is to be counted as ten warheads (with the exception noted below), but it may actually be equipped for up to 20 such missiles. Each current and future Soviet heavy bomber so equipped is to be counted as eight warheads (with the exception noted below), but it may actually be equipped for up to 16 missiles.

The United States may apply the above rule to 150 heavy bombers, whereas the Soviet Union may apply it to 180 heavy bombers. For any heavy bombers equipped for long-range nuclear ALCMs in excess of these levels, the number of attributable warheads would be the number for which the bombers were actually equipped. Multiple-warhead long-range nuclear ALCMs are banned.

Mobile Missiles. The Soviet SS-24 and SS-25 are mobile missiles. For the sake of reciprocity, the US MX missile was also to be treated as mobile.

Neither party may keep more than 250 non-deployed ICBMs for mobile launchers; of those retained, no more than 125 may be non-deployed ICBMs for rail-mobile launchers. There is also a numerical limit of 110 on non-deployed launchers for mobile ICBMs, of which no more than 18 may be non-deployed launchers for rail-mobile ICBMs.

The treaty contains provisions designed to inhibit the rapid reloading of ICBM launchers. There are no limits on the number of non-deployed cruise missiles and other heavy bomber weapons, but some restrictions are placed on the location of long-range nuclear ALCMs.

SLCMs. During the negotiations the Soviet Union sought legally binding limits on nuclear sea-launched cruise missiles (SLCMs). However, the United States saw insurmountable difficulties in verifying such limits, and the Treaty left SLCMs virtually unconstrained. Nevertheless, in separate statements, the two sides agreed to provide each other with a politically binding but not verified annual declaration concerning the deployments of long-range nuclear SLCMs. For the duration of the Treaty, the parties are also to provide each other annually with 'confidential' information regarding their nuclear SLCMs with a range of between 300 and 600 kilometres. SLCMs with multiple warheads must not be produced or deployed.

Exemptions. The START I rules exempt certain test equipment from counting. Included in this category are: 75 non-modern heavy bombers equipped for non-nuclear arms, former heavy bombers and heavy bombers for training; 20 test heavy bombers; 25 test silo launchers; and 20 test mobile launchers at test ranges.

Elimination. Deployed SLBMs and most deployed ICBMs may be removed from accountability either by destroying their launchers – silos for fixed ICBMs, mobile launchers for mobile ICBMs and launcher sections of submarines for SLBMs – or by converting those launchers so that they could carry only another type of permitted missile. However, the requirement to eliminate 154 deployed Soviet/Russian SS-18s must be met exclusively through silo destruction.

Non-Circumvention. Strategic offensive arms may not be transferred to third countries. Nor is permanent basing of such arms outside national territory permitted. Temporary stationing of heavy bombers in other countries is permitted subject to notifications, and port calls for strategic submarines are allowed.

Duration. The START I Treaty has a duration of 15 years, unless superseded by another agreement. The parties may agree to extend the Treaty for successive five-year periods, but each party has the right to withdraw from it at any time if it decides that extraordinary events have jeopardized its supreme interests.

The Soviet side stated that the START I Treaty may be effective and viable only under conditions of compliance with the 1972 ABM Treaty and that the extraordinary events referred to above include events related to withdrawal by one of the parties from the ABM Treaty or to a material breach of the ABM Treaty. In other words, the Soviet Union considered that territorial defence against ballistic missiles and significant reductions in ballistic missiles are mutually incompatible.

The 1992 Lisbon Protocol

The dissolution of the Soviet Union in December 1991, leaving nuclear arms deployed in several former Soviet republics, gave rise to fears that new nuclear-weapon powers would emerge and make it impossible for the START I Treaty to become effective. The fears were somewhat allayed when, in January 1992, the Russian Federation formally declared itself the 'legal successor of the Soviet Union from the standpoint of responsibility for the fulfilment of international obligations', covering obligations 'under bilateral and multilateral agreements in the field of arms limitation and disarmament'; these agreements include the 1968 Non-Proliferation Treaty (NPT) prohibiting Russia from transferring control over nuclear weapons to any country, 'directly or indirectly'.

The Russian statement, of which the international community had taken note, was not challenged by the non-Russian republics when it was made. Nonetheless, in a Protocol signed in Lisbon on 23 May 1992, Belarus, Kazakhstan and Ukraine – which at that time had nearly one-third of the total ex-Soviet inventory of strategic nuclear weapons stationed on their territories – were recognized by the Russian Federation and the United States as 'successor states' of the Soviet Union 'in connection' with the START I Treaty. The original bilateral agreement was thus converted into a multilateral one, and the three republics and the Russian Federation undertook to make arrangements among themselves for the implementation of its provisions. Since the Lisbon Protocol became an integral part of the START I Treaty, it had to be ratified together with it.

In separate formal letters addressed to the President of the United States, the leaders of Belarus, Kazakhstan and Ukraine pledged to 'guarantee' the elimination of all the nuclear weapons located on their territories. They further pledged that their countries would accede to the NPT as non-nuclear-weapon states 'in the shortest possible time'. Indeed, the presence of nuclear weapons on the territory of a state does not prevent that state from becoming a non-nuclear-weapon party to the NPT as long as the weapons are controlled by a nuclear-weapon state. No deadline was set for accession to the NPT. In ratifying the START I Treaty, the US Senate required that the letters from the leaders of Belarus, Kazakhstan and Ukraine that accompanied the Lisbon Protocol be regarded as legally binding obligations with the same effect as the provisions of the Treaty.

Assessment

The START I Treaty was the first arms control agreement to significantly reduce strategic nuclear forces. Its accomplishments can be summarized as follows.

By reducing the number of the most threatening ballistic missile warheads, and by substantially cutting the aggregate missile throw-weight, the START I Treaty reduced the nuclear attack potential of the superpowers. Since it contained incentives to decrease, through downloading, the number of warheads on deployed MIRVed missiles, and since it promoted a shift from missiles to slower-flying bombers, it rendered the nuclear forces of either side less capable of threatening a first strike. The Treaty institutionalized unprecedentedly extensive and intrusive measures of verification. It provided each side with transparency and predictability with regard to the strategic nuclear programmes of the other side and could bring significant savings in military spending. With the START I agreed counting rules and definitions, as well as its notification, elimination and verification procedures, deeper reductions in strategic weapons became easier to negotiate.

Nevertheless, the START I Treaty fell short of the envisaged ambitious goal of a 50% reduction of US and Soviet strategic forces. Even after the Treaty had been fully implemented, the United States and the Soviet Union were still permitted to have more weapons than they had in 1972, when the SALT I Interim Agreement was signed. In emphasizing reductions in long-range missiles, warheads and throw-weight, the START I Treaty discounted nuclear-armed gravity bombs; it limited air-launched missiles only partially; and it left sea-launched cruise missiles practically unconstrained. Moreover, the parties were allowed to make qualitative improvements to their strategic weapons arsenals as older weapons were retired. As in the case of the INF Treaty, they were permitted to reuse the removed nuclear warheads.

Post-START I Initiatives

START I was used as a 'springboard' for additional stabilizing changes. A few months after the signing of the Treaty, President Bush directed that all US strategic bombers as well as all ICBMs scheduled for deactivation be removed from their alert posture; that the development of the mobile MX/Peacekeeper ICBM as well as the mobile portions of the small ICBM programme be terminated; and that the programme to build a replacement for the nuclear short-range attack missile for strategic bombers be cancelled. Shortly thereafter, President Gorbachev reciprocated by announcing that Soviet heavy bombers would be taken off alert; that work on a modified short-range nuclear missile for heavy bombers would be halted; that work on a small mobile ICBM would be stopped; that plans to build new launchers for ICBMs on railway cars and to modernize these ICBMs would be abandoned; that all

Soviet ICBMs on railway cars would be returned to their permanent storage sites; that 503 Soviet ICBMs, including 134 MIRVed ICBMs, would be removed from day-to-day alert status; and that several SLBM-carrying submarines would soon be removed from active service.

In January 1992, in his State of the Union Address, President Bush announced that, after the United States had completed 20 planes for which the procurement had begun, it would stop further production of the B-2 bomber. It would also cancel the small ICBM programme, halt the production of new warheads for sea-based ballistic missiles, stop all new production of the MX/Peacekeeper missile and cease purchasing any more advanced cruise missiles. At the same time, President Yeltsin stated that Russia's programmes for the development or modernization of several types of strategic weapon had been cancelled; that the production of heavy Tu-160 and Tu-95 bombers would stop; that the production of airborne long-range cruise missiles as well as long-range sea-based cruise missiles of existing types would be halted; and that the number of SLBM-carrying submarines on combat patrol, which had been halved, would be further reduced.

These declarations by the United States and the Soviet Union/Russia may have demonstrated that under the prevailing political circumstances a surprise nuclear attack was not considered a real threat to either party.

5.10 The 1993 START II Treaty

The main shortcoming of the START I Treaty was insufficient arms reductions. This was to be remedied by the US–Russian Joint Understanding reached by Presidents Bush and Yeltsin on 17 June 1992. According to this understanding, the levels projected for START I had to be more than halved. The most outstanding feature of the new arms control agreement was the elimination of all MIRVed ICBMs; hence its name, the De-MIRVing Agreement. At the same time, several US–Russian agreements were signed to assist Russia in the safe and secure transportation and storage of nuclear weapons in connection with its planned destruction of these weapons.

The De-MIRVing Agreement was codified in the US–Russian Treaty on Further Reduction and Limitation of Strategic Offensive Arms, known as the START II Treaty. This treaty, signed on 3 January 1993, included two Protocols and a Memorandum of Understanding.

Main Provisions

The START II Treaty set equal numerical ceilings for the strategic nuclear weapons that might be deployed by either side. The agreed ceilings were to be reached in two stages: stage one was to be completed seven years after entry into force of the START I Treaty; stage two, by the year 2003. Stage two could be completed even earlier, by the end of the year 2000, if the parties concluded, within one year after entry into force of the START II Treaty, an agreement on a programme of assistance to promote the fulfilment of the relevant provisions. (Russia pointed out that it would otherwise bear a disproportionate economic cost burden in implementing the Treaty.)

Stage One. By the end of the first stage, each side must have reduced the total number of its deployed strategic nuclear warheads to 3,800–4,250. These figures include the number of warheads on deployed ICBMs and SLBMs as well as the number of warheads for which heavy bombers with nuclear missions are equipped.

Of the total of 3,800–4,250 warheads, no more than 1,200 may be on deployed MIRVed ICBMs, no more than 2,160 on deployed SLBMs and no more than 650 on deployed heavy ICBMs.

Stage Two. By the end of the second and final stage, each side must have reduced the total number of its deployed strategic nuclear warheads to 3,000–3,500. Within this numerical band, the parties are free to choose the level they wish to settle at.

Of the retained warheads, none may be on MIRVed ICBMs, including heavy ICBMs; only ICBMs carrying a single warhead will be allowed. Russia has thus given up the most threatening component of its nuclear panoply. No more than 1,700–1,750 deployed warheads may be on SLBMs. This, in turn, was a concession on the part of the United States, which had planned to deploy considerably more warheads under the START I Treaty. MIRVed SLBMs are not prohibited, however.

Downloading. The START II Treaty permits the United States to download its Minuteman III ICBMs and Russia its SS-N-18 SLBMs, as well as two additional existing types of ballistic missile, by up to four warheads per missile. However, unlike the START I Treaty, the START II Treaty does not limit the aggregate number of warheads that may be downloaded. The US MX/Peacekeeper ICBMs, as well as the Russian SS-18 heavy ICBMs and SS-24 ICBMs, each of which carry ten warheads, and the Russian six-warhead SS-19 ICBMs must be eliminated.

Elimination. START I rules for missile system elimination apply to START II with one exception regarding the SS-18 – a concession to Russia. As many as 90 SS-18 silos may be converted to carry a single-warhead missile which Russia stated would be of the SS-25 type. The START II Treaty stipulates special procedures to ensure that those converted silos would never again be able to launch a heavy ICBM. The remaining SS-18 silos will have to be destroyed.

All SS-18 missiles and their launch canisters, both deployed and non-deployed, must be eliminated no later than by 1 January 2003 in accordance with the agreed procedures, or by using such missiles for the delivery of objects into the upper atmosphere or space. There may be no transfer of heavy ICBMs to any recipient, including any other party to the START I Treaty. As in the START I Treaty, elimination of retired warheads is not required.

Heavy Bombers. According to the START II Treaty, heavy bombers are to be counted using the number of nuclear weapons – whether long-range nuclear ALCMs, short-range missiles or gravity bombs – for which they are actually equipped. The numbers are specified in the Memorandum of Understanding on Warhead Attribution and are subject to confirmation by a one-time exhibition and by routine on-site inspections. This change in the START I counting rules was introduced at the insistence of Russia.

The START II Treaty provides that up to 100 heavy bombers, not accountable under the START I Treaty as long-range nuclear ALCM-carrying heavy bombers, may be 'reoriented' to a conventional role. Such bombers must be based separately from heavy bombers equipped for nuclear roles. They may be used only for non-

Table 5.1   *US and Soviet/Russian Strategic Nuclear Forces, 1990 and 2001*

Deployed weapons	As of September 1990	As of July 2001
US warheads [a]		
ICBMs	2,450	2,079
SLBMs	5,760	3,616
Bombers	2,353	1,318
Total US	*10,563*	*7,013*
Soviet/Russian warheads [a,b]		
ICBMs	6,612	3,364
SLBMs	2,804	1,868
Bombers	855	626
Total Soviet/Russian	*10,271*	*5,858*

[a] Warhead attributions are based on the START I counting rules.

[b] The figures for 1990 include weapons in Russia, Belarus, Kazakhstan and Ukraine; the figures for 2001 include only weapons in Russia.

nuclear missions, and must have observable differences from other heavy bombers of the same type that have not been reoriented to a conventional role. Reoriented heavy bombers may be returned to a nuclear role after 90 months' notification, but thereafter they may not be reoriented again to a conventional role.

Entry into Force and Duration. The START II Treaty is to enter into force on the date of the exchange of instruments of ratification. The provision banning the transfer of heavy ICBMs was to be applied provisionally by the parties from the date of signature of the Treaty.

Each party has the right to withdraw from the START II Treaty if it decides that extraordinary events have jeopardized its supreme interests. According to the law on the ratification of the START II Treaty, adopted by the Russian Duma in 2000, such extraordinary events include a breach of the START II Treaty; withdrawal by the United States from the ABM Treaty; build-up of strategic offensive arms by states that are not parties to the START II Treaty in a way posing a threat to the national security of Russia; deployment of nuclear weapons on the territory of states which joined NATO after the date of the START II Treaty signature (NATO stated that it had no plans to do so, but refused to give a formal assurance of non-deployment); deployment by the United States or any other state of armaments preventing the normal functioning of the Russian system of early warning of missile attacks; and events of economic or technical nature making it impossible for Russia to fulfil its obligations under the START II Treaty, or jeopardizing the environmental security of Russia.

5.11   Agreements Complementary to START II

In a Joint Statement on Parameters on Future Reductions in Nuclear Forces, issued on 21 March 1997 by the summit meeting held at Helsinki, Russia and the United

States reached an understanding that, once the START II Treaty entered into force, they would immediately begin negotiations on a START III agreement. The new agreement would include, among other things, the following basic components: the establishment, by 31 December 2007, of lower aggregate levels of 2,000–2,500 strategic nuclear warheads for each of the parties; measures relating to the transparency of strategic nuclear warhead inventories and the destruction of warheads, as well as other jointly agreed measures to promote the irreversibility of reductions; resolving issues related to the goal of making the current START treaties unlimited in duration; and placement in a deactivated status of all strategic delivery vehicles to be eliminated under the START II Treaty by 31 December 2003 by removing their nuclear warheads or taking other agreed steps.

Subsequently, Russia expressed its readiness to reduce its strategic offensive arms – on the basis of reciprocity with the United States – to a lower level than that provided for in the 1997 Helsinki statement, namely, to 1,500 warheads. Given the poor state of the Russian economy, many experts doubted whether the country could afford keeping nuclear forces even at lower levels.

To speed up the entry into force of the START II Treaty and the commencement of negotiations for further reductions of strategic arms, several agreements were reached on 26 September 1997, at the same time as the set of agreements related to the implementation of the ABM Treaty were signed (see section 3 above).

START II Protocol

This Protocol extended the date by which the START II limitations and reductions must be completed from 1 January 2003 to 31 December 2007. It also extended the date by which the interim limitations and reductions of the START II Treaty must be carried out from seven years after entry into force of the START I Treaty (5 December 2001) to 31 December 2004.

The Protocol also stated that the parties might conclude an agreement on a programme of assistance for the purpose of facilitating and accelerating implementation of the START II reductions and limitations. This provision replaced the START II provision that required early implementation of the treaty obligations if the parties concluded, within one year of the START II Treaty entry into force, an agreement on the programme of assistance.

The START II Protocol is to enter into force upon the exchange of the instruments of ratification.

Joint Agreed Statement

This statement recorded the agreement between the parties that reductions in the number of warheads attributed to Minuteman III ICBMs under the START II Treaty might be carried out at any time before 31 December 2007, the deadline for completing all treaty-mandated reductions. This provision was to ensure that de-MIRVing under START II would take place in a stable and equivalent manner.

The Joint Agreed Statement had no effect on the downloading provisions of the START I Treaty, which remained unchanged.

Exchange of Letters on Early Deactivation

This exchange between the Russian Foreign Minister and the US Secretary of State codified the previous commitment that the United States and Russia would deactivate by 31 December 2003 all strategic nuclear delivery vehicles which, under the START II Treaty, were to be eliminated by 31 December 2007. Deactivation was to be achieved either by removing the nuclear re-entry vehicles from the missiles or by taking other jointly agreed steps.

The letters would enter into force when the START II Treaty entered into force.

Assessment

The START II Treaty was meant to improve strategic stability through the agreed elimination of MIRVed ICBMs, which – because of their lethality and vulnerability – were most likely to be launched in a pre-emptive attack. Its implementation would have resulted in a two-thirds reduction in the strategic nuclear weapons that the Soviet Union and the United States maintained at the height of the Cold War. However, the Treaty did not enter into force because of the United States' refusal to ratify it.

New negotiations started in 2002 on the lowering of the START II ceilings for US and Russian strategic warheads to between 1,500 or 1,700 and 2,200, and on the adoption of a bilateral declaration on a new strategic relationship. The main problems encountered in these negotiations concerned the way in which deployed warheads should be counted, the irreversibility and verifiability of the cuts, and the format of the planned accord.

5.12 The 2002 Treaty on Strategic Offensive Reductions

Contrary to the predictions that the US withdrawal from the 1972 ABM Treaty would make it impossible for Russia to continue its nuclear arms control transactions with the United States, on 24 May 2002, in Moscow, the two powers signed a new nuclear arms control agreement – the Treaty on Strategic Offensive Reductions. The Treaty is subject to ratification.

Undertakings

Russia and the United States undertook to reduce their respective inventories of nuclear warheads so as not to exceed the aggregate limit of 1,700–2,200 warheads by 31 December 2012. (The pace of the reductions was left to their discretion.) This legally binding commitment codified the reductions announced in unilateral statements by the presidents of Russia and the United States in 2001. As stated by the US spokesman, the limitations are to apply only to warheads operationally deployed on launchers. The Treaty does not spell out measures to verify compliance, but Russia and the United States will meet at least twice a year in a Bilateral Implementation Commission (BIC) to discuss issues related to the Treaty. It is understood that the verification regime of the START I Treaty, which remains valid until 2009, will provide a foundation for transparency and predictability regarding the implementation of the Strategic Offensive Reductions Treaty. It is not clear, however, to what extent the former treaty, which deals exclusively with means of delivery, could help in controlling the observance of the latter treaty, which deals exclusively with warheads.

The Treaty is to remain in force until 31 December 2012, but it may be extended by agreement of the parties or superseded earlier by a subsequent agreement. Withdrawal is allowed upon three months' notice.

Assessment

By drastically reducing the number of warheads that can be launched instantaneously (by two-thirds, from the 2002 levels of 5,000–7,000), the Strategic Offensive Reductions Treaty may diminish the likelihood of unauthorized or accidental nuclear war between Russia and the United States, but its arms control benefit is meagre. It falls far short of the nuclear powers' obligations under the 1968 Non-Proliferation Treaty (see Chapter 6).

The 2002 Treaty is remarkable for what it allows rather than for what it prohibits. The parties remain free to produce both warheads and means of delivery (missiles, including ICBMs equipped with MIRVs, and bombers) without any restriction. Non-deployed warheads possessed in excess of the agreed limits do not have to be decommissioned and destroyed; they may be stored without being subject to external controls. The constraints imposed by the Treaty can thus be easily reversed. As stated in the text, each party will determine for itself the composition and structure of its strategic offensive arms. Instead of actually cutting their nuclear arsenals, Russia and the United States could simply rearrange them, qualitatively and quantitatively, or even increase them.

Nevertheless, the Treaty reinforces the political rapprochement between the two nuclear superpowers. In the Joint Declaration, issued on the same day they signed the Treaty, the United States and Russia agreed that a new strategic relationship between them, 'based on the principles of mutual security, trust, openness, cooperation, and predictability', required substantive consultation across a broad range of international security issues. They therefore decided to establish a Consultative Group for Strategic Security, to be chaired by their foreign ministers and defence ministers. This group is to be the principal mechanism through which the sides should strengthen mutual confidence, expand transparency, share information and plans, and discuss strategic issues of mutual interest.

5.13 Tactical Nuclear Forces

Even before the agreed substantial cuts in strategic nuclear arsenals were made, the United States and the Soviet Union took a series of measures to reduce their tactical nuclear forces, which consist of short-range systems for use in battlefield or theatre-level operations. These measures were not embodied in a formal treaty but were announced separately and unilaterally by each power at the highest political level. They were clearly made in the expectation of reciprocity by the other side.

US Undertakings

On 27 September 1991, President Bush announced that he was directing the elimination of the entire US inventory of ground-launched short-range nuclear weapons.

All US nuclear artillery shells and short-range ballistic missile warheads were to be brought back to the United States and destroyed. Air-delivered nuclear capability in Europe was, however, to be preserved. In return, the Soviet Union was asked to destroy not only its nuclear artillery and nuclear warheads for short-range ballistic missiles but also those weapons which the United States no longer possessed, namely, nuclear warheads for air-defence missiles and nuclear landmines.

The US announcement also contained a commitment to withdraw all tactical nuclear weapons from US surface ships and attack submarines as well as nuclear weapons associated with land-based naval aircraft. This entailed removing all nuclear Tomahawk cruise missiles from ships and submarines as well as nuclear bombs aboard aircraft carriers. Many of these land- and sea-based warheads were to be dismantled and destroyed. Those remaining were to be placed in secure storage, to be made available if necessary should a crisis arise. Again, the Soviet Union was invited to match US actions – by removing all tactical nuclear weapons from its surface ships and attack submarines, by withdrawing nuclear weapons for land-based naval aircraft, and by dismantling or destroying many of these weapons and consolidating the rest at central locations.

Soviet Undertakings

On 5 October 1991, in response to the US undertakings, President Gorbachev announced the following steps regarding tactical nuclear weapons. All nuclear artillery ammunition and nuclear warheads for tactical missiles would be destroyed. Nuclear warheads of anti-aircraft missiles would be removed from the army and stored in central bases; some of them would be destroyed. All nuclear mines would be eliminated. All tactical nuclear weapons would be removed from surface ships and multi-purpose submarines. These weapons, as well as weapons from ground-based naval aviation, would be stored; some of them would be destroyed.

Moreover, President Gorbachev proposed that the United States and the Soviet Union remove all tactical nuclear weapons from their naval forces and destroy them. Also – on a reciprocal basis – all nuclear ammunition (bombs and aircraft missiles) should be removed from active units of forward-based tactical aviation and stored. On 29 January 1992 the Soviet commitments were confirmed by President Yeltsin on behalf of Russia.

French Undertakings

In June 1992, the French government decided to cancel the production of a tactical nuclear missile known as the Hadès. This missile, originally meant as a replacement for the Pluton missile designed for use against a massive attack by Warsaw Treaty Organization forces, had long been a point of friction between Germany and France. Indeed, since the range of the Hadès was to be shorter than 500 kilometres, it would land on German soil even if fired from the easternmost regions of French territory.

British Undertakings

The British Secretary of State for Defence announced, also in June 1992, that Royal Navy ships and aircraft and the Royal Air Force maritime patrol aircraft would no longer have the capability to deploy tactical nuclear weapons. By August 1998 all British free-fall bombs had been dismantled.

Assessment

Because of their relatively small size, large numbers and widespread dispersal, tactical nuclear weapons cannot be kept under strict supervision. Many, especially those of older types, are not equipped with electronic locks to prevent their unauthorized employment. Maintaining command and control over such weapons in a wartime situation would be particularly difficult: the fear that they may be overrun by an enemy early in a conventional armed conflict could prompt local military commanders to resort to their early use and start a nuclear war unintended by political leaders. In this respect, short-range tactical weapons are even more dangerous than long-range strategic weapons.

The unilateral undertakings to reduce or eliminate tactical nuclear weapons, especially those assumed by the United States and the Soviet Union, marked a change in the official policies of both powers. They could be understood as an indirect recognition that nuclear weapons were no longer useful for war-fighting, even though the possibility of using tactical nuclear weapons remained a component of the military doctrines of the nuclear-weapon powers. A limited number of US air-delivered nuclear bombs continued to be stationed in several NATO countries of Western Europe, but the underlying rationale was less military than political – to demonstrate the US commitment to the defence of Europe, and perhaps also to accentuate the burden-sharing of the nuclear risk among NATO allies.

The declaratory form of the new obligations was a departure from the generally accepted requirement that arms reductions must be effectively verified. However, given the high levels of the remaining nuclear weapons, none of the great powers was running a serious risk to its security by not verifying compliance. Detailed and time-consuming negotiations would have certainly delayed the removal of short-range nuclear weapons from regions of ethnic and political strife in the former Soviet Union, where they could have been taken over by sub-national units or terrorists. Nonetheless, it would appear desirable to codify these undertakings – which were assumed under special circumstances – in a formal treaty, check their implementation and make them more difficult to reverse than unilateral statements.

According to a statement made by NATO in July 1992, the United States had, by that time, removed its land-based stockpile of nuclear artillery shells, short-range missiles and naval nuclear depth bombs from Europe. It was then also announced that tactical nuclear weapons had been taken off US surface ships and attack submarines. The withdrawal of US nuclear weapons from South Korea – although never officially acknowledged by the United States – made it possible for the two Korean states to sign the 1992 Joint Declaration on the Denuclearization of the Korean Peninsula.

The Russian authorities stated that by the end of 1992 all tactical nuclear weapons stationed on the territory of the former Soviet republics had been withdrawn to Russia. By March 2000 – as subsequently stated by the Russian representative to the Conference on Disarmament – Russia had removed all tactical nuclear weapons from surface ships and multipurpose submarines, as well as from naval land-based aircraft, and moved them to central storage facilities; it had eliminated one-third of the total number of nuclear munitions for tactical sea-based missiles and naval air-craft; and it had destroyed almost all nuclear warheads of tactical missiles, artillery shells and nuclear mines. Half of the total number of nuclear warheads for anti-air-craft missiles and half of the total number of nuclear air-bombs had been destroyed.

In order to significantly reduce the risk of nuclear war, all tactical nuclear weapons – not only those possessed by the two nuclear superpowers – should be drastically reduced and eventually eliminated, as they were built to fight such a war. This appears to be urgent in view of the reported renewed interest of the military, in both Russia and the United States, in tactical weapons. However, a verifiable formal multilateral agreement to this effect would require an unambiguous definition of the term 'tactical'; the explosive yield and the geographic range do not suffice as criteria. This may present certain difficulties, for a weapon categorized as tactical or sub-strategic by the United States and Russia may be viewed as strategic by other states.

6

Nuclear-Weapon Proliferation

From the beginning of the nuclear age there has been an awareness that the spread of nuclear weapons to additional countries – referred to as 'horizontal proliferation', as distinct from the growth of the nuclear arsenals of the nuclear-weapon powers, referred to as 'vertical proliferation' – would increase the danger to world security. This awareness has led to the development of the nuclear-weapon non-proliferation regime, which encompasses various restrictive rules as well as specialized institutions, both national and international.

6.1   The 1968 Non-Proliferation Treaty

The pivotal role in the non-proliferation regime belongs to the Treaty on the Non-Proliferation of Nuclear Weapons (NPT), signed on 1 July 1968. The NPT – in force since 1970 – is a unique international document in that it prohibits possession of the most destructive weapons yet invented, by an overwhelming majority of states, while tolerating possession of the same weapons, for an undefined period, by a handful of states. In addition to retaining their nuclear arsenals, the nuclear-weapon powers are free to assist each other in developing nuclear warheads and in testing them, to receive from any state the material necessary to pursue their nuclear-weapon programmes, to deploy nuclear weapons on the territories of other states and to decide by themselves whether, and to what extent, to accept international controls over their peaceful nuclear activities. The non-nuclear-weapon states have thus assumed the main burden of obligation. However, the Treaty is not an end in itself: the declared aim of the parties is to pave the way towards nuclear disarmament.

Main Provisions

The essential non-proliferation obligations are contained in the first three articles of the NPT.

Non-Transfer and Non-Acquisition of Nuclear Weapons. The nuclear-weapon states have undertaken not to transfer 'to any recipient whatsoever' nuclear weapons or other nuclear explosive devices or control over them, and not in any way to 'assist, encourage, or induce' any non-nuclear-weapon state to manufacture or acquire such weapons or devices. The non-nuclear-weapon states have pledged not to receive nuclear weapons or other nuclear explosive devices or control over them, as well as not to manufacture them or receive assistance in their manufacture.

'Nuclear weapons or other nuclear explosive devices', the proliferation of which the NPT was meant to prevent, were not defined by the Treaty. A nuclear-weapon state was defined as one that had exploded a nuclear explosive device prior to 1 January 1967. The effect of setting this date was to limit the number of nuclear-weapon states to five, namely, the United States, the Soviet Union, the United Kingdom,

France and China, but it later proved difficult to maintain that a state exploding such a device after the set time limit should continue to be classified as non-nuclear. This question first arose in 1974, when India conducted a nuclear explosion and thereby crossed the formal threshold separating nuclear-weapon from non-nuclear-weapon states. The reiterated assurances by successive Indian governments that they were pursuing only peaceful ends put India in the intermediate class of nuclear-threshold states until 1998, when both India and Pakistan tested nuclear explosive devices. However, none of these states was formally recognized as a nuclear-weapon state.

Nor is it clear what is meant by the NPT ban on the 'manufacture' of nuclear weapons. The unchallenged US interpretation, given in the course of the negotiation of the Treaty, was that facts indicating that the purpose of a particular activity is to acquire a nuclear explosive device would tend to indicate non-compliance. Thus, according to the negotiating record, the construction of an experimental or prototype nuclear explosive device would be covered by the term 'manufacture', as would the production of components relevant only to a nuclear explosive device. However, the NPT does not provide for means to verify whether parties are engaged in developing prototype nuclear devices or weapon components. Research relevant to nuclear weapons and their components is not explicitly prohibited.

Another deficiency is the lack of an explicit ban on the provision of assistance in the manufacture of nuclear weapons by the non-nuclear-weapon parties to the NPT to non-nuclear-weapon states not party to the NPT. This omission, if taken advantage of, could enhance proliferation. However, as early as 1968, the Soviet Union and the United States, the powers responsible for the formulation of the relevant clauses of the NPT, expressed the opinion that such assistance would constitute a violation of the Treaty. This interpretation appears to have been accepted by all parties.

In the process of ratification of the NPT by the US Congress, the US government made a declaration of interpretation, according to which the Treaty would cease to be valid in time of war. In other words, from the start of hostilities, transfer of nuclear weapons or of control over them, as well as their acquisition by non-nuclear-weapon states by other means, would cease to be prohibited. This so-called 'war reservation' is highly controversial, as it contradicts the essential provisions of the NPT. Nevertheless, the 'nuclear sharing' arrangements for participation and cooperation by NATO allies in the use of nuclear weapons in case of war, as developed in the late 1960s, remain in force. War does cancel *ipso facto* certain treaties previously concluded between the belligerents, especially treaties of a political nature. It should, however, stand to reason that an arms control treaty that imposes restrictions on the possession of a certain type of weapon with a view to minimizing the risk of its use must remain in force during armed conflict, even if the verification and certain other provisions of the treaty have ceased to function. The NPT clearly belongs to this category of treaties.

In ratifying the NPT, several states placed on record their understanding that the Treaty should not impede unification of Western Europe. In other words, they wanted to keep open the possibility of a united Europe sharing the nuclear weapons of France and the United Kingdom. However, since Article I of the NPT prohibits transfer of nuclear weapons to 'any recipient whatsoever', sharing the possession of, and control over, such weapons among the sovereign members of the European Union must be ruled out. Only a Europe fully integrated in a federated state could qualify as a successor to the nuclear status of the present European nuclear-weapon powers without causing an increase in the number of nuclear-weapon states. This prospect is rather remote.

On the other hand, the drafters of the NPT did not foresee the disintegration of a nuclear-weapon power, and yet this occurred. The breakup of the Soviet Union gave rise to claims by some of the newly independent states to those portions of the Soviet nuclear arsenal which were stationed on their territories. These claims were eventually abandoned and the integrity of the NPT was maintained.

Nuclear Safeguards. Should a non-nuclear-weapon state decide to produce a nuclear weapon, it would need the requisite quantity of weapon-grade fissile material. The availability of this material is of crucial significance; hence the need for international control. Safeguards which have been devised to meet this need must be able to detect in a timely fashion the diversion of 'significant' quantities of nuclear material from peaceful nuclear activities to the manufacture of nuclear explosive devices as well as deter diversion by creating the risk of early detection. Subject to safeguards are plutonium and uranium, the fissionable materials defined in the Statute of the International Atomic Energy Agency (IAEA), as well as the equipment for their processing, use or production. Neptunium and americium, which could also be used in a nuclear explosive device if they were available in separated form and in sufficient quantities, are not covered by that definition.

The verification functions are performed by the IAEA, which is an autonomous intergovernmental organization founded in 1957 to promote peaceful uses of nuclear energy. The IAEA safeguards adopted before the conclusion of the NPT were intended to ensure that nuclear items obtained by non-nuclear-weapon states, with the help of the IAEA or under its supervision, were not used for *any* military purpose. The safeguards adopted for the NPT made an allowance for the withdrawal from international control of nuclear material destined for non-explosive military purposes. This allowance could be misused because enriched uranium used for the propulsion of ships, especially submarines, is often the same as that used in nuclear weapons. To prevent abuses, special arrangements would have to be made between the state withdrawing the nuclear material in question and the IAEA in order to identify the circumstances under which safeguards would not be applied. The state would have to make it clear that the unsafeguarded material (the quantity and composition of which would have to be known to the IAEA) would not be used for the production of nuclear weapons or other nuclear explosive devices. Safeguards would apply again as soon as the nuclear material was re-introduced into a peaceful nuclear activity. Such verification, however, could be thwarted by claims of military secrecy.

When in 1987 Canada decided to take advantage of the above-mentioned exemption provision – never applied before – and acquire a fleet of 10–12 nuclear-powered (but conventionally armed) attack submarines in order to assert its claims to sovereignty in Arctic waters, doubts arose about the compatibility of such an acquisition with Canada's commitment to the cause of non-proliferation. These plans were subsequently cancelled. If Canada had come into possession of nuclear-powered submarines, the letter of the NPT would not have been affected, but an unfortunate precedent would have been set for non-application of safeguards by the parties to the NPT.

Precise time limits are stipulated in the NPT for the initiation of negotiations for, and the entry into force of, safeguards agreements between the parties and the

IAEA. Several dozen states, mostly those without substantial nuclear activities, failed to conclude such agreements in time. In a few cases, when the relevant treaty provision had been ignored, suspicions arose that the basic non-proliferation obligations were being ignored as well. Thus, when North Korea, which was engaged in significant nuclear activities, refused, under varying pretexts, first to negotiate and later to agree to comprehensive controls over these activities, its refusal was interpreted by many as an attempt to conceal a nuclear-weapon development programme. North Korea eventually concluded the required agreement, but doubts persisted as to whether it had taken advantage of the several years' long delay to extract a significant amount of plutonium from the nuclear fuel irradiated in one of its reactors and to hide it away for weapon purposes. The IAEA was unable to conclude that there had been no such diversion. There is no specific clause in the NPT to deal with such a situation, but the additional protocol to the safeguards agreements, approved by the IAEA in 1997, will provide greater transparency of, and better access to, the pertinent nuclear facilities.

The NPT requires safeguards to be implemented in such a manner as to avoid hampering the economic or technological development of the parties or international cooperation in the field of peaceful nuclear activities. This requirement seems to have been met, although there have been occasional complaints that controls complicate the production process or are a burden for enterprises because of the cost and the threat to industrial secrets.

The accumulation of large quantities of readily accessible weapon-usable nuclear material is difficult to safeguard because of measurement uncertainties: the margin of error is dangerously high. In addition to plutonium separated by certain states from spent nuclear power reactor fuel, hundreds of tons of weapon-grade fissile material will be released as a result of the envisaged dismantlement of Russian and US nuclear weapons. The IAEA Statute requires that any special fissionable material in excess of the amount needed for peaceful purposes by member-states be deposited with the Agency. However, proposals for setting up an international plutonium storage (IPS), in compliance with this provision, have not materialized, mainly because of different opinions regarding the procedures for withdrawing the stored material.

For many years, the NPT clause which sets forth the safeguards requirement had been applied in a way that led to an absurd situation: the non-nuclear-weapon parties to the NPT, those that have formally undertaken not to acquire nuclear weapons, were subject to safeguards covering all their nuclear activities, both current and future, whereas the nuclear activities of states refusing to join the NPT and keeping their nuclear-weapon option open were controlled only partially, by safeguards applying exclusively to imported nuclear material or equipment. A significant part of the nuclear fuel cycle of non-parties could therefore remain unsafeguarded. Several countries concerned about the dangers of nuclear proliferation inherent in this unjustified distinction between foreign and indigenous technology had been seeking to impose on non-parties full-scope safeguards, as comprehensive as NPT-type safeguards, as a condition for nuclear trade. A few suppliers, however, in pursuit of commercial interests, continued providing nuclear material and equipment to countries accepting safeguards only on imported items. They may have thereby contributed, consciously or unconsciously, to the recipients' capabilities to produce nuclear weapons.

In April 1992, the Nuclear Suppliers Group adopted a common export policy. They agreed that transfer to a non-nuclear-weapon state of nuclear facilities, equipment, components, material and technology, as specified in the so-called trigger list, should not be authorized unless that state had brought into force an agreement with the IAEA requiring the application of safeguards on all source and special fissionable material in its current and future peaceful activities. In 1993 this agreement was formally recorded, but not all exporters of nuclear items subscribed to it.

Nuclear-weapon states are not obligated by the NPT to accept international control. They may, however, do so upon request of the suppliers of nuclear materials wanting to ensure that their materials are not used for the manufacture of nuclear weapons. A certain number of facilities in the nuclear-weapon states have been submitted to IAEA safeguards on a voluntary basis. Moreover, in the late 1990s Russia and the United States agreed to submit to IAEA safeguards weapon-origin fissile material designated as no longer required for defence purposes.

What is clearly missing is an international body to which complaints of non-compliance with the NPT, other than those related to nuclear safeguards, could be directed for investigation. The absence of such a body led to the application by some states of unilateral sanctions against suspected but not proven violators.

Peaceful Uses of Nuclear Energy. The NPT affirms the right of the parties to develop and use nuclear energy for peaceful purposes and obligates the parties in a position to do so to contribute to such efforts in non-nuclear-weapon states with due consideration for the needs of the developing areas of the world. The implementation of this provision of the NPT was affected by the slowdown in the growth of civilian nuclear power owing to safety factors, especially after the 1979 Three Mile Island accident in the United States, the 1986 Chernobyl accident in Ukraine and the 1999 Tokai-mura accident in Japan. It was also affected by economic factors, which included a weak increase in electricity demand, high initial investment and shortage of capital, as well as by the belief that spent fuel and high-level radioactive waste cannot be safely managed. In many countries, nuclear energy did not appear to be an economically competitive means to generate electricity. In Canada, France, Germany, the United Kingdom, the United States and a few other Western countries, no new nuclear power plants had been ordered for many years. Some industrialized European states decided to abandon nuclear energy altogether and started decommissioning their power reactors. Japan, the Republic of Korea and Taiwan did pursue nuclear plant construction, but they were able to do so without financial assistance from the nuclear-weapon powers. In Central and Eastern Europe there is a debate over the need to complete the construction of partially built plants; a few will be completed, while ageing units will be shut down. Assistance in non-power applications of nuclear energy – in medicine, biological research and agriculture – continues to be provided to several countries, mainly through the IAEA. The NPT did not eliminate the sovereign right of states to choose their trading partners and to judge themselves whether or not certain requested supplies were consistent with the basic objectives of the Treaty.

Under the NPT, the potential benefits of peaceful applications of nuclear explosions were to be made available by the nuclear-weapon parties to non-nuclear-

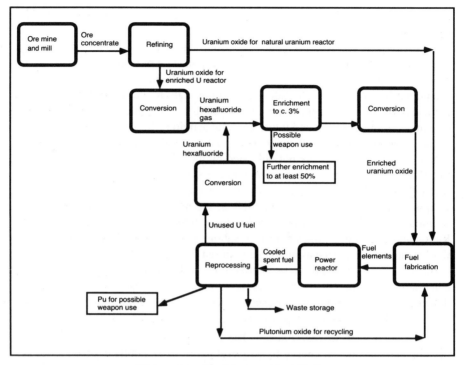

Figure 6.1 *The Nuclear Fuel Cycle*

weapon parties under appropriate international observation. This promise was made
in exchange for the renunciation by the latter states of the right to conduct *any*
nuclear explosions, as there is no way that a nuclear explosion can be carried out
with assurance that it performs no military function. Indeed, 'peaceful' nuclear
explosive devices, which can be used for industrial ends, could also be used as
weapons. They are transportable, and the amount of energy they are able to release
could cause mass destruction. Consequently, any of the non-nuclear-weapon countries which exploded such devices would de facto become a nuclear-weapon power.

It is now recognized that conventional explosives can achieve results equivalent to
those of nuclear explosives. Moreover, health and environmental risks would make
nuclear explosions unacceptable to the public in many countries. The prevailing
opinion is that peaceful uses of nuclear explosions would entail more risks than benefits. For this reason, the NPT clause which calls for the conclusion of a special
international agreement or agreements to provide nuclear explosion services to non-
nuclear-weapon states was not implemented. The Comprehensive Nuclear Test-Ban
Treaty (CTBT), signed in 1996, prohibited nuclear explosions for both military and
non-military purposes.

Disarmament Obligations. In one of the most important articles of the NPT
(Article VI), the parties undertook to pursue negotiations 'in good faith' on measures relating to cessation of the nuclear arms race 'at an early date' and to nuclear
disarmament, and on a treaty on general and complete disarmament. The NPT

negotiating history suggests that the clause requiring the cessation of the nuclear arms race was understood by the signatories as denoting a package of measures, which included the termination of nuclear-weapon test explosions and a ban on further production of fissile material for nuclear explosive purposes.

However, the NPT clause providing for nuclear disarmament has given rise to sharp controversies. Most non-nuclear-weapon states interpret it as an obligation to negotiate the abolition of nuclear armaments. They argue that the NPT was a bargain between the non-nuclear-weapon and nuclear-weapon states: the self-imposed nuclear arms denial of the former was to be matched by corresponding acts of the latter. They also refer to the Advisory Opinion of the International Court of Justice (ICJ) of 8 July 1996, which acknowledged that there existed an international obligation to achieve nuclear disarmament in all its aspects. Most nuclear-weapon states treat the relevant NPT clause as an obligation to negotiate only reductions or limitations of nuclear weapons and to preserve, thereby, what they consider to be strategic stability. They do envisage the elimination of nuclear arsenals, but only within the framework of general and complete disarmament. However, the undertaking to negotiate a treaty on general and complete disarmament – a remnant of the international debate conducted in the Cold War spirit of the early 1960s (see Chapter 3) – is little more than a 'ritual' formula appearing as a desideratum in certain UN resolutions or preambles to multilateral arms control agreements.

Amendments. The NPT is subject to amendments, but an amendment requires the consent of the nuclear-weapon parties, as well as those other parties that are members of the IAEA Board of Governors on the day the amendment is circulated. Whereas the nuclear-weapon powers might agree on certain changes in the text of the NPT, it is not likely that in the Board of Governors – a large and heterogeneous group – unanimity could be obtained among NPT parties on any significant modification of the Treaty. Moreover, even if an amendment were adopted by the required majority, it might fail to enter into force if the parties decided not to ratify it. This is why, in their endeavours to clarify ambiguities and to strengthen the NPT, the parties consider it safer and simpler to resort to agreed understandings, formal or informal, rather than tampering with the language of the Treaty.

Entry into Force and Duration. The initial duration of the NPT was set at 25 years. The decision concerning the extension of the Treaty for an indefinite period of time or for an additional fixed period or periods was to be taken by a majority of the parties at a specially convened conference. This conference was convened in April 1995. Since the same conference was charged with reviewing the operation of the NPT, it was called the NPT Review and Extension Conference.

On 11 May 1995, when it was obvious that a majority of the parties, as required by the NPT, supported an indefinite extension of the Treaty, the conference decided without a vote (although not unanimously or by consensus) that the Treaty would continue in force 'indefinitely'. Two documents closely linked with the Decision on Extension and with each other were adopted on the same day, also without a vote. One was about the revised arrangements for reviewing the implementation of the NPT, and the other about the principles and objectives of non-proliferation.

The Decision on Strengthening the Review Process for the Treaty provided that review conferences should be held every five years, as had been the case during the preceding 25 years. A preparatory committee was to meet several times prior to each review conference to 'consider principles, objectives and ways . . . to promote the full implementation of the NPT as well as its universality, and to make recommendations thereon'. The review conferences themselves had to look forward as well as back, evaluate the results of the period under review, including the implementation of the parties' undertakings under the Treaty, and identify the areas in which, and the means through which, further progress should be sought in the future.

The Decision on Principles and Objectives for Nuclear Non-Proliferation and Disarmament was intended as a 'yardstick' to measure progress in the fulfilment of the obligations under the NPT. It required that the parties' programme of action should include: completion of the negotiations on a nuclear test ban treaty; 'immediate' commencement and early conclusion of negotiations on a convention banning the production of fissile material for nuclear weapons or other nuclear explosive devices; pursuit of systematic and progressive efforts by the nuclear-weapon states to reduce nuclear weapons globally, with the ultimate goal of eliminating those weapons; and pursuit by all states of general and complete disarmament under strict and effective international control. In addition, a resolution sponsored by Russia, the United Kingdom and the United States called upon all states of the Middle East that have not yet done so to accede to the NPT as soon as possible and to place their nuclear facilities under full-scope IAEA safeguards.

Withdrawal. Each party to the NPT has the right to withdraw from it if 'extraordinary events, related to the subject matter of this Treaty, have jeopardized the supreme interests of its country'. A party decides for itself whether such events have occurred and does not need to justify its action to any external authority. A notice addressed by it, three months in advance, to all other parties to the NPT as well as to the UN Security Council, with a statement of the events regarded as jeopardizing its security, should suffice. It is not clear from the language of the NPT which extraordinary events the drafters of the Treaty had in mind other than the acquisition of nuclear weapons by a potential adversary, and what action they expected from the Security Council.

In more than three decades of the Treaty's existence, North Korea was the only country to take advantage of the withdrawal clause. When in March 1993 the North Korean government gave notice of its withdrawal from the NPT, it referred to the joint US–South Korean military manoeuvres, which it considered threatening, and to the IAEA request to conduct a special inspection of North Korean facilities, which it considered unjustified. In June 1993, one day before the expiration of the three months' notice period, the North Korean government suspended the 'effectuation' of its withdrawal from the NPT. It did, however, withdraw from the IAEA.

Assessment

The NPT established a norm of international behaviour in the nuclear field. It is therefore of paramount importance for arms control: it constitutes an obstacle to nuclear anarchy and makes it possible for the nuclear-weapon powers to engage in significant reductions of their arsenals. Despite the asymmetry of the rights and obligations of the nuclear-weapon and non-nuclear-weapon parties, the NPT has attracted a record number of adherents; by the year 2002 only Cuba, India, Israel and Pakistan had remained outside the NPT.

The nuclear test explosions conducted in 1998 by India and Pakistan dealt a blow to the international nuclear non-proliferation regime. They did not, however, directly impair the integrity of the NPT, as neither country was a party to the Treaty. In spite of being de facto nuclear-weapon powers, India and Pakistan cannot join the NPT as nuclear-weapon states; this would be contrary to the letter of the Treaty. Nor are they likely to follow the example of South Africa in destroying the nuclear weapons they have manufactured and in joining the NPT as non-nuclear-weapon states. However, they appear willing to formally commit themselves to behave like nuclear-weapon parties to the NPT in not transferring nuclear weapons to any recipient whatsoever and in not assisting anyone in acquiring such weapons. (This was also the position of France before it joined the NPT.)

The 2000 NPT Review Conference agreed on a plan of action, consisting of steps for the 'systematic and progressive' efforts to implement Article VI of the NPT. These steps included: achieving an early entry into force of the CTBT; declaring a moratorium on nuclear-weapon-test explosions pending the entry into force of the CTBT; concluding, within five years, a treaty banning the production of fissile material for nuclear weapons or other nuclear explosive devices; establishing a subsidiary body of the Conference on Disarmament with a mandate to deal with nuclear disarmament; recognizing the principle of irreversibility of nuclear arms control and disarmament; unequivocal undertaking by the nuclear-weapon states to accomplish the total elimination of their nuclear arsenals; accelerating the entry into force and full implementation of the START II Treaty and concluding the START III Treaty as soon as possible, while preserving and strengthening the ABM Treaty; implementing the 'trilateral initiative' between Russia, the United States and the IAEA regarding nuclear safeguards to be applied to fissile material that is surplus to military requirements; taking measures leading to nuclear disarmament of all nuclear-weapon states; making arrangements to place the fissile material of all nuclear-weapon states that is no longer required for military purposes under international verification to ensure that such material remains permanently outside military programmes; reaffirming that general and complete disarmament is the 'ultimate' objective of the disarmament process; submitting regular reports on the implementation of the NPT Article VI and of the 1995 Decision on Principles and Objectives for Nuclear Non-Proliferation and Disarmament; and further developing the verification capabilities.

The future viability of the NPT will depend on whether the nuclear-weapon powers live up to the above postulates. If they do not, some non-nuclear-weapon states or a group of such states might decide to withdraw from the NPT. States could use this exit clause to demonstrate their disappointment and disapproval, even without an intention to 'go nuclear', but such a demonstration could start the unravelling of the Treaty.

6.2 Security Assurances for Non-Nuclear-Weapon States

Except for a reference to the obligation of all states under the UN Charter to refrain in their international relations from the threat or use of force, no specific obligation has been laid down in the NPT to ensure the security of non-nuclear-weapon states. However, states which have renounced their claims to nuclear weapons, including those enjoying the protection of nuclear-weapon powers, have all along insisted on obtaining security assurances, considered by many to be an essential component of an effective nuclear non-proliferation regime.

Positive Assurances

As early as 1968, under the pressure of non-nuclear-weapon states, the UN Security Council adopted Resolution 255, by which the Soviet Union, the United Kingdom and the United States pledged immediate assistance, in accordance with the UN Charter, to any non-nuclear-weapon state party to the NPT that is a 'victim of an act or an object of a threat of aggression in which nuclear weapons are used'. These pledges, usually referred to as 'positive assurances', were clearly insufficient, as they merely reaffirmed the duty of UN members to provide assistance to a country which has been aggressed, irrespective of the type of weapon used in aggression. Moreover, China and France, the remaining nuclear-weapon powers, which at that time were not parties to the NPT, were not bound by this resolution, adopted by a majority vote.

At the 1990 NPT Review Conference, Egypt submitted a proposal for a new Security Council resolution. It envisaged a collective commitment, instead of a mere tripartite commitment of the depositaries of the NPT, to provide assistance to the affected states, as well as an obligation of the Security Council to decide immediately upon measures to be taken in response to a threat of use or actual use of nuclear weapons against a non-nuclear-weapon state party to the NPT. The measures in question would be adopted in conformity with Chapter VII of the UN Charter, which deals with 'action with respect to threats to the peace, breaches of the peace, and acts of aggression'. They could include technical, financial and humanitarian assistance to the victims as well as sanctions against any state, party or non-party to the NPT which had used nuclear weapons against a non-nuclear-weapon party to the NPT. The proposal was not taken up for discussion.

Negative Assurances

Both Resolution 255 and that proposed by Egypt provided for action only when a threat of nuclear attack had been made or an attack had occurred. Therefore, states which have forsworn nuclear weapons under the NPT have also demanded formal assurances that nuclear weapons would not be used against them. Such assurances – usually called 'negative' because they amount to a non-use obligation, as distinct from assurances containing an obligation to assist, as described above – were given to states establishing nuclear-weapon-free zones. Negative security assurances were also contained in statements made by the nuclear-weapon powers in connection with the 1978 and 1982 Special Sessions of the UN General Assembly devoted to disarmament, as well as on other occasions. However, they were conditional, phrased in a different way by different countries, and merely declaratory.

For years, efforts have been made in various forums, including the Conference on Disarmament (CD), to develop negative security assurances that would be uniform, unconditional and legally binding. The UN General Assembly adopted several resolutions recommending the conclusion of an international convention on the non-use of nuclear weapons against non-nuclear-weapon states. In 1990 the following proposal was put forward by Nigeria. The nuclear-weapon states would undertake, under an international agreement, not to use or threaten to use nuclear weapons against any non-nuclear-weapon state party to the NPT which does not belong to a military alliance and does not have other security arrangements with a nuclear-weapon state, as well as against any non-nuclear-weapon state party to the NPT which belongs to a military alliance or has other security arrangements with a nuclear-weapon state but has no nuclear weapons stationed on its territory. Non-nuclear-weapon states in the latter category would, for their part, undertake not to participate in, or contribute to, a military attack against any nuclear-weapon state or its allies parties to the NPT, except in self-defence. A special conference would be convened to conclude such an agreement in the form of a protocol to the NPT. The Nigerian proposal was re-submitted in 1995, but was not subject to international consideration.

In 1992, at the CD, France formulated what it considered to be the basic elements of a possible agreement on negative security assurances. These elements included a pledge by the nuclear-weapon powers to refrain from the threat or use of nuclear weapons against non-nuclear-weapon states party to the NPT or regional denucle-arization treaties or against states not party to these treaties which have concluded with the IAEA an agreement for the application of full-scope safeguards. Neither states belonging to military alliances nor states having nuclear weapons stationed on their territory but considered as non-nuclear-weapon states under the NPT would be a priori excluded from such assurances. Under certain circumstances, namely to repel aggression, nuclear weapons could be used against any non-nuclear-weapon state.

In 1994, 11 non-aligned members of the Conference on Disarmament submitted a draft protocol on security assurances. The nuclear-weapon states would pledge not to use or threaten to use nuclear weapons against non-nuclear-weapon states, the latter being defined as all states other than those falling under the NPT definition of a nuclear-weapon state. In the case of nuclear aggression or threat of such aggression against a non-nuclear-weapon state, the necessary help and assistance would be provided by a conference of the parties to the NPT and the UN Security Council. The proposed protocol was to become an integral part of the NPT. In fact, it would have been only indirectly related to this Treaty, as it would provide negative security assurances to non-parties to the NPT as well. The protocol was to enter into force under the same conditions as the NPT, that is, even before China and France had ratified it.

Combined Assurances

The proposals described above did not prove generally acceptable, and the nuclear-weapon states refused to enter into negotiations on any one of them. Only in 1995, a few days before the NPT Review and Extension Conference, did the great powers decide to jointly sponsor UN Security Council Resolution 984, which combined positive and negative security assurances. This resolution was adopted unanimously.

The new positive assurances, now given by all five declared nuclear-weapon states, are more specific than those included in Resolution 255. They provide that, in response to a request from a state victim of an act of nuclear aggression, or object of a threat of such aggression, the Security Council members would help to settle the dispute and restore international peace and security, as well as take 'appropriate' measures, individually or collectively, for technical, medical, scientific or humanitarian assistance. In addition, 'appropriate' procedures might be recommended by the Security Council regarding compensation under international law from the aggressor for loss, damage or injury sustained as a result of the aggression. Thus, at least certain postulates put forward by Egypt in 1990 (see above) were met.

With regard to negative assurances, no progress whatsoever was achieved. Upon declaring the obvious, namely, that an aggression with the use of nuclear weapons would endanger international peace and security, Resolution 984 simply took note of the statements made by the nuclear-weapon states, in which the conditions for non-use of such weapons were reiterated. France, Russia, the United States and the United Kingdom reaffirmed that they would not use nuclear weapons against non-nuclear-weapon states parties to the NPT, except in the case of an invasion or any other attack on them, their territories, their armed forces or other troops, their allies, or on a state to which they have a security commitment, carried out or sustained by such a non-nuclear-weapon state in 'association or alliance' with a nuclear-weapon state. For Russia, the above statement confirmed the reversal of the policy of no first use of nuclear weapons, advocated until 1993, and the official adoption of the doctrine of nuclear deterrence.

Only China undertook not to use or threaten to use nuclear weapons against non-nuclear-weapon states or nuclear-weapon-free zones at any time and under any circumstance. This commitment applies to non-nuclear-weapon parties to the NPT or non-nuclear-weapon states that have undertaken comparable internationally binding commitments not to manufacture or acquire nuclear explosive devices.

Resolution 984 refers (as Resolution 255 did) to Article 51 of the UN Charter dealing with the right of self-defence. This Charter provision does not have direct relevance to the issue of providing security assurances to non-nuclear-weapon states, but a reference to it may serve to legitimize the use or the threat of use of nuclear weapons in countering *any* armed attack, including one carried out solely with conventional means of warfare, as if the right of self-defence were unlimited. Thus, by Resolution 984 the nuclear-weapon powers did not enter into any new international commitments.

Assessment

It is doubtful whether at any time during the Cold War the nuclear-weapon powers had seriously contemplated the possibility of renouncing all use of nuclear weapons. It is surprising, however, that after the termination of the Cold War confrontation, the elimination of the US and Soviet intermediate-range nuclear forces, the withdrawal of most tactical nuclear weapons to central locations and the beginning of the process of strategic weapons dismantlement, the nuclear postures have remained unchanged. Each nuclear-weapon state possesses conventional armed forces quantitatively and/or qualitatively superior to those of its potential non-nuclear-weapon adversaries and would not need to resort to nuclear weapons to stop an aggression launched by the latter. The argument that the option of using nuclear weapons against non-nuclear-weapon states must be retained to react to a possible use of chemical or biological weapons is not convincing. Should a chemical or biological threat emerge, a massive response with sophisticated conventional weapons would suffice, as was convincingly demonstrated by the UN coalition forces during the

1991 Gulf War. The residual role of nuclear weapons amounts now to nothing more than deterring – through a threat of reprisal in kind – the first use of these weapons. All nuclear-weapon states have declared that their nuclear weapons are not targeted at any state, and yet the nuclear security assurances they have given to non-nuclear-weapon states are still neither unconditional nor legally binding. The results of the US Department of Defense's 2001 Nuclear Posture Review were understood by many as undermining even the conditional security assurances of the United States.

Resolution 984 (1995) pointed out that the issues raised in its provisions remained of continuing concern to the Security Council. This statement may serve as a point of departure for negotiating a more meaningful international instrument.

6.3 Protection of Nuclear Material

The following measures to protect nuclear material form part of the non-proliferation regime.

Protection in International Transport

A major step towards reducing the risks of diversion of nuclear material to non-peaceful purposes was made in 1980 with the signing of the Convention on the Physical Protection of Nuclear Material. This convention, in force since 1987, obliges the parties to ensure that, during international transport across their territory or on ships or aircraft under their jurisdiction, nuclear material for peaceful purposes, as categorized in an annex, is protected at the agreed level. (It does not apply to the physical protection of nuclear material for military purposes or to the protection of other radioactive sources.) Thus, for example, transportation of 2 kilograms or more of plutonium or of 5 kilograms or more of enriched uranium must take place under constant surveillance by escorts and under conditions which ensure close communications with 'appropriate response forces'.

Furthermore, the parties undertook not to export or import nuclear material or allow its transit through their territory, unless they had received assurances that this material would be protected during international transport in accordance with the levels of protection determined by the convention. The parties also agreed to share information on missing nuclear material to facilitate recovery operations. Robbery, embezzlement or extortion in relation to nuclear material, and acts without lawful authority involving nuclear material which cause or are likely to cause death or serious injury to any person or substantial damage to property, are to be treated as punishable offences. Each party must inform the depositary of its laws and regulations giving effect to the convention. In 1997 the International Maritime Organization (IMO) decided to incorporate the Code for the Safe Carriage of Irradiated Nuclear Fuel, Plutonium and High-Level Radioactive Wastes in Flasks on Board Ships (INF Code) into the International Convention for the Safety of Life at Sea.

In 1998 a group of experts convened by the IAEA Director General to review all Agency programmes urged that consideration be given to the possible revision of the Convention on the Physical Protection of Nuclear Material. Subsequently, the IAEA experts recommended that the scope of the Convention be expanded by requiring member-states to pass legislation implementing IAEA Guidelines on a range of issues, including how nuclear materials and facilities can be protected from sabotage. A formal conference is needed to amend the Convention.

Protection in Domestic Activities

Within states, the responsibility for physical protection of nuclear material and facilities rests with the governments of these states. However, such protection is a matter of worldwide concern. Since the effectiveness of physical protection in one state may depend on measures taken by another state, there is a need for international cooperation. Theft of plutonium or highly enriched uranium could lead to the construction of an explosive device capable of causing mass destruction. Moreover, an act of sabotage against a nuclear facility – nuclear reactor, separate irradiated fuel storage site, reprocessing plant or fuel fabrication facility utilizing plutonium – or against a shipment of nuclear material within one country could create a radiological hazard to the populations of other countries. To deal with these problems, the IAEA published recommendations for what member-states can do to establish national systems for the protection of nuclear facilities and of nuclear material in use, transport and storage, or to improve the quality and the effectiveness of the existing systems. The IAEA International Physical Protection Advisory Service (IPPAS) provides advice and assistance to member-states in translating these recommendations into specific requirements.

The 1997 Guidelines for the Management of Plutonium set out the policies that a number of states, including the nuclear-weapon states, decided to follow with regard to plutonium. In accordance with these guidelines, annual statements of national holdings of civil unirradiated plutonium and of plutonium contained in spent civil reactor fuel are submitted to the IAEA.

Under the 1994 Convention on Nuclear Safety, in force since 24 October 1996, the contracting parties agreed to achieve a high level of nuclear safety worldwide through the enhancement of national measures and international cooperation, including safety-related technical cooperation; to establish effective defences in nuclear installations against potential radiological hazards in order to protect individuals, society and the environment from harmful effects of ionizing radiation from such installations; to prevent accidents with radiological consequences and to mitigate such consequences should they occur. The parties must submit reports at periodic review meetings on measures taken to implement their obligations. The Convention covers only civilian nuclear power plants.

Protection of Weapons

There are no means to prevent nuclear weapons from falling into the hands of subnational political groups or foreign governments when law and order in a nuclear-weapon state break down. Tactical weapons, which exist in greater numbers than any other type of nuclear weapon, present a particular danger; they are relatively small and therefore easy to conceal and transport. Some are not equipped with a protective mechanism and may be directly usable. Full awareness of the dangers of nuclear terrorism or of an accidental nuclear explosion should lead to the abolition of tactical nuclear weapons. In the meantime, all nuclear weapons, both deployed and non-deployed, must be fitted with use-denial mechanisms that disable the weapons when unauthorized persons attempt to use them.

The Problem of Smuggling

The disintegration of the Soviet Union weakened the security of nuclear installations and storage facilities and brought about the loosening of nuclear export controls in the new independent states. In 1995, the nuclear material inventory in Russia consisted of some 1,100 to 1,300 tons of highly enriched uranium and some 165 tons of separated plutonium, distributed over more than 50 sites. Russian officials acknowledged that there had been many violations of the Russian regulations for securing and accounting for nuclear materials. Proliferation-significant quantities of these materials were also stocked in Belarus, Kazakhstan and Ukraine. These circumstances, as well as the fact that weapon-grade fissile materials can be safely handled and transported and cannot be easily detected by law-enforcement authorities, created conditions that facilitated theft and smuggling.

The smugglers first offered very small quantities of plutonium or low-enriched uranium, probably as samples for possible customers, but since mid-1994 the police of several countries have intercepted substantial quantities of plutonium and weapon-usable uranium. (Other intercepted radioactive materials proved unfit for weapon purposes.) In none of the known cases was the amount of the confiscated material enough for an industrially underdeveloped country to manufacture a nuclear explosive device, and the probability that a terrorist group would have the capability to construct such a device is low. Obtaining the necessary materials is only the first step in building a nuclear bomb; its production requires highly qualified personnel in the fields of physics, chemistry, metallurgy and electronics.

Russia was not the only country to have difficulties with fissile material management. Other countries encountered problems as well, but no buyer of smuggled nuclear-weapon-usable material was identified. None of the states considered to be potential nuclear proliferators appeared to show interest in the material offered. Because smuggling on a massive scale is highly unlikely, those aspiring to nuclear-weapon status would most probably try to acquire a weapon-producing potential rather than a limited amount of material for only one or two weapons. Although a real black market of plutonium and highly enriched uranium does not, as yet, seem to exist, surveillance of nuclear facilities must be reinforced and border controls rendered more effective to prevent its emergence.

Anti-smuggling efforts ought to be coordinated internationally. At the request of its Board, the IAEA developed a database of incidents of illicit trafficking in nuclear material and in other radioactive sources. The data include open information voluntarily provided by states as well as information obtained from the media and other unofficial sources. States may also provide information that they consider confidential. For the purposes of reporting, an 'illicit trafficking incident' is a situation in which the movement or sale of nuclear material or other radioactive sources is not in conformity with national law and involves a quantity or quality of material which is of interest from either a proliferation or radiation protection perspective. The analysis of the data on confirmed cases of illicit trafficking made available by March 2000 indicated that the majority of seizures of nuclear materials had been made in Europe and that most material had been stolen while in domestic use or storage rather than in transit.

In their joint statement of 28 September 1994, Presidents Clinton and Yeltsin agreed to cooperate in combating the illegal trade in nuclear material and enhance transparency in nuclear matters. Moreover, the participants in the April 1996 Nuclear Safety and Security Summit agreed to ensure increased cooperation among their governments in all aspects of prevention, detection, exchange of information, investigation and prosecution in cases of illicit nuclear trafficking. In this context, the 1 September 2000 US–Russian Agreement Concerning the Management and Disposition of Plutonium Designated as no Longer Required for Defense Purposes and Related Cooperation is significant. This so-called Plutonium Management and Disposition Agreement (PMDA) requires that 68 metric tons of weapon-grade plutonium – 34 tons for each party (enough for thousands of nuclear weapons) – be disposed by irradiating it as fuel in reactors or by immobilizing it with high-level radioactive waste. The PMDA establishes conditions for ensuring that this plutonium can never be used for weapons or any other military purposes.

The United States has assisted the authorities of several former Soviet republics in developing and installing modern surveillance and monitoring equipment for use at sites where sensitive nuclear material is stored. Complete material accountancy in both the nuclear- and non-nuclear-weapon states would further help to enhance the safety of fissile material.

It was proposed that an international anti-smuggling convention be negotiated to complement the existing rules dealing with the threat of diversion of nuclear weapons or nuclear-weapon material. Such a convention could also strengthen export controls.

Action against Nuclear Terrorism

On 19 March 2002 the IAEA Board of Governors agreed on an 'action plan' designed to upgrade worldwide protection against acts of nuclear terrorism. The plan covers the following areas: physical protection of nuclear material and nuclear facilities; detection of malicious activities (such as illicit trafficking) involving nuclear and other radioactive materials; strengthening of state systems for nuclear material accountancy and control; security of radioactive sources; assessment of safety- and security-related vulnerabilities at nuclear facilities; response to malicious acts or threats thereof; adherence to international agreements and guidelines; and enhancement of programme coordination and information management for nuclear safety-related matters.

A number of states pledged specific sums of money for a special fund set up to support the plan. Several other states announced in-kind support.

6.4   Nuclear Supplies

From the political perspective, the threat of nuclear-weapon proliferation has diminished since the entry into force of the NPT, but from a technical perspective it may have increased because it has become easier for states to develop nuclear weapons. Nuclear-weapon technology is no longer a secret shared by a few, and most non-nuclear components of the weapons are available in international commerce. Hence the need for ever stricter measures of control over nuclear supplies.

Guidelines for Nuclear Transfers

In 1977 a group of nuclear material and equipment exporters, the so-called 'London Club', adopted a set of principles for safeguards and export controls. The group included France, which for the first time participated in the formulation of international nuclear export controls, even though it was not yet party to the NPT. The Guidelines for Nuclear Transfers, agreed by what was subsequently called the Nuclear Suppliers Group (NSG), were several times revised taking account of the advances in technology, the proliferation sensitivity and the changes occurring in procurement practices. Unanimous consent of the NSG members is needed for modification of the Guidelines.

The Guidelines apply to nuclear transfers for peaceful purposes to any non-nuclear-weapon state and, in the case of controls on retransfer, to transfers to any state. An export 'trigger list' was defined. Suppliers may authorize the transfer of items or related technology identified in this list only upon formal governmental assurances from recipients explicitly excluding uses which would result in a nuclear explosive device. All listed nuclear materials and facilities should be placed under physical protection to prevent unauthorized use and handling. Arrangements should be made for a clear definition of responsibilities for the transport of the trigger list items. Suppliers may transfer trigger list items or related technology to a non-nuclear-weapon state only when the receiving state has brought into force an agreement with the IAEA requiring the application of safeguards on all source and special fissionable material in its current and future peaceful activities. Transfers to a non-nuclear-weapon state without such a safeguards agreement may be authorized only in exceptional cases, when they are deemed essential for the safe operation of existing facilities, and only if safeguards are applied to those facilities. The above policy does not apply to agreements or contracts drawn up on or prior to 3 April 1992. Suppliers reserve the right to apply additional conditions of supply as a matter of national policy. All these requirements also apply to facilities for reprocessing, enrichment, or heavy-water production, utilizing technology directly transferred by the supplier or derived from transferred facilities. Transfers of such facilities, or major critical components thereof or related technology, require an undertaking that IAEA safeguards apply to any facility of the same type (that is, if the design, construction or operating processes are based on the same or similar physical or chemical processes, as defined in the trigger list) constructed during an agreed period in the recipient country and that there is at all times in effect a safeguards agreement permitting the IAEA to apply Agency safeguards with respect to such facilities identified as using transferred technology.

Suppliers must exercise restraint in the transfer of sensitive facilities, technology and weapons-usable materials. If enrichment or reprocessing facilities, equipment or technology are to be transferred, suppliers should encourage recipients to accept, as an alternative to national plants, supplier involvement and/or other appropriate multinational participation in resulting facilities. For a transfer of an enrichment facility, or technology therefor, the recipient nation must agree that neither the transferred facility, nor any facility based on such technology, will be designed or operated for the production of greater than 20% enriched uranium without the consent of the supplier nation. Transfer of trigger list items or related technology may take place only upon the recipient's assurance that in the case of retransfer of such items

or technology, or transfer of trigger list items derived from facilities originally transferred by the supplier, or with the help of equipment or technology originally transferred by the supplier, the recipient of the transfer or retransfer has provided the
same assurances as those required by the supplier for the original transfer. In addition, the supplier's consent is required for certain specified transfers and retransfers.
In general, suppliers may authorize transfer of the trigger list items or related technology only when they are satisfied that the transfers will not contribute to the proliferation of nuclear weapons or other nuclear explosive devices. Suppliers should
promote international exchange of physical security information, protection of
nuclear materials in transit and recovery of stolen nuclear materials and equipment.
They should also encourage the designers and producers of sensitive equipment to
construct it in such a way as to facilitate the application of safeguards. In the event
that one or more suppliers believe that there has been a violation of sup-
plier/recipient understandings resulting from the Guidelines, particularly in the case
of an explosion of a nuclear device, or illegal termination or violation of IAEA safeguards by a recipient, suppliers should consult promptly through diplomatic channels in order to determine and assess the reality and extent of the alleged violation.
Upon the findings of such consultations, the suppliers should agree on an appropriate response and possible action which could include the termination of nuclear
transfers to that recipient.

Nuclear Dual-Use Guidelines

In March–April 1992 the NSG meeting in Warsaw adopted the Guidelines for
Transfers of Nuclear-Related Dual-Use Equipment, Material and Related Technology (the so-called Warsaw Guidelines), which became effective in January 1993.
According to these guidelines, the suppliers may not authorize transfers of equipment, materials, software or related technology, identified in the Annex, for use in a
non-nuclear-weapon state in a nuclear explosive activity or an unsafeguarded
nuclear fuel cycle activity or, in general, when there is an unacceptable risk of diversion to such an activity, or when the transfers are contrary to the objective of averting the proliferation of nuclear weapons. (A number of states notified the IAEA that,
in the light of developments in nuclear-related technology, they have updated parts
of the list of items incorporated in the Annex.)

Export licensing procedures for the transfer of relevant items, which are to be
established by the suppliers, should include enforcement measures for violations. In
considering whether transfers should be authorized, the most important factor to be
taken into account is whether the recipient state is a party to the NPT or to a similar
international, legally binding nuclear non-proliferation agreement, and has an IAEA
safeguards agreement in force applicable to all its peaceful nuclear activities. Before
authorizing a transfer, the supplier should obtain a statement from the end-user specifying the uses and end-use locations of the proposed transfer, as well as an assurance that the proposed transfer or any replica thereof will not be used in any nuclear
explosive activity or unsafeguarded nuclear fuel cycle activity. In case of transfer to
a non-adherent to the Warsaw Guidelines, suppliers should obtain an assurance that
their consent will be secured prior to any retransfer of the relevant items or replica
thereof to a third country. Each supplier country may apply the Guidelines to other
items of significance in addition to those specified in the Annex. It may also apply

other conditions for transfer in addition to those provided for in the Guidelines. Suppliers should exchange information and consult with other states adhering to the Guidelines.

Members of the NSG stated that in adopting the Nuclear Dual-Use Guidelines they were aware of the need to contribute to economic development while avoiding contributing in any way to the dangers of proliferation of nuclear weapons or other nuclear explosive devices, and of the need to remove non-proliferation assurances from the field of commercial competition. Japan serves as a point of contact for administering the transfer control arrangements through its Permanent Mission to the IAEA in Vienna.

The Zangger Committee

Since 1971 another intergovernmental group, the Nuclear Exporters Committee, known as the Zangger Committee (after its first chairman, the representative of Switzerland to the IAEA), has been active in establishing the conditions and procedures to govern exports of nuclear equipment or material in accordance with the obligations set out in the NPT as well as on the basis of fair commercial competition.

The Zangger Committee is engaged in the exchange of information about exports, or licences for exports, to any non-nuclear-weapon state not party to the NPT, through a system of annual returns that are circulated on a confidential basis among the members. Understandings reached in the Committee are communicated by individual countries to the IAEA and are carried into effect through domestic export control legislation. An agreed trigger list specifies items which, when exported, must be subject to safeguards under an agreement with the IAEA. The list is continuously reviewed and updated following the developments in nuclear technology. The Zangger Committee is an informal body; its understandings have no status in international law but are arrangements unilaterally entered into by member-states.

Since the NSG, which comprises the members of the Zangger Committee, has adopted strict guidelines for nuclear supplies, including a detailed trigger list, and since the major suppliers have committed themselves not to export nuclear material or equipment to states which are not covered by full-scope safeguards, the Zangger Committee may appear superfluous. It continues, nevertheless, as a technical body complementary to the NSG, to develop and clarify the trigger list.

Assessment

Spokesmen of certain developing countries have criticized the restrictive measures taken by suppliers as an infringement of the right to nuclear supplies implied in the NPT. Their argument is that, since governments have accepted the safeguards required by the Treaty, no further limitation should be placed on their peaceful nuclear programmes. However, under the NPT, the right of parties to obtain equipment, material and technology for peaceful uses of nuclear energy is not unlimited: any such supplies are subordinated to non-proliferation goals. This means that they must not in any way facilitate the acquisition of nuclear weapons.

The nuclear export controls may have slowed the pursuit of nuclear weapons by certain non-NPT states. There were cases of illegal exports, but these have been prosecuted by the authorities of the countries concerned as criminal offences.

6.5   Fissile Material Production Cut-Off

Following 1993 UN General Assembly Resolution 48/75L, which called for a cut-off of production of fissile materials for nuclear explosive purposes – the first such resolution adopted by consensus – it was widely expected that the matter would soon become the subject of negotiations at the CD. However, the CD encountered difficulties in defining a mandate of the ad hoc committee to be entrusted with such negotiations. Some delegations were of the view that the mandate should permit consideration only of the future production of fissile material, that is, production after an agreed cut-off date. Other delegations insisted that it should permit consideration of past production as well, so as to eliminate the asymmetry in the possession of fissile material stockpiles by various states. Still others proposed that, in addition to the question of production of fissile material, consideration be given to the management of such material. In March 1995 agreement was reached to negotiate a 'non-discriminatory, multilateral and internationally and effectively verifiable treaty banning the production of fissile material for nuclear weapons or other nuclear explosive devices'. A proviso was made that no delegation would be precluded from raising for consideration any of the above-mentioned controversial issues. However, the cut-off negotiating committee could not start work in spite of its formal establishment, because several delegations demanded that other measures be simultaneously negotiated, in particular, the prevention of an arms race in outer space and the elimination of nuclear weapons.

Importance of the Cut-Off Measure

Already at the beginning of the 1960s, in the wake of the conclusion of the Partial Test Ban Treaty, the United States was prepared to cut the production of fissile material for nuclear-weapon purposes down to its actual needs, on condition that the Soviet Union acted likewise. On 20 April 1964 President Johnson announced a substantial reduction in US production of enriched uranium to be carried out over a four-year period. Simultaneously with this announcement, Chairman Khrushchev made public the decision of the Soviet government to stop the construction of two new large atomic reactors for the production of plutonium; to reduce substantially the production of uranium-235 for nuclear weapons; and to allocate accordingly more fissile materials for peaceful uses. On 21 April 1964 Prime Minister Douglas-Home stated that plutonium production in the United Kingdom was being gradually terminated and that the plutonium produced by civil reactors would not be used in the weapons programme.

The above measures were largely understood as the start of a process leading to an internationally agreed complete cessation of production of fissile materials for weapons, but the nuclear-weapon powers continued producing these materials. In the course of several decades they accumulated such enormous quantities of weapon-grade uranium and plutonium that they could, without risk, stop their production unilaterally, without a formal treaty. Most of them did so. Russia and the United States went even further. In addition to the undertaking to stop the operations of plutonium production reactors, they agreed, bilaterally, to definitively dispose of large quantities of weapon-grade plutonium withdrawn from their respective nuclear-weapon programmes. To render it irreversibly unusable for nuclear weapons, the plutonium is to be irradiated in nuclear power reactors as so-called mixed-oxide (MOX) fuel and/or 'immobilized' (in glass or ceramic forms) in high-level radioactive waste and buried.

Nevertheless, a global treaty banning the production of fissile materials for nuclear explosive devices would strengthen the non-proliferation regime, even if it did not affect the existing stockpiles. Plutonium or highly enriched uranium is the basic component of all nuclear weapons. These materials are also the most complicated and expensive parts of nuclear weapons to produce. A halt to their production would limit the size of potential nuclear arsenals. Moreover, in depriving the nuclear-weapon powers of their right to produce unsafeguarded fissile materials, a verified cut-off measure would also attenuate the present inequality of the NPT parties with regard to nuclear safeguards. For, whereas the non-nuclear-weapon states are obliged under the NPT to apply safeguards to all their nuclear activities, the nuclear-weapon states are not; they have submitted to international controls only a certain number of nuclear facilities and have done so only on a voluntary basis. Under a multilateral cut-off treaty there would be no mandatory verification of stocks of weapon-usable materials from past production, but verification of future non-production of weapon-usable materials could be the same for all parties. Particularly, the enrichment and reprocessing plants in the territories of the parties would have to be subject to undifferentiated international verification. Production of highly enriched uranium for naval reactors and certain research reactors would have to be addressed separately. (It is becoming increasingly feasible to use low enriched uranium for the propulsion of ships.)

Suggestions have been made to extend the fissile material cut-off measure to include tritium production. Tritium, produced in reactors, is an important constituent of many nuclear warheads, where it 'boosts' the yield of the fission explosion, but it also has civilian uses.

Prospects

Over 180 non-nuclear-weapon parties to the NPT are under the obligation not to produce nuclear-weapon-usable materials and are subject to full-scope IAEA safeguards; they are not expected to assume additional non-proliferation obligations or controls. Cessation of production of the materials in question directly concerns only states which conduct significant nuclear activities but are not subject to full-scope IAEA safeguards. It would, therefore, be more expedient to negotiate the proposed cut-off measure in a forum composed of these countries, whether parties or non-parties to the NPT, rather than at the CD, composed of 66 countries. Other states could be involved, through the IAEA, in verifying compliance with the reached agreement, but only states affected by the agreement should bear the additional costs.

Certain opponents of a cut-off treaty argue that it would amount to indirectly recognizing the freedom of India, Israel and Pakistan, in addition to the five recognized nuclear-weapon states, to retain their unsafeguarded stocks of fissile materials, use these materials for the production of nuclear weapons and retain the weapons already manufactured. This freedom cannot be taken away by a cut-off treaty alone, but it may be significantly curtailed.

To avoid interpreting the cut-off as an arrangement legitimizing the nuclear status of the three above-mentioned states, the measure should be unambiguously recognized as a temporary, transitional step in the process of nuclear disarmament. The cut-off treaty should provide for, or be followed by, the establishment of a comprehensive, regularly updated global register of stocks of plutonium (both weapon-grade and reactor-grade, as the latter, too, can be used to make nuclear weapons) and highly enriched uranium. Such transparency would facilitate a possible future, internationally verified prohibition on the use of any fissile material, including the material extracted from dismantled weapons, for the production of new weapons.

6.6   The Missile Technology Control Regime

A recommendation frequently made to strengthen the non-proliferation regime was to complement the existing restraints on supplies of nuclear material and equipment by restraints on supplies of dual-capable weapon systems, that is, systems capable of delivering both conventional and nuclear weapons. This recommendation was partly put into practice when, in April 1987, seven governments – those of Canada, France, the FRG, Italy, Japan, the United Kingdom and the United States – established the Missile Technology Control Regime (MTCR).

Guidelines for Sensitive Missile-Relevant Transfers

The agreed Guidelines for Sensitive Missile-Relevant Transfers were originally meant to cover only transfers of equipment and technology which could make a contribution to missile systems capable of delivering a nuclear weapon, and were to be applied only to missiles exceeding certain specified thresholds for the range (300 kilometres) and weight of payload (500 kilograms). In 1992, they were amended to cover missiles capable of delivering not only nuclear but also biological and chemical weapons, regardless of range and payload. They must not impede space programmes, as long as such programmes could not contribute to the delivery systems (other than manned aircraft) for weapons of mass destruction.

The revised MTCR Guidelines – in effect since January 1993 – are accompanied by an Annex specifying two categories of item, which term includes equipment and technology. Category I items, all of which are in Annex items 1 and 2, are those of greatest sensitivity. If a Category I item is included in a system, that system will also be considered as Category I, except when the incorporated item cannot be separated, removed or duplicated. Particular restraint is to be exercised in the consideration of Category I transfers regardless of their purpose, and there is a strong presumption to deny such transfers. Particular restraint is to be exercised also in the consideration of transfers of any items in the Annex, or of any missiles (whether or not figuring in the Annex), if the supplier government itself judges, on the basis of available, 'persuasive' information, that they are intended to be used for the delivery of weapons of mass destruction; there is a strong presumption to deny such transfers. Until further notice, transfer of Category I production facilities is not to be authorized.

The remaining 18 items in the Annex are Category II items. They are not on a denial list; their transfers are to be considered on a case-by-case basis. Concern about the proliferation of weapons of mass destruction occupies a prominent place among the factors that must be taken into account in the evaluation of all transfer applications. Decisions concerning membership, like all other MTCR decisions are made only by consensus. Membership of the MTCR does not involve an entitlement to obtain technology from another partner or an obligation to supply it. A country can choose to adhere to the MTCR Guidelines without being obligated to join the group. An office in the French Ministry for Foreign Affairs acts as a point of contact for coordinating the schedule of MTCR meetings and their agendas.

By 2001 the MTCR had attracted over 30 states. Most of them possess either ballistic missiles or advanced ballistic missile-related technological capacities.

Assessment

Although missiles can carry all kinds of weapon, the acquisition of missiles in regions of tension may engender pressure for the acquisition of weapons of mass destruction, in particular nuclear weapons, or arouse suspicion that the country importing or producing missiles is planning to acquire such weapons. Indeed, if a nuclear-capable country which possesses missiles decided to 'go nuclear', it would have readily available nuclear delivery vehicles which are more threatening than aircraft: the time of travel from the missile launch pad to target is measured in minutes instead of hours and, once launched, missiles cannot be recalled and are very difficult to intercept. Most missiles that have so far been acquired by developing countries are known to be relatively inaccurate, as exemplified by the Scud missile used by Iraq in the 1991 Gulf War. They would be militarily more effective if they were equipped with weapons of mass destruction rather than with conventional weapons. Hence the importance of the MTCR.

By introducing export licensing requirements for rocket systems (including ballistic missiles) and unmanned air vehicle systems (including cruise missiles and drones) as well as related equipment, material and technology, the MTCR has contributed to stopping or slowing down the missile programmes pursued by several countries, even though it is not embodied in a formal treaty. To make it even more complicated and more costly for countries to acquire sensitive missile technology, the MTCR should be joined by all missile-producing states and its proceedings should cease to be secretive. Furthermore, the MTCR rules, which lend themselves to different interpretations, should be tightened: the restrictions must be made legally binding and an international body must be entrusted with monitoring compliance.

The draft International Code of Conduct against Ballistic Missile Proliferation, agreed in September 2001 by the members of the MTCR as the basis for further consultation and elaboration, is intended to complement and reinforce the missile non-proliferation regime. It does not, however, cover cruise missiles. It contains a set of principles, commitments and confidence-building measures to be implemented via a multilateral instrument open to all states. An extensive exchange of information and reporting are to ensure transparency. A mechanism is to be established for the voluntary resolution of questions arising from national declarations and/or questions pertaining to space-launch vehicle and ballistic missile activities. In fact, however, the Code adds little to the basic provisions of the MTCR Guidelines. If it were to be adopted universally, this could be interpreted as legitimizing, at least indirectly, the indigenous production and deployment of all types of missile.

In the long run, it will certainly prove untenable to enforce regulations aimed at denying certain missiles or missile technology to the majority of nations, while reserving them for a few. There is, moreover, no way to clearly separate the peaceful uses of outer space from the pursuit of long-range missiles. The Russian proposal for a Global Control System (GCS) for the Non-Proliferation of Missiles and Missile Technology, presented by the Russian President in 1999 and reiterated by the Russian Foreign Minister at the 2000 NPT Review Conference, would increase transparency and reduce the risk of miscalculation or misunderstanding by requiring governments to provide notification of ballistic missile launches. It would thus complement the relevant bilateral US–Russian agreements (see Chapter 18). It would not, however, remove or even attenuate the patently unequal treatment of states under the MTCR. Transparency cannot do much to promote the cause of non-proliferation, whereas the security assurances against attacks with missiles carrying weapons of mass destruction, and/or assistance in the peaceful uses of space, which would be offered under the GCS to nations that had voluntarily renounced the use of missiles as delivery vehicles for weapons of mass destruction, might not suffice as positive incentives.

Only a universal renunciation of the missiles covered by the MTCR could significantly reduce the armaments asymmetry between the missile 'haves' and 'have-nots'. It could pave the way towards a general ban on all ballistic missiles – as proposed by President Reagan at the 1986 Reykjavik summit meeting – and on all nuclear-capable cruise missiles. Aircraft, the nuclear-weapon delivery vehicle which would still be left in the possession of states if the 'zero missile' idea were realized, are slower and more vulnerable than missiles and, therefore, somewhat less threatening. Before such ambitious initiatives could be contemplated, the arms control negotiators should devote more attention to stopping and preventing the spread of nuclear explosives and chemical and biological warfare agents than to the spread of the means of their delivery. Strict compliance with the multilateral treaties which ban the proliferation of weapons of mass destruction will inevitably degrade the military value of missiles as carriers of these weapons, especially the utility of missiles of intercontinental range.

7

Proposals for the Abolition of Nuclear Weapons

In April 1995, during the Non-Proliferation Treaty (NPT) Review and Extension Conference, peace activists from a number of countries produced a statement urging that a world free of nuclear weapons be achieved and that environmental degradation and human suffering – a legacy of several decades of nuclear-weapon testing and production – be redressed. This statement became the founding document of the movement called 'Abolition 2000, A Global Network to Eliminate Nuclear Weapons'. It was signed by hundreds of non-governmental organizations (NGOs), many of which had been working for the abolition of nuclear weapons since the 1950s.

In December 1996, in what was generally viewed as a surprise move, retired General Lee Butler, former Commander-in-Chief of the US Strategic Air Command, and retired General Andrew Goodpaster, former Supreme Allied Commander in Europe, released a joint statement in favour of the elimination of nuclear weapons worldwide. The next day 61 retired generals and admirals from 17 states, including 18 from Russia and 19 from the United States, issued a statement claiming that the continuing existence of nuclear weapons constituted a peril to global peace and security and to the safety and survival of the people 'we are dedicated to protect'. The signers of the statement concluded that the creation of a nuclear-weapon-free world was a challenge of the highest possible historic significance and that the dangers of proliferation, of terrorism, and of a new nuclear arms race rendered it necessary.

A statement made by international civilian leaders in February 1998 also had considerable impact on world opinion regarding nuclear armaments. This statement, signed within a month by well over 100 outstanding individuals from dozens of nations, including former heads of states or government, called for specific steps to reduce the dangers inherent in nuclear weapons and urged that the nuclear powers declare unambiguously that their goal is the abolition of nuclear weapons.

Subsequently, in June 1998, the Foreign Ministers of Brazil, Egypt, Ireland, Mexico, New Zealand, Slovenia, South Africa and Sweden made a joint declaration calling upon the governments of the five nuclear-weapon states and of the three nuclear-capable states to commit themselves unequivocally to the elimination of their nuclear weapons and nuclear-weapon capability. They also requested that negotiations begin to achieve the sought goal.

Following the above initiatives, or in parallel with them, detailed proposals for accomplishing nuclear disarmament were submitted by groups of states, panels of independent experts and research institutions. They are summarized here in the chronological order of their presentation.

7.1   The Stimson Center's Report

In December 1995 the Washington-based Henry L. Stimson Center published a report on *An Evolving US Nuclear Posture*. The report, adopted by a commission chaired by General Goodpaster, stated that the possession of nuclear weapons and reliance on nuclear deterrence entailed significant economic and political costs and that the very existence of these weapons entailed a risk that they would be used one day with devastating consequences for the United States and other nations. In the view of the authors of the report, US nuclear weapons were of declining military and political utility in both addressing the residual threats of the Cold War and in encountering emerging threats to the security of the United States.

The postulated 'evolutionary' nuclear posture would establish a long-term objective, that of eliminating all nuclear weapons of all states, but would enable the United States to undertake changes in the size and operational status of its nuclear forces in a gradual manner. The essential prerequisites for progress towards the above objective were increased openness and access to information regarding the activities, facilities and materials related to national defence postures and weapons of mass destruction, as well as arms control regimes making reductions of nuclear weapons and of weapon materials irreversible. In the long term, effective regional and collective security regimes were likely to be necessary if states were to be persuaded to forgo acquisition of all weapons of mass destruction.

The following four phases were suggested as guidelines for US policy. During the first phase, the United States and Russia would work to shift the foundation of their relationship away from mutual assured destruction and would reduce their nuclear arsenals to roughly 2,000 warheads each. During the second phase, nuclear deterrence would become far less central to maintaining stable relations among the declared nuclear-weapon powers, which would allow them to reduce their arsenals to hundreds of nuclear weapons each. During the third phase, nuclear weapons would be further marginalized in national policies and interstate relations through the establishment of reliable cooperative security and verification regimes, and all remaining nuclear powers would reduce their arsenals to tens of weapons. At this point, the international community could evaluate the relative costs and benefits of eliminating all nuclear weapons from all nations. During the fourth and final phase, the international community would have to have effective and reliable security alternatives to the threat of mass violence and sufficiently stringent verification regimes to allow for the complete elimination of nuclear weapons from all countries.

The Stimson Center's Report dealt almost exclusively with US security interests. The reason given was that the United States, the leading military and political power in the world, bore a special responsibility to 'spearhead the movement' to gradually decrease and, if possible, eliminate the dangers associated with nuclear weapons. Adoption of an evolutionary nuclear posture and a revitalized commitment to the long-term objective of eliminating all nuclear weapons could bring important national security benefits to the United States while entailing minimal risks. However, under current political conditions, the authors of the report considered the elimination of nuclear weapons as 'infeasible'. It was deemed achievable only after far-reaching changes had occurred in the principles that guide state policies and actions.

7.2   Programme of Action Proposed at the CD

In August 1996 a group of 21 non-aligned countries participating in the Conference on Disarmament (CD) proposed a Programme of Action for the elimination of nuclear weapons in three phases.

In the first phase – from 1996 to 2000 – multilateral negotiations would commence with a view to the early conclusion of a legally binding instrument to assure non-nuclear-weapon states against the use or threat of use of nuclear weapons; of a convention prohibiting the use or threat of use of nuclear weapons; of a treaty to eliminate nuclear weapons; and of a treaty banning the production of fissile material for nuclear weapons. Agreements had to be reached to end the qualitative improvement of nuclear weapons by stopping all nuclear-weapon tests and closing all nuclear-weapon test sites, as well as by preventing the use of new technologies for the upgrading of nuclear-weapon systems, including the prohibition of relevant research and development. Additional nuclear-weapon-free zones were to be established and declarations made of the stocks of nuclear weapons and of nuclear-weapon-usable material. Moreover, nuclear-weapon systems were to be taken down from the state of operational readiness; the ABM Treaty preserved; the testing of outer space weapon systems suspended and then prohibited; the START II Treaty ratified; negotiations on further reductions of nuclear arsenals initiated and concluded; the fissile material transferred from military to peaceful uses placed under IAEA safeguards; negotiations for nuclear disarmament, including the cessation of production of nuclear warheads by all nuclear-weapon states, continued; and the decade 2000–2010 declared as the 'Decade for Nuclear Disarmament'.

In the second phase – from 2000 to 2010 – the treaty eliminating nuclear weapons was to enter into force and a single integrated multilateral comprehensive verification system established, including such measures as: the separation of nuclear warheads from their delivery vehicles; the placement of nuclear warheads in secure storage under international supervision; and transfer of fissile materials and delivery vehicles to peaceful purposes. Moreover, an inventory of nuclear arsenals was to be prepared under international auspices, missiles intended to carry nuclear warheads reduced in a balanced manner, and the decade 2010–2020 declared as the 'Decade for the Total Elimination of Nuclear Weapons'.

In the third phase – from 2010 to 2020 – principles and mechanisms for a global cooperative security system would be adopted, and the treaty eliminating all nuclear weapons would be fully implemented. All facilities devoted to the production of nuclear weapons would be converted to peaceful purposes; safeguards on nuclear facilities universally applied; and all nuclear weapons eliminated.

The above programme of action never became the subject of negotiation or of detailed multilateral examination, because of the opposition of the great powers and their allies.

7.3   The Canberra Report

In 1995 the Australian government established an independent commission, composed of internationally known personalities, to propose practical steps towards a nuclear-weapon-free world, 'including the related problem of maintaining stability and security during the transitional period and after this goal is achieved'. This commission, called the Canberra Commission on the Elimination of Nuclear Weapons, submitted its report in August 1996.

In presenting the case for a nuclear-weapon-free world, the commission used the following major arguments: the destructiveness of nuclear weapons is so great that they have no military utility against a comparably equipped opponent, other than the belief that they deter the opponent from using nuclear weapons, whereas the use of nuclear weapons against a non-nuclear-weapon opponent is politically and morally indefensible; the indefinite deployment of nuclear weapons carries a high risk of their ultimate use through accident or inadvertence; and the possession of nuclear weapons by some states stimulates other states to acquire them, reducing the security of all. Consequently, the elimination of nuclear weapons must be an endeavour of all states. The process of elimination should ensure that no state feels, at any stage, that further nuclear disarmament would threaten its security. The elimination should, therefore, be conducted as a series of verified phased reductions, in order to allow states to satisfy themselves, at each stage of the process, that further movement towards elimination can be made safely.

The following 'immediate steps' were recommended to demonstrate the intent of the nuclear-weapon states to reduce the role of nuclear weapons in their security postures: taking the nuclear forces off alert; removing the warheads from delivery vehicles; ending the deployment of non-strategic nuclear weapons; ending nuclear testing; initiating negotiations for further reductions of US and Russian nuclear arsenals; assuming reciprocal no-first-use of nuclear weapons undertakings among the nuclear-weapon states and a non-use undertaking by them with respect to non-nuclear-weapon states.

As 'reinforcing steps', the commission recommended: action to prevent further horizontal proliferation of nuclear weapons; development of verification arrangements for a nuclear-weapon-free world; and cessation of the production of fissile material for nuclear explosive purposes. It pointed out that the political commitment to eliminate nuclear weapons must be matched by a willingness to make available the resources needed for nuclear disarmament. States must be confident that detected violations would be acted upon.

Concurrent with the central disarmament process, there would be a need for activities supported by all states to build an environment conducive to nuclear disarmament and non-proliferation. The integrity of the ABM Treaty would have to be protected and the nuclear-weapon-free zones supported. The commission concluded that the world would be a much more secure place for everyone if there were no nuclear weapons, but it refrained from setting out a precise time frame for the elimination of these weapons.

The Canberra report contained an exhaustive and – to many – persuasive argumentation in favour of the abolition of nuclear armaments. It lacked, however, a coherent programme of action to reach the pursued objective. Nor did it satisfactorily answer the question of how world security would be maintained in a nuclear-weapon-free environment. An annex to the report claimed that the elimination of nuclear weapons could be checked to an acceptable degree of certainty, but no blueprint for a verification system was produced.

7.4   The CISAC Study

In 1997 the Committee on International Security and Arms Control (CISAC), a standing committee of the US National Academy of Sciences, completed a study on *The Future of US Nuclear Weapons Policy*. The study proposed that the United States should pursue a two-part programme of change in its nuclear-weapon policies. The first part was to be a near- and mid-term set of force reductions to diminish further confrontational and potentially destabilizing aspects of force postures; to reduce the risks of erroneous, unauthorized or accidental nuclear-weapon use; and to help curb the threat of further nuclear proliferation. In their early phases these measures would be largely bilateral, between the United States and Russia. The second part was to be a long-term effort to foster international conditions in which the possession of nuclear weapons would no longer be seen as necessary or legitimate for the preservation of national and global security.

In the view of the authors of the CISAC study, the benefits of comprehensive nuclear disarmament would be as follows: it would virtually eliminate the possibility of the use of nuclear weapons – whether authorized and deliberate or not – by states now possessing them; it would reduce the likelihood that additional states would acquire nuclear weapons; and it would deal decisively with the moral and legal status of nuclear weapons.

The study also discussed the risks of nuclear disarmament. It warned that the prohibition on nuclear weapons might break down via cheating or overt withdrawal from the disarmament regime. It therefore suggested that the regime be built within a larger international security system capable not only of deterring or punishing the acquisition or use of nuclear weapons but also of responding to major aggression. The study further referred to the argument that, if the major powers believed that the risk of nuclear war had been eliminated, they might initiate or intensify conflicts that could otherwise have been avoided or limited. It pointed out, however, that there had been changes in the structure of the international order that were acting to reduce the probability of major war independent of nuclear deterrence and that, moreover, the inherent capacities to rebuild nuclear weapons could act as a deterrent to the outbreak of major wars.

To manage the transition to comprehensive nuclear disarmament, the CISAC saw the possibility of an international agency assuming custody of the arsenals remaining during the transition to prohibition. Alternatively, nations might find it preferable to bypass the intermediate step involving an international agency and proceed directly to negotiations for the prohibition of nuclear weapons either globally, in a single agreement, or in steps involving successive expansions in the number and the geographical scope of nuclear-weapon-free zones. Whatever path was chosen, complete nuclear disarmament would require continued evolution of the international system towards collective action, transparency and the rule of law. A comprehensive system of verification, as well as safeguards to protect against cheating or rapid breakout, would also be required.

The CISAC report used the word 'prohibit' rather than 'eliminate' or 'abolish' because, in the opinion of its drafters, the world could never truly be free from the potential reappearance of nuclear weapons and their effects on international politics. The knowledge of how to build nuclear weapons could not be erased from human minds. Even if every nuclear warhead were destroyed, the current nuclear-weapon states and a growing number of other technologically advanced states would be able to build nuclear weapons within a few months or years of a national decision to do so.

7.5   The Model Nuclear Weapons Convention

In 1997 the International Network of Engineers and Scientists Against Proliferation (INESAP) published a model nuclear weapons convention (NWC). The convention would prohibit the development, testing, production, stockpiling, transfer, use and threat of use of nuclear weapons. States possessing nuclear weapons would be required to destroy their arsenals. The NWC would also prohibit the production of weapon-usable fissile material and would require that nuclear delivery vehicles be destroyed or converted for non-prohibited purposes.

The elimination of nuclear weapons would take place in phases. Each phase would have a deadline for the completion of specific activities. Verification would include declarations and reports from states, routine inspections, challenge inspections, on-site sensors, satellite photography, radionuclide sampling and other remote sensors, information sharing and citizen reporting. Persons reporting suspected violations of the convention would be provided protection, including the right of asylum. An international monitoring system would be established to gather information and make it available through a registry. Information that might jeopardize commercial secrets or national security would be kept confidential.

States parties to the NWC would be required to adopt legislative measures to implement their obligations. An agency would be set up to deal with verification and to ensure compliance. As in certain existing arms control conventions, the agency would comprise a conference of the parties, an executive council and a technical secretariat. The model NWC provides for graduated responses to non-compliance, beginning with consultation, clarification and negotiation, and, if necessary, recourse to the UN General Assembly and Security Council. The nuclear-weapon states would cover the costs of the elimination of their nuclear arsenals. However, an international fund could be established to assist states that might have financial difficulties in meeting their obligations.

The main purpose of the model NWC was to encourage governments to engage in nuclear disarmament talks. Its text was not included in the agenda of interstate negotiations, but became the subject of international discussions at different levels. Some of the many questions that arose in the course of these discussions were as follows: What would be the incentives for states to join a NWC? Could a NWC be enforced? Would it require the establishment of a new international security order? How could a sudden breakout from the NWC be prevented? How could a threat of terrorists acquiring and possibly using nuclear weapons be dealt with? How could the health and environmental challenges of nuclear-weapon dismantlement and destruction be met? How would research related to nuclear weapons be treated? The answers to these questions, given by the drafters of the model NWC, were only partly convincing.

7.6   The Tokyo Forum Report

In August 1998, in the wake of the Indian and Pakistani nuclear-weapon test explosions, an independent panel of international disarmament experts, diplomats, government officials and military strategists was organized at the initiative of the Prime Minister of Japan. In July 1999 this panel, called the Tokyo Forum for Nuclear Non-Proliferation and Disarmament, released a report entitled *Facing Nuclear Dangers: An Action Plan for the 21st Century*. The report addressed four areas: new nuclear dangers; mending strategic relations to reduce nuclear dangers; stopping and reversing nuclear proliferation; and achieving nuclear disarmament. A number of recommendations were made. The most important were as follows.

The reversing and unravelling of the NPT regime must be stopped by a reaffirmation of the Treaty's requirements for both disarmament and non-proliferation. A permanent secretariat and consultative commission should be created to deal with questions of compliance and to consider measures that would strengthen the Treaty.

The nuclear-weapon states must reaffirm the goal of eliminating nuclear weapons and take sustained, concrete steps to this end. No other city must experience the fate of Hiroshima and Nagasaki.

The CTBT must be ratified urgently by the states still holding out, including India, Pakistan, North Korea and Israel. All states must respect a moratorium on nuclear testing.

The United States and Russia should pursue the reductions of their nuclear arms to the level of 1,000 deployed strategic warheads. Verifiable reductions and elimination should cover non-deployed and non-strategic nuclear weapons. China should join the United Kingdom and France in reducing and, in the first instance, not increasing the nuclear-weapon inventory.

Transparency regarding the numbers and types of nuclear weapons and the amounts of fissile material should be encouraged.

All states with nuclear weapons should endorse and achieve the goal of zero nuclear weapons on hair-trigger alert.

A fissile material cut-off treaty should be promptly concluded. China, India, Pakistan and Israel should declare moratoria on the production of fissile material for nuclear weapons. Nuclear-weapon states should put all excess military stocks of fissile materials and civil fissile materials under IAEA safeguards.

Regional and global cooperative efforts should be made to prevent weapons of mass destruction from falling into the hands of extremist, fanatical or criminal groups.

The international community should explore ways to control and reverse missile proliferation, including global and regional agreements which would draw upon the provisions of the 1987 INF Treaty. A special conference of concerned states should be convened to deal with the growing problem of missile proliferation.

All states contemplating the deployment of advanced missile defences should proceed with caution, in concert with other initiatives to reduce the salience of nuclear weapons.

India and Pakistan should – in the near term – maintain moratoria on nuclear testing, sign and ratify the CTBT, support prompt negotiation of a fissile material cut-off treaty, adopt and implement nuclear risk-reduction measures, suspend missile flight-tests, and confirm pledges to restrain nuclear and missile-related exports. In the long term, both countries should accede to the NPT as non-nuclear-weapon states.

Weapons of mass destruction must be eliminated in the Middle East. All nuclear-weapon and missile-related activities in North Korea must cease, and the goal of a denuclearized Korean Peninsula must be achieved as soon as possible.

The permanent members of the UN Security Council should refrain from exercising their veto against efforts to assist or defend UN member-states that have become victims of the use or of the threat of use of weapons of mass destruction. All current and prospective permanent members of the UN Security Council should have exemplary non-proliferation credentials.

The CD should revise its procedures, update its work programme and carry out purposeful work or suspend its operations. Consensus among CD members should not be necessary to begin or conclude negotiations on a multilateral convention.

The scope of verification of nuclear disarmament should be expanded to cover non-deployed nuclear weapons and the dismantling of nuclear weapons.

The international community must be united and unequivocal in its response to would-be violators, based on a broad consensus, including possible recourse to Chapter VII of the UN Charter. A revitalized United Nations with a reformed and authoritative Security Council is essential for building and maintaining the support of the international community for the effective enforcement of compliance.

There were many disagreements among the members of the Tokyo Forum. The most controversial issue was that of ballistic missile defences. The views ranged from those completely opposed to such defences to those favouring both national and theatre missile defences. Nevertheless, the participants agreed that the issue required further multilateral debate and that all implications of possible deployment of missiles defences should be considered.

Many people expected that the Tokyo Forum Report would complement the Canberra Report by providing a blueprint for nuclear disarmament. In fact, however, the recommended measures – many of which had been proposed in several other forums – were not organically linked with each other. Moreover, the advocacy of the elimination of nuclear weapons was considerably weaker than in the Canberra Report.

7.7   Assessment

Since complete nuclear disarmament is intended to reform the world security architecture, it can hardly be achieved through a single international treaty. A series of measures would have to be negotiated and carried into effect in the course of what is bound to be a complex process of unpredictable length. The required negotiations need not be conducted in one forum. It would be more efficient to use several forums – open-ended or composed of states directly concerned – functioning simultaneously, without time constraints.

In order to start the disarmament process leading to the abolition of nuclear weapons, it would be necessary, in the first place, to render the nuclear non-proliferation regime universal and to ensure the enforcement of the non-proliferation norms. Nuclear-weapon tests would have to be definitively and universally banned, the production of fissile materials for explosive devices stopped, further proliferation of nuclear delivery vehicles prevented, the establishment of new nuclear-weapon-free zones encouraged, and the use of nuclear weapons prohibited. Nuclear energy systems lending themselves readily to nuclear-weapon production would have to be placed under international management. Tactical nuclear weapons would have to be eliminated prior to, or simultaneously with, drastic reductions of strategic nuclear weapons, and compliance with the prohibitions on other weapons of mass destruction – chemical and biological – ensured.

As a result of the series of incremental steps specified above, the existing nuclear forces could be brought down to low or even very low levels. Given the inequalities of states in conventional armaments, a problem would then arise how to proceed to the final elimination of nuclear weapons, for nuclear forces, even relatively small forces, are considered by some nations as a counterbalance to the superior conventional forces of their adversaries. A fully equitable solution to this dilemma might require the abolition of conventional weapons as well. However, resuscitating the utopian idea of general and complete disarmament would lead nowhere. A more realistic approach would be to bring about radical overall reductions of conventional armed forces and armaments, coupled with deep cuts in military production and spending, so as to achieve at least rough regional military balances. Such measures – to be based on generally agreed criteria – should result in force structures significantly minimizing the offensive capabilities of states.

Among the obstacles to nuclear disarmament which are most often referred to are the difficulty to verify compliance with the obligation to eliminate all nuclear weapons and nuclear-weapon components, as well as the impossibility to 'disinvent' these weapons. It is true that no verification can be absolutely foolproof, but full transparency and sophisticated technical means of supervision could render the probability of a nuclear disarmament treaty being violated very small. In particular, strict international verification of all stocks of fissile material usable in nuclear weapons, and of all facilities producing these materials for peaceful uses, would make clandestine development of nuclear-weapon capabilities practically impossible. The effectiveness of a technical control system could be significantly enhanced by using the so-called societal verification, as proposed by Professor Joseph Rotblat, a Nobel Laureate. This would mean that all citizens, not only experts, would be called upon to ensure the integrity of the Treaty, and each member of the community would become its custodian. Signatory states would be required to pass national laws making it the right and duty of their citizens to notify an international verification authority of any preparation for a breakout from the Treaty. Societal verification would, of course, be possible only in democracies tolerating transparency in military affairs, open discussion of security issues and unhampered activities of the mass media. Democratization of the political systems of at least the most powerful states is an indispensable requirement for general and complete nuclear disarmament.

It is also true that nuclear weapons cannot be disinvented. Indeed, the know-how and the capability to rebuild them cannot be eliminated. However, this is not a reason for them not to be outlawed. Chemical and biological weapons – much easier to manufacture than nuclear weapons – cannot be disinvented either, and yet they are banned under international conventions.

Nuclear disarmament could not take place in a political vacuum. The deep-rooted suspicions of bad faith among nations would have to be dissipated through confidence building. This is a condition for creating a cooperative relationship among the great powers, a relationship necessary for common action against emerging proliferators of nuclear and other weapons of mass destruction, and also against one of their own number that may secretly retain or reconstruct such weapons. This is also a condition for avoiding nuclear powers' involvement in regional disputes that should be settled by regional security organizations. The UN's conflict-resolution and peacekeeping capabilities would have to be considerably strengthened. States must become persuaded that the possession of nuclear weapons is a liability rather than an asset and that a nuclear-weapon-free world will be safer than a world with nuclear weapons.

8

Chemical and Biological Weapons

During World War I, the extensive use of poisonous gases resulted in many casualties, over 90,000 of which were fatal. Although the death toll from these chemical weapons was relatively low in comparison with the number of deaths caused by conventional weapons, the extreme suffering which they caused reinforced popular demands for a ban on this method of warfare. This led to the signing in Geneva, on 17 June 1925, of the Protocol for the Prohibition of the Use in War of Asphyxiating, Poisonous or Other Gases, and of Bacteriological Methods of Warfare.

8.1   The 1925 Geneva Protocol

The Geneva Protocol – originally a protocol to the 1925 Convention for the Supervision of the International Trade in Arms and Ammunition and in Implements of War (see Chapter 2.3) – entered into force in 1928. In the part dealing with gases and all analogous liquids, materials or devices, the Protocol only reaffirmed the ban which was already in existence and had been declared in several previously signed international documents.

Scope of the Obligations

For many years, the interpretation of the scope of obligations under the Geneva Protocol was a matter of dispute. In 1969, a majority of UN member-states adopted Resolution 2603 A(XXIV) expressing the view that the Protocol embodied the generally recognized rules of international law prohibiting the use in international armed conflicts of all biological and chemical methods of warfare, regardless of any technical developments. In particular, the resolution declared as contrary to the rules of international law the use in international armed conflicts of: (a) any chemical agents of warfare – chemical substances, whether gaseous, liquid or solid – which might be employed because of their direct toxic effects on man, animals or plants; and (b) any biological agents of warfare – living organisms, whatever their nature, or infective material derived from them – which are intended to cause disease or death in man, animals or plants, and which depend for their effects on their ability to multiply in the attacked person, animal or plant. The 1925 Geneva Protocol is now understood to cover not only bacteria but also other micro-organisms, such as viruses or rickettsiae (unknown at the time the Protocol was signed) – hence the use of the term 'biological'.

The United States had been in the forefront of the group of states which gave the Geneva Protocol a narrow interpretation and which contended that the use of irritants (such as tear gas) and anti-plant chemicals was not covered by the Protocol. In 1975, after the Indo-China War, in which such substances were used on a large scale, the United States decided to renounce, as a matter of national policy, the first use of herbicides in war, except for control of vegetation within US bases and installations or around their immediate defensive parameters. It also decided to renounce the first use of riot control agents in war except in defensive military methods of saving lives, such as: (a) to control rioting prisoners of war in areas under US military control; (b) to reduce or avoid civilian casualties when civilians are used to mask or screen attacks; (c) to recover downed aircrews and passengers, as well as escaping prisoners, in rescue missions in remote isolated areas; or (d) to protect convoys outside the combat zone from civilian disturbances, terrorists and paramilitary organizations. This interpretation was more liberal than the one previously advocated, but it still fell short of the understanding of the scope of the Geneva Protocol as formulated in UN Resolution 2603.

Weaknesses

The Geneva Protocol restricts its non-use obligation to the conditions of 'war'. Therefore it is, strictly speaking, not applicable to internal conflicts. It might also be argued that the Protocol does not cover those international conflicts in which the belligerents do not consider themselves to be formally at war. The Protocol does not ban the threat of use of the prohibited weapons and applies only to relations 'as between' the parties.

There is no mechanism to verify compliance with the Protocol prohibitions or to clarify ambiguous situations. This shortcoming created a number of problems. Since the 1980s, however, this gap has been filled by the UN resolutions empowering the UN Secretary-General to investigate reports on possible violations of the Geneva Protocol.

Reservations

In joining the Geneva Protocol, over 40 states entered reservations. These reservations upheld the right of the reserving states to employ the banned weapons against non-parties to the Protocol, or in response to the use of these weapons by a violating party, or even against the allies of the violator that have not committed a violation. For many states, the Protocol was essentially a no-first-use treaty. Proposals have frequently been put forward that those who made a reservation should withdraw it and give up the right to use chemical and biological weapons under any circumstance. A number of states did so after the conclusion of the 1972 Biological Weapons Convention and, especially, after the conclusion of the 1993 Chemical Weapons Convention, which prohibited the very possession of the weapons in question. (See sections 8.2 and 8.5 below.)

During the 1991 Gulf War, the French President stated that France would not respond with chemical or bacteriological weapons should Iraq employ such weapons against the forces of the anti-Iraq coalition. The United States, for its part, did not formally rule out such a response.

Assessment

The Geneva Protocol is a document of historic significance. Its importance lies in the fact that an international legal constraint, 'binding alike the conscience and the practice of nations', was imposed on acts which were generally held in abhorrence and which had been condemned by the opinion of the civilized world. According to a widely shared opinion, the Protocol is part of customary international law, to be complied with by parties and non-parties alike.

8.2   The 1972 Biological Weapons Convention

Since the signing of the 1925 Geneva Protocol, the prevailing opinion had been that the possession of chemical and biological weapons should be prohibited simultaneously. Both categories of weapon are usually referred to as weapons of mass destruction (along with nuclear weapons), a usage in line with the definition formulated in 1948 by the UN Commission for Conventional Armaments. The issue of banning their development, production and stockpiling was placed on the international disarmament agenda in 1968.

In 1969, the UN Secretary-General issued the report on *Chemical and Bacteriological (Biological) Weapons and the Effects of Their Possible Use*, which concluded that certain chemical and biological weapons cannot be confined in their effects and might have grave, irreversible consequences for man and nature. This would apply to both the attacking and the attacked nations. The report on *Health Aspects of Chemical and Biological Weapons*, published a year later by the World Health Organization (WHO), pointed out that chemical and biological weapons pose a special threat to civilians and that the effects of their use are subject to a high degree of uncertainty and unpredictability. Nevertheless, several Western countries proposed a treaty banning only biological weapons. The main reason for separate treatment of these two categories of weapon, as put forward by the sponsors of the proposal, was that a ban on biological weapons did not require intrusive verification and could therefore be concluded quickly, without serious risks, and that this was not the case with chemical weapons. After a period of hesitation, especially on the part of the non-aligned states, the Western countries' approach was adopted by the Eighteen-Nation Committee on Disarmament (ENDC) and its successor, the Conference of the Committee on Disarmament (CCD), where the negotiations were taking place.

A factor that had facilitated this development was the unilateral renunciation of biological weapons by the United States, announced in 1969, and the decision by the US government to destroy its stockpile of these weapons irrespective of the possible conclusion of an international agreement. In 1970 the United States also formally renounced the production, stockpiling and use of toxins for war purposes. It stated that military programmes for biological agents and toxins would be confined to research and development for defensive purposes.

On 10 April 1972 the Convention on the Prohibition of the Development, Production and Stockpiling of Bacteriological (Biological) and Toxin Weapons and on Their Destruction was opened for signature. This convention, generally known as the Biological Weapons (BW) Convention and often also referred to as the Biological and Toxin Weapons (BTW) Convention, entered into force in 1975.

Scope of the Obligations

The BW Convention prohibits the development, production, stockpiling, or acquisition by other means, or retention of microbial or other biological agents, or toxins, as well as of weapons, equipment or means of delivery designed to use such agents or toxins for hostile purposes or in armed conflict.

Definitions. The Convention did not define the prohibited items or the targets to which the prohibitions relate. There exists, however, an authoritative definition of biological agents formulated by the WHO. In its 1970 report, mentioned above, the WHO described 'biological agents' as agents that depend for their effects on multiplication within the target organism. Toxins are poisonous products of organisms; unlike biological agents, they are inanimate and not capable of reproducing themselves. The Convention applies to all natural or artificially created toxins 'whatever their origin or method of production', that is, it covers toxins produced biologically, as well as those produced by chemical synthesis. Since toxins are chemicals by nature, their inclusion in the BW Convention was a step towards a ban on chemical weapons. All biological agents and toxins intended to be used for hostile purposes or in armed conflict are thus covered by the BW Convention. This implies that the prohibitions under the convention relate to all possible targets.

Whereas there were no disputes among the parties regarding the definition of biological agents or toxins, the lack of definition of 'weapons, equipment or means of delivery' led to a controversy. In ratifying the BW Convention, Switzerland reserved the right to decide for itself which items fall within the definition of weapons, equipment or means of delivery designed to use biological agents or toxins. The United States entered an objection to this reservation, claiming that it would not be appropriate for states to unilaterally reserve the right to take such decisions. In its opinion, the prohibited items are those whose design indicates that they could have no other use than that specified in the Convention, or that they are intended to be capable of the use specified. There are, however, few weapons, equipment or means of delivery which would meet such criteria.

Permitted Activities. The prohibition on developing, producing, stockpiling or otherwise acquiring or retaining biological agents and toxins is not absolute. It applies only to types and to quantities that have no justification for 'prophylactic', 'protective' or 'other peaceful purposes'. Retention, production or acquisition by other means of certain quantities of biological agents and toxins may thus continue, and there may be testing in laboratories and even in the field. In the course of negotiations, a clarification was given that the term 'prophylactic' encompasses medical activities such as diagnosis, therapy and immunization. The term 'protective' covers the development of protective masks and clothing, air and water filtration systems, detection and warning devices, and decontamination equipment; it must not be interpreted as permitting possession of biological agents and toxins for defence, retaliation or deterrence. The term 'other peaceful purposes' has remained unclear. One can assume that it includes all types of scientific experimentation.

There are no provisions in the BW Convention restricting research activities. One reason for this omission may be that research aimed at developing agents for civilian purposes may be difficult to distinguish from research serving military purposes, whether defensive or offensive. Moreover, in the biological field it is difficult to draw a dividing line between research and development. A country can develop warfare agents in research facilities; once developed, these agents can be rapidly produced in significant quantities. This circumstance and the express authorization to engage in some production of biological warfare agents and toxins create a risk that the provisions of the Convention may be circumvented. The stipulation that any development, production, stockpiling or retention of biological agents or toxins must be justified does not carry sufficient weight. There are no agreed standards or criteria for the quantities of biological agents or toxins that may be needed by individual states for the different purposes recognized by the Convention. The parties are not even obliged to declare the types and amounts of agents or toxins they possess or the use they make of them. The system of material accountancy that is useful in the verification of certain measures of arms control is not practicable in the case of biological agents or toxins. It is thus not evident how much of a certain prohibited substance stocked by a given country would constitute a violation of the Convention. The secrecy surrounding biological research activities and, in particular, the maintenance of defensive preparations which at certain stages may be indiscernible from offensive preparations could generate suspicions leading to allegations of breaches.

Transfers. A separate article of the Convention prohibits the transfer of the agents, toxins, weapons, equipment or means of delivery specified above to 'any recipient whatsoever', that is, to any state or group of states or international organizations, as well as sub-national groups or individuals. The provision of assistance, encouragement or inducement in the acquisition of the banned weapons is likewise forbidden. These non-proliferation clauses may appear hard to reconcile with the commitment of the parties to engage in the 'fullest possible' exchange of biological agents and toxins and of equipment for the processing, use or production of such agents and toxins for peaceful ends because such materials and technologies, as well as expertise, are dual-use and therefore widespread. To meet the concerns of the developing countries, the parties have undertaken to cooperate in the further development and application of scientific discoveries in the field of biology for peaceful purposes. However, since the BW Convention is essentially a disarmament treaty, it cannot serve as an effective instrument for such cooperation. There exist specialized bodies for this purpose – intergovernmental agencies or non-governmental scientific associations – which function irrespective of the BW Convention.

Destruction. The most remarkable feature of the BW Convention is the disarmament obligation of the parties: to destroy or divert to peaceful purposes all biological agents, toxins, weapons, equipment and means of delivery. The envisaged destruction or diversion was to take place not later than nine months after entry into force of the Convention, it being understood that for states acceding to the Convention after its entry into force the destruction or diversion would be completed upon accession. All the necessary safety precautions must be observed during the destruction operations to protect 'populations' (that is, not only the population of the country carrying out these operations) as well as the environment in general. However, states joining the Convention are not required to declare the possession or non-possession of the banned weapons. Nor are they obligated to prove that they have fulfilled the commitment to destroy the stocks of these weapons or to divert them to peaceful purposes.

After the BW Convention entered into force, the United States announced that its stockpile of biological and toxin agents and all associated munitions had been destroyed, except for small quantities for laboratory defensive research purposes. It also made it known that its former biological warfare facilities had been converted to medical research centres. The United Kingdom said that it had no stocks of biological weapons. The Soviet Union stated that it did not possess any biological agents or toxins, weapons, equipment or means of delivery, as prohibited by the Convention, but this statement was proved to be false. In 1992 Russia admitted that it had not destroyed its stockpiles.

Relationship to the 1925 Geneva Protocol

The BW Convention does not expressly prohibit the use of biological or toxin weapons, but It does make it clear that the obligations assumed under the 1925 Geneva Protocol, which prohibits such use, remain valid. It also refers to UN General Assembly resolutions condemning actions contrary to the principles and objectives of that Protocol. However, adherents to the BW Convention do not need to be parties to the Geneva Protocol. Moreover, the Convention stipulates that nothing in its provisions shall be interpreted as in any way limiting or detracting from the obligations assumed by states under the Geneva Protocol. This implies that the reservations to the Protocol, which form part of the obligations contracted by the parties, continue to subsist. To the extent that the reservations concern the right to employ the banned weapons against non-parties or in retaliation against a party violating the protocol, they are incompatible with the obligation of the parties to the Convention never 'in any circumstances' to acquire biological weapons. They also contradict the parties' expressed determination to exclude 'completely' the possibility of biological agents and toxins being used as weapons. Over the years, a number of states have withdrawn their reservations to the Geneva Protocol, either with regard to biological weapons alone or with regard to both biological and chemical weapons. They have thereby recognized that, since the retention and production of biological weapons are banned, so must, by implication, be the use, because use presupposes possession.

Nonetheless, in 1996 Iran proposed that the Convention be amended so as to make the ban on use explicit rather than implicit. The Iranian proposal was opposed by many states, which feared the risks of having other provisions of the Convention opened up for renegotiation as well.

Measures of Implementation

Each party is obligated to take measures, in accordance with its constitutional processes, to prohibit and prevent the activities banned by the BW Convention from taking place anywhere within its territory and under its jurisdiction or control. The term 'measures' covers legislative, administrative or regulatory measures, whereas the term 'under its jurisdiction or control' extends the bans to non-self-governing territories administered by states parties and to territories under military occupation. 'Anywhere' implies that even transnational corporations operating in the territories of non-parties to the Convention are covered by the prohibitions if they remain under the jurisdiction or control of the parties. The parties have undertaken to consult one another and to cooperate in solving problems relating to the objective or the application of the provisions of the Convention.

Entry into Force, Duration, Amendments and Reviews

As stipulated in the BW Convention, it entered into force after the deposit of the instruments of ratification by 22 signatory governments, including the governments of the Soviet Union, the United Kingdom and the United States, which had been designated as depositaries of the Convention.

The Convention is of unlimited duration. However, each party has the right to withdraw if it decides that extraordinary events, related to the subject matter of the Convention, have jeopardized its supreme interests. A notice of withdrawal should be given to all other parties and to the UN Security Council three months in advance. It should include a statement about the 'extraordinary events' justifying the withdrawal.

Changes in the text of the Convention can be brought about through amendments. Amendments may be proposed by any party, but they enter into force for each state accepting them only upon their acceptance by a majority of the parties.

The BW Convention provides for conferences of the parties to review its operation with a view to ensuring that its purposes and provisions are being realized.

Assessment

The BW Convention was the first international agreement after World War II to provide for the elimination of an entire class of weapon. Many considered it regrettable that chemical weapons, which are associated in the public mind with biological weapons and which – unlike biological weapons – have already been used on a large scale in war, were not prohibited at the same time. Nevertheless, the parties to the BW Convention recognized that it was a step towards an agreement effectively prohibiting chemical weapons as well as providing for their destruction. Without a formal commitment included in the Convention that such an agreement should be reached at an 'early' date, many countries would have probably refrained from joining the Convention.

The aim of the BW Convention was not so much to remove an immediate peril as to eliminate the possibility that scientific and technological advances, modifying the conditions of production, storage or use of biological weapons, would render these weapons militarily attractive. Indeed, the discoveries of recent years have made it possible to develop and mass-produce agents and toxins which would be more lethal and easier to stockpile than those already in existence. Moreover, normally harmless organisms that do not cause diseases can be modified so as to produce diseases for which there is no known treatment. As repeatedly emphasized by the review conferences of the parties to the Convention, the prohibitions are comprehensive enough to cover all relevant scientific and technological developments, including biological agents and toxins that could result from genetic engineering processes.

The admitted violation of the BW Convention by Russia (see Chapter 19.6), followed by the disclosure of an offensive biological weapons programme in Iraq, as well as reports that several other nations also have or are seeking to acquire a biological weapon capability, indicated that the threat of biological warfare is real. Since biological weapons can be produced relatively easily and cheaply, they may also prove useful to terrorists. The fear of bioterrorism grew considerably when – following the September 2001 attacks on the United States – anthrax spores were sent to a number of places through the US mail. However, no specific measures were set forth in the BW Convention to verify compliance with the obligation not to develop, produce, stockpile or otherwise acquire or retain biological agents or toxins for hostile purposes. Toxic substances may be stored in inconspicuous repositories and eventually 'weaponized', that is, filled into missiles or bombs or spray systems similar to those used to deliver pesticides. Consequently, a violator could relatively easily break out from the Convention. The lack of verification machinery in the BW Convention is a serious lacuna, which must be filled to increase the effectiveness of the ban on biological warfare agents and toxins.

Efforts to Strengthen the BW Convention

The first steps to strengthen the BW Convention were taken at the second review conference of the parties, held in 1986. The parties then agreed on a set of confidence-building measures which included: exchanging data on research laboratories that meet very high national or international safety standards; sharing information on all outbreaks of unusual diseases; encouraging publication of results of biological defence research in scientific journals; and promoting scientific contact related to the Convention. At the third review conference, in 1991, the parties added two measures: declaration of past activities in offensive and/or defensive biological research and development programmes, and declaration of vaccine production facilities. All these politically (but not legally) binding undertakings proved insufficient, as many governments supplied incomplete data or did not supply any data at all.

In 1991 the states parties to the BW Convention decided to set up an Ad Hoc Group of Governmental Experts (also known as the group of 'verification experts' or VEREX group) to identify and examine potential verification measures from a scientific and technical standpoint. The report produced by this group in 1993 stated that certain measures, used singly or in combination, could strengthen the BW Convention regime by helping to distinguish prohibited activities from those permitted and thereby reduce ambiguities about compliance. In 1994 this report was considered at a special conference which decided to establish another Ad Hoc Group, open to all states parties to the BW Convention, to negotiate a legally binding protocol to the Convention in order to improve its implementation. The negotiations started in 1995. By 2001 many provisions of the draft protocol had been agreed upon, but a number of controversial issues, some of them of crucial importance, were still unresolved.

The protocol would, of course, have to be ratified by the parties to the BW Convention to be applicable to them. Non-ratification by some states would result in a two-tier regime.

Transfer/Export Control. Under this rubric, the controversy was about the kind of export control regime that should be established to facilitate peaceful technical cooperation among states, and about how the regime should regulate transfers between protocol parties and transfers from protocol parties to protocol non-parties, both signatories and non-signatories to the BW Convention. The developing countries were opposed to any discriminatory transfer control regime that would preclude their access to biotechnology, whereas the developed countries were opposed to lowering their current export control standards. In fact, the disagreement was over whether the so-called Australia Group – an informal arrangement set up in the mid-1980s (largely because of the use of CW in the Iraq–Iran War) by a number of industrialized countries to ensure that dual-use items or technologies are not transferred to chemical or biological armament programmes – should continue to operate as a safeguard after entry into force of the protocol.

Declaration Triggers. While broadly agreeing that facilities and activities of particular relevance to the Convention must be declared to initiate the process of verification, the negotiators disagreed on the precise criteria that would 'trigger' the declaration of a given facility or activity. Again, the disagreement was mainly between the developed and developing countries. The former insisted that all relevant facilities be declared in all countries, without placing a disproportionate burden on some of them, whereas the latter wanted the burden to be placed on the developed countries, as those deemed to have the most facilities of concern to the BW Convention.

Follow-Up of Declarations. While agreeing that submission of declarations by the parties should be followed by well-defined verification procedures, the negotiators disagreed on some of these procedures. Thus, many non-aligned states suggested that only biodefence and maximum containment facilities should receive randomly selected 'transparency visits' to confirm that the declaration for a declared site was consistent with the obligations under the protocol; thereby, mainly facilities in developed countries would be subject to such visits. The Western Group of states argued that all declared facilities should be subject to these short and infrequent visits. There was also a dispute over whether 'clarification visits', those intended to address concerns at the low level of controversy (ambiguities or omissions identified in declarations), should be voluntary or mandatory, and whether they should apply to facilities that appeared to meet the declaration requirements but had not been declared.

Investigations. A large group of Western nations favoured the so-called 'red light' procedure, according to which an investigation requested to check compliance should take place unless a majority of the executive council of the envisaged international implementing organization voted to stop it. Some non-aligned countries and the United States preferred the so-called 'green light' procedure, according to which a requested investigation should not take place unless a majority of the parties voted for it to take place. The declared aim of the latter approach was to deter frivolous challenges. The degree of intrusiveness of investigations was also a matter of dispute.

Definitions. In order to enhance the accuracy of what should be subject to verification, some states proposed to include in the protocol definitions of the key terms used in the Convention, namely, 'biological weapon', 'biological agent' and 'hostile purposes'. Others preferred keeping the broad formulations of the Convention, especially the 'general purpose criterion', fearing that adoption of precise definitions – which would require an amendment to the Convention – might restrict the scope of the prohibitions and create undesirable loopholes.

Thresholds. A proposal was made to set quantitative thresholds below which the possession of listed agents would be justified. This, in the view of the proponents, would establish universal guidelines of what was a permissible amount of agent.

The opponents contended that the nature of biological agents would make it possible for a state to easily and rapidly grow them to levels exceeding the agreed thresholds (hence, there was no need to maintain large stockpiles) and that a quantity of agent justified for one state may not be justified for another.

Entry into Force. The most popular formula for the entry into force of the protocol appeared to be the setting of a numerical target of 60–70 ratifications. Those opposed to this solution argued that it could result in some biotechnologically developed countries remaining outside the protocol. They preferred (by analogy with the 1996 Comprehensive Nuclear Test-Ban Treaty) to specify the countries whose ratification would be indispensable for the protocol to become effective – in the first place the depositaries of the BW Convention (Russia, the United Kingdom and the United States). Others suggested that a certain agreed number of countries from each geographical area be required to ratify, so as to deny to any country the right to veto the bringing of the protocol into effect.

Assessment. The differences of opinion on the basic provisions of the negotiated protocol were substantial. Consequently, the required consensus could not be reached – as planned – before the fifth review conference of the parties to the BW Convention. At the July 2001 meeting of the Ad Hoc Group, the United States rejected the draft protocol submitted by the Group's chairman as a compromise between divergent positions as well as the very approach that it represented. The draft had the support of many states – including the European Union member-states – as a basis for reaching agreement.

At the review conference, which took place in November 2001, the United States argued that the chairman's text posed a risk to national security and to proprietary commercial information. It then put forward several proposals. The parties should: agree to enact national criminal legislation to enhance bilateral extradition agreements with respect to BW offences and to make it a criminal offence for any person to engage in activities prohibited by the BW Convention; adopt strict standards for the security of pathogenic micro-organisms; establish a mechanism for international investigations of suspect disease outbreaks and/or alleged BW incidents; set up a voluntary cooperative mechanism for clarifying and resolving compliance concerns by mutual consent; and adopt and implement strict biosafety procedures, based on World Health Organization or equivalent national guidelines.

On the last day of the review conference, the United States unexpectedly called for a formal end of the Ad Hoc Group's work. After more than six years of efforts, the negotiating process broke down. The review conference adjourned until November 2002.

8.3   The 1990 US–Soviet Chemical Weapons Agreement

On 1 June 1990, the United States and the Soviet Union signed the Agreement on Destruction and Non-Production of Chemical Weapons and on Measures to Facilitate the Multilateral Convention on Banning Chemical Weapons (US–Soviet Chemical Weapons Agreement). This bilateral accord crowned several years of US–Soviet talks conducted in parallel with multilateral negotiations at the Conference on Disarmament, aimed at reaching a comprehensive and worldwide chemical disarmament treaty.

Scope of the Obligations

The most essential obligations assumed by the United States and the Soviet Union under the Chemical Weapons Agreement were: to halt the production of chemical weapons; to reduce chemical-weapon stockpiles to equal, low levels; and to accept measures necessary to verify compliance. The clauses dealing with the projected multilateral chemical weapons convention had the form of proposals to be considered by all the negotiators.

Cessation of Production. The parties undertook to stop the production of chemical weapons upon entry into force of the US–Soviet Chemical Weapons Agreement. This undertaking was a concession on the part of the United States, which for several years had insisted on the right to continue manufacturing chemical weapons. Indeed, in 1987, when the Soviet Union announced that it had ceased producing chemical weapons, the United States decided to terminate its unilateral, 18-year moratorium on the production of such weapons in order to replace the ageing stocks with so-called binary munitions. Binary munitions are filled with two chemicals of relatively low toxicity which mix and react (the reaction product being a super-toxic agent) only when the munition is being delivered to the target; they are easier and safer to store and employ than 'traditional' chemical weapons.

However, in its endeavours to modernize its chemical-weapon stockpile, the US government encountered apparently insurmountable political and technical obstacles. One of these was the inability to find an US-based company willing to supply a component or a precursor chemical for the nerve agent to be used in artillery shells. The Soviet government, on its part, did not at the time seem interested in renewing its stockpile.

Reduction of Stockpiles. Each party to the US–Soviet Chemical Weapons Agreement undertook to reduce the aggregate quantity of its chemical weapons to the level of 5,000 metric tons of chemical agents. This meant that the Soviet Union would have to destroy about 90% and the United States about 80% of their respective stockpiles. The total to be eliminated amounted to approximately 65,000–70,000 tons of agents, but the composition of the stocks to be retained was not constrained. The aggregate capacity of empty chemical munitions and devices was to be reduced as well. The destruction operations – which, in addition to the chemicals themselves, were to cover munitions, devices and containers from which the chemicals had been removed – were to begin no later than 31 December 1992, and by 31 December 2002 each party should have destroyed the total stocks subject to elimination. It might be noted that the US Administration was already committed by a congressional decision unilaterally to destroy a major part of its stocks by the mid-1990s.

The parties to the Agreement were to be allowed to retain the technical capacity to manufacture chemical weapons without restriction. This allowance, which weakened the credibility of the obligation assumed by the two powers not to produce chemical weapons, should be seen in conjunction with the reference made to the US and Soviet 'rights' under the 1925 Geneva Protocol. The reference confirmed that the United States and the Soviet Union were at that time not prepared to give up the option to use chemical weapons as a means of retaliation in kind, as formulated in their reservations to the Geneva Protocol.

Support for the Projected Multilateral Convention

The United States and the Soviet Union stated that they would make every effort to have the multilateral convention, then under negotiation, concluded at the earliest date and that they would accord it precedence over their bilateral agreement. They also undertook to take practical steps to encourage all states capable of producing chemical weapons to join the multilateral convention. To reach these goals, they agreed on the following measures.

Each side would reduce its stockpile to 500 agent tons within eight years after the entry into force of the multilateral convention. The remaining stocks would be eliminated during the subsequent two years only if a decision to this effect had been taken by a special conference of states parties to the multilateral convention. The conference was to be convened at the end of the eighth year of the convention's operation, and its decision was to be based on assessment of 'whether the participation in the multilateral convention is sufficient' for taking the envisaged action. In a joint statement accompanying the agreement, the United States and the Soviet Union specified that an affirmative decision of the conference would require the consent of a majority of the parties attending it. This majority would have to include states that had submitted, before 31 December 1991, a written declaration to the Conference on Disarmament that they possessed chemical weapons, had signed the convention within 30 days after its opening for signature and had become parties to the convention not later than one year after its entry into force. The proposed voting mechanism was meant to induce countries to declare the possession of chemical weapons even before the conclusion of the convention and to sign and ratify the convention soon after it had been agreed.

The proposals concerning the projected multilateral convention proved the most controversial part of the bilateral agreement. As pointed out above, the two signatory powers were still unwilling to commit themselves unconditionally to the destruction of all their chemical-weapon arsenals and chemical-weapon production facilities. Consequently, those joining the multilateral convention could not be certain that its ultimate goal – the complete destruction of chemical-weapon stockpiles and production facilities by all states – would ever be reached. Under such circumstances, many would hesitate to forswear the chemical-weapon option, especially since the chemical-weapon powers were to be accorded the privilege of veto at the envisaged special conference of states parties.

Entry into Force and Duration

The US–Soviet Chemical Weapons Agreement was to enter into force upon the exchange of instruments stating its acceptance by each party. The two powers thus chose a procedure which is simpler and quicker than ratification, often required for major arms control treaties.

The agreement was to be of unlimited duration, unless the two sides agreed to terminate it after the envisaged multilateral convention had become effective. A document with detailed provisions for the implementation of inspection measures was to be completed by 31 December 1990, but it was not. The agreement never entered into force.

Assessment

The decision of the two superpowers – the possessors of the largest stockpiles of chemical weapons – to cease chemical-weapon production and eliminate a major part of their stockpiles before a multilateral chemical weapons convention had been signed was generally considered an important event. However, in claiming the right to retain a certain quantity of chemical weapons and to use them in retaliation even after the conclusion of the projected comprehensive multilateral convention, the United States and the Soviet Union were trying to usurp the privilege of deciding whether and when the convention could be fully implemented. The fact that the United States and the Soviet Union, which also possessed the strongest nuclear and conventional forces, considered it necessary to keep 500 tons of chemical weapons for their defence was difficult to comprehend. The importance which the two states appeared to attach to such a relatively small amount of chemical weapons may have conveyed the wrong impression about the actual value of these weapons for national security; it may have even encouraged their proliferation.

A convention for the total elimination of chemical weapons incorporating the conditions proposed by the two powers did not attract broad adherence. Many countries would have refused to accept permanent status as a non-possessor of chemical weapons while a handful of parties retained for an indefinite period of time certain quantities of such weapons as well as certain facilities capable of producing more weapons.

8.4   Negotiations and Initiatives for a Multilateral Ban on Chemical Weapons

Several proposals for a comprehensive ban on chemical weapons were made after the conclusion of the BW Convention.

US–Soviet Reports

In 1979 and 1980 joint reports were submitted by the United States and the Soviet Union on their bilateral negotiations for a general prohibition of chemical weapons.

The two powers expressed the opinion that the parties to a multilateral convention should assume the obligation never to develop, produce, otherwise acquire, stockpile or retain super-toxic lethal, other lethal or other harmful chemicals, or precursors of such chemicals, unless these were intended for non-hostile purposes or for military purposes not involving the use of chemical weapons, and unless the types and quantities of the chemicals were consistent with such purposes. The prohibition was also to apply to munitions and devices specifically designed to cause death or other harm through the toxic properties of chemicals released as a result of the employment of such munitions or devices. The reports did not envisage the possibility of concluding a partial ban.

In a future convention the scope of the prohibition was to be determined on the basis of the general-purpose criterion complemented by toxicity criteria, which would serve as a basis for identifying lethal and harmful chemicals and facilitate verification. The dispute regarding the extent to which irritants, toxins and precursors should be covered could not be settled, however.

Agreement was also reached that the destruction of declared stocks or their diversion to non-weapon ends should be completed not later than ten years after a state had become party to the convention. The declared means of production were to be destroyed or dismantled. The fulfilment of obligations would have to be subject to 'adequate' verification based on a combination of national and international measures. However, important issues relating to international verification measures remained unresolved.

The bilateral talks were broken off in 1980 as a consequence of deteriorating East–West relations, but multilateral discussions at the Conference on Disarmament continued.

The 1984 US Draft Convention

In April 1984 the United States proposed a draft convention for a comprehensive ban on chemical weapons. The central feature of this proposal was the requirement that each party must consent, at 24 hours' notice, to a 'special inspection' (permitting 'unimpeded access') of one of the sites for which systematic international on-site inspection was authorized – namely, facilities for 'permitted' activities – as well as of chemical-weapon stockpiles and production plants destined for destruction, or of any location or facility owned or controlled by the government of a party, including military facilities. The purpose of such an inspection would be to clarify and resolve any matter which might cause doubts about compliance. The United States later explained that 'controlled by the government' meant controlled through contract or regulatory requirements. For locations and facilities not subject to the above provisions, requests for 'ad hoc on-site inspections' might be refused, but the party in question would have to explain its refusal and suggest alternative methods for resolving the compliance concerns.

This novel US approach to verification, termed an 'open invitation' to inspect all suspect sites, was categorically rejected by the Soviet Union on the grounds that its adoption would result in the disclosure of state secrets unrelated to the production or storage of chemical weapons. (Only in 1987 did the Soviet Union accept, in principle, the idea of mandatory on-site inspections.) Nevertheless, many of the provisions of the US draft were incorporated in the draft convention subsequently developed by the Conference on Disarmament.

The 1989 Paris Conference

The lack of progress in negotiations on a multilateral ban on chemical weapons at a time when these weapons were being used by Iraq in the 1980–88 Iraq–Iran War induced the French government, the depositary of the 1925 Geneva Protocol, to convene a special conference in Paris.

In the Final Declaration of the Paris conference, adopted on 11 January 1989, the representatives of nearly 150 states expressed their determination to prevent any recourse to chemical weapons by completely eliminating them. They recognized the continuing validity of the 1925 Geneva Protocol and recalled their concern at violations of the Protocol, as established and condemned by the United Nations; they further reaffirmed the necessity of concluding, at an early date, a convention prohibiting the development, production, stockpiling and use of chemical weapons, and providing for their destruction; and they emphasized that the convention must be global, non-discriminatory, comprehensive, effectively verifiable and of unlimited duration. They also stated that they wished to strengthen UN procedures related to investigations of alleged violations of the Geneva Protocol.

Some representatives of Arab countries at the Paris conference intimated that, as long as Israel had not formally renounced nuclear weapons, its neighbours could not be expected to renounce their chemical-weapon option. This linkage found scant support among states outside the Middle East region. Most participants held the view that chemical weapons – although repulsive and inhumane – are not comparable to nuclear weapons in terms of destructiveness or perceived usefulness for deterrence against aggression. The Arab position did not prevent the Final Declaration from being adopted by consensus.

The Paris conference was an important political event in that it highlighted the risk of repeated use of chemical weapons as long as these weapons remained in stockpiles and proliferated to new countries. In practical terms, however, it had little impact on the negotiations for a comprehensive ban on chemical weapons.

Initiatives Preceding the Global Ban

Pending the conclusion of a chemical weapons convention, Australia initiated informal multilateral consultations aimed at curbing the proliferation of chemical weapons by restricting the export of precursors of these weapons. The Australia Group drew up a list of chemicals subject to export controls and agreed on means to prevent circumvention of these controls by companies or individuals. (It subsequently also agreed to control exports of biological agents and toxins and of the relevant equipment – see section 2 above.) The obligations assumed by the members of the group, although not legally binding, have raised the cost of acquiring chemical as well as biological means of warfare.

In an agreement signed on 5 September 1991 at Mendoza (Argentina), Argentina, Brazil and Chile reaffirmed their unilateral statements on non-possession of chemical weapons. They also pledged not to develop, produce or acquire these weapons and expressed the intention to establish appropriate inspection mechanisms in their respective countries with regard to the precursors of chemical warfare agents. The Mendoza Agreement was acceded to by several other South American countries.

On 4 December 1991, the countries of the Andean Group – Bolivia, Colombia, Ecuador, Peru and Venezuela – signed at Cartagena de Indias (Colombia) a declaration on the renunciation of weapons of mass destruction. They proclaimed a commitment not to possess, produce, develop, use, test or transfer nuclear, biological, toxin or chemical weapons, and to refrain from storing, acquiring or holding such weapons in any circumstances. The Cartagena Declaration expressed the determination of its signers to promote the transformation of Latin America and the Caribbean into an area free of weapons of mass destruction.

In a Joint Declaration signed at New Delhi on 19 August 1992, India and Pakistan undertook not to develop, produce or otherwise acquire chemical weapons, not to use these weapons, and not to assist, encourage or induce anyone to engage in such activities.

8.5   The 1993 Chemical Weapons Convention

In May 1991 the United States retreated from its position that it must be allowed to keep a chemical-weapon stockpile of 500 tons until all chemical weapon-capable states had joined the projected multilateral convention. It was thus ready to commit itself unconditionally to the destruction of all its chemical-weapon stocks and chemical-weapon production facilities. It stated that once the convention became effective it would give up the right to retaliate with chemical weapons.

This US renunciation of the postulates of the 1990 US–Soviet Chemical Weapons Agreement narrowed the gap between the positions of the chemical-weapon 'haves' and 'have-nots'. It was welcomed by the Soviet Union and many other states and gave a new impulse to the multilateral talks. In September 1992, the Conference on Disarmament finalized the text of a Convention on the Prohibition of the Development, Production, Stockpiling and Use of Chemical Weapons and on Their Destruction. This text – which includes the Annex on Chemicals, the Annex on Implementation and Verification (Verification Annex) and the Annex on the Protection of Confidential Information (Confidentiality Annex) – was forwarded to the UN General Assembly, which endorsed it. In January 1993 the Chemical Weapons (CW) Convention was opened for signature, and on 29 April 1997 it entered into force.

Definitions

For the purposes of the CW Convention the following definitions were adopted.

Chemical Weapons. Toxic chemicals and their precursors, except where intended for not-prohibited purposes as long as the types and quantities are consistent with such purposes; munitions and devices specifically designed to cause death or other harm through the toxic properties of the toxic chemicals referred to above and which would be released as a result of the employment of such munitions and devices; and any equipment specifically designed for use directly in connection with the employment of the mentioned munitions and devices.

Toxic Chemical. Any chemical which, through its chemical action on life processes, can cause death, temporary incapacitation or permanent harm to humans or animals; all such chemicals are covered, regardless of origin or method of production.

Precursor. Any chemical reactant which takes part at any stage in the production of a toxic chemical; this includes any key component of a binary or multi-component chemical system, the key component being a precursor which plays the most important role in determining the toxic properties of the final product.

Riot Control Agent. Any chemical which rapidly produces in humans sensory irritation or disabling physical effects that disappear within a short time following termination of exposure.

Chemical Weapon Production Facility. Any equipment, as well as any building housing such equipment, designed, constructed or used at any time since 1946 as part of the final technological stage in the production of the banned chemicals, or for filling chemical weapons into munitions or bulk storage containers, or for loading chemical sub-munitions, such as containers of binary components, into chemical munitions; not included is the single small-scale facility permitted to each party for production of toxic chemicals for purposes not prohibited under the Convention.

Purposes not Prohibited. Industrial, agricultural, research, medical, pharmaceutical or other peaceful purposes; protective purposes: those directly related to protection against chemical weapons; military purposes not connected with the use of chemical weapons and not dependent on the use of the toxic properties of chemicals as a method of warfare; and law enforcement, including domestic riot control.

Scope of the Obligations

To exclude the possibility of chemical warfare, the CW Convention prohibits the development, production, acquisition by other means, stockpiling or retention of chemical weapons, or their transfer, directly or indirectly, to anyone. It prohibits unconditionally the use of chemical weapons as well as military preparations for such use. Parties are under the obligation not to assist, encourage or induce anyone to engage in the prohibited activities.

Riot control agents must not be used as a 'method of warfare', but may be employed for law enforcement. Tighter restrictions equivalent to those imposed on other chemical agents could not be incorporated in the Convention because several countries, especially the United States, saw some legitimate uses of these agents to save lives in wartime situations. Certain countries interpret the relevant provision as permitting the use of tear gas and similar incapacitants *only* for domestic law enforcement.

Herbicides, the use of which during the Indo-China War prompted the negotiations on chemical weapons, have not been satisfactorily covered either, because the definition of chemical weapons, as formulated in the Convention, does not cover toxic chemicals causing harm to plants. The preamble does mention the internationally recognized prohibition on the use of herbicides as a method of warfare, but this is considered by many as insufficient. Indeed, the applicability of the 1925 Geneva Protocol to herbicides is not universally accepted (see section 1 above), whereas the 1977 Convention on the Prohibition of Military or Any Other Hostile Use of Environmental Modification Techniques (the so-called Enmod Convention), which is also referred to in this context, has few adherents and bans exclusively those uses of herbicides that produce 'widespread, long-lasting or severe' effects in upsetting the ecological balance of a region. (See Chapter 9.1.)

Within 30 days of the Convention's entry into force, declaration had to be made of the kind and quantity of chemical weapons and chemical-weapon production facilities located on the party's territory or in any other place under its jurisdiction or control, as well as of plans for destroying these weapons and facilities.

National Implementation Measures and Reservations

Each party must: (a) prohibit natural and legal persons anywhere on its territory or elsewhere under its jurisdiction as recognized by international law from undertaking any activity prohibited to a state party under the convention, including enacting penal legislation with respect to such activity; (b) not permit in any place under its control any activity prohibited to a state party under the convention; and (c) extend its penal legislation enacted under (a) above to any prohibited activity undertaken anywhere by natural persons possessing its nationality. A national authority is to be designated or established by each party to fulfil its obligations under the Convention.

The resolution of advice and consent to ratification of the CW Convention adopted by the US Senate contains a number of conditions, some of which contravene the letter or the spirit of the Convention. For example, inspection of US facilities may be refused and, if allowed, the collected laboratory samples may not be transferred for analysis to a laboratory outside the territory of the United States. These conditions are formulated like reservations to the Convention, but they have no international legal force. Articles of the CW Convention are not subject to reservations, while the annexes to it are subject only to those reservations which are not incompatible with the object and purpose of the Convention.

Schedules

To facilitate implementation of the CW Convention and application of verification measures, toxic chemicals and their precursors are listed in three schedules corresponding to the degree of concern they give rise to.

Schedule 1. Chemicals that have been developed, produced, stockpiled or used as chemical weapons; that otherwise pose a high risk to the object and purpose of the convention by virtue of their high potential for use in prohibited activities; and that have little or no use for purposes that are not prohibited.

Schedule 1 chemicals must be destroyed, except for a small quantity – no more than one metric ton annually – that may be produced in a single small-scale facility for not-prohibited purposes.

Schedule 2. Chemicals that possess such lethal or incapacitating toxicity as well as other properties that could enable them to be used as chemical weapons; that may be used as precursors in chemical reactions at the final stage of formation of chemicals listed in schedule 1 and partly in schedule 2; that pose a significant risk by virtue of their importance in the production of chemicals listed in schedule 1 and partly in schedule 2; and that are not produced in large commercial quantities for not-prohibited purposes.

Schedule 2 chemicals may not be transferred to non-parties after the Convention has been in force for three years, that is, after April 2000. In the meantime, such transfers required 'end-use certificates' containing recipients' pledges not to use the chemicals for prohibited purposes.

Schedule 3. Chemicals not listed in other schedules and that have been produced, stockpiled or used as chemical weapons; that otherwise pose a risk because they possess such lethal or incapacitating toxicity as well as other properties that might enable them to be used as chemical weapons; that pose a risk by virtue of their importance in the production of chemicals listed in schedule 1 and partly in schedule 2; and that may be produced in large commercial quantities for not-prohibited purposes.

Schedule 3 chemicals and facilities must be declared if more than 30 metric tons are produced annually. Facilities producing more than 200 tons are subject to routine inspection. Reports on schedule 3 chemicals must include data for the previous calendar year on quantities produced, imported and exported. When transferring schedule 3 chemicals to non-parties, each party must adopt measures to ensure that the transferred chemicals will be used only for not-prohibited purposes. Five years after entry into force of the Convention, that is, in 2002, other measures may be adopted regarding transfers to non-parties.

Destruction

The destruction of chemical weapons is to begin not later than two years after the Convention enters into force for a given party. Within three years 1% of the stocks should be destroyed; within five years, 20%; within seven years, 45%; and the remainder within ten years. The order of destruction does not take into account the qualitative aspects of chemical weapons. This was criticized by states which would have preferred to see the most toxic chemical agents destroyed first.

If a party believes that it will be unable to ensure the destruction of all schedule 1 chemical weapons and their components within the above time limit, it may request an extension of the deadline. Such a request, to be made no later than nine years after the entry into force of the convention, should specify the duration of the proposed extension, explain the reasons and contain a detailed plan of destruction during the remaining portion of the original ten-year period and the proposed extension. A decision on the request is to be taken by the Conference of States Parties. In no case may the deadline be extended beyond 15 years. An extension granted to one party would not automatically entitle another party to obtain an extension; a special application would have to be submitted by the state concerned.

A state party that has on its territory chemical weapons belonging to another state must make efforts to ensure that the weapons are removed from its territory no later than one year after the Convention enters into force. If these weapons are not removed, the party is obliged to destroy them; it may request other states to provide assistance in the destruction. Old chemical weapons (defined as those produced before 1925 or between 1925 and 1946 but no longer usable) as well as abandoned chemical weapons (defined as those left by a state after 1 January 1925 on the territory of another state without the consent of the latter) do not pose a significant threat to the object and purpose of the Convention but constitute a threat to the environment. (In 1999 a bilateral agreement was reached between China and Japan on the destruction of the chemical weapons abandoned by the former Imperial Japanese Army on the territory of China.)

All chemical-weapon production facilities must cease production immediately after the Convention has entered into force and be closed within 90 days thereafter. Destruction of the facilities should begin not later than one year after the Convention's entry into force and be completed in the course of the subsequent nine years.

Chemical-weapon production facilities may be temporarily converted for destruction of chemical weapons. Such a converted facility must be destroyed as soon as it is no longer in use for destruction and in any case no later than ten years after entry into force of the Convention. In exceptional cases of 'compelling need', permission may be granted to convert a chemical-weapon production facility for purposes not prohibited under the Convention rather than destroy it. Conversion should be carried out in such a manner as to make the converted facility no more capable of being reconverted into a chemical-weapon production facility than any other facility used for peaceful purposes.

Each party must meet the costs of destruction and assign the highest priority to ensuring the safety of people and to protecting the environment during the destruction processes. Each may determine how it will destroy the weapons, except that dumping in any body of water, land burial or open-pit burning are not allowed. Each party must also meet the costs of verification of storage and destruction of weapons, unless decided otherwise by the Executive Council of the Organisation for the Prohibition of Chemical Weapons (OPCW).

At the time the Convention was concluded, Russia said that it would not be able to meet the ten-year destruction deadline; it had neither an operating chemical destruction facility nor a destruction plan. It also insisted that the costs of verification of destruction should be met by all parties on the basis of the UN scale of assessment rather than by the verified party alone. In July 1992, the United States and Russia signed an Agreement concerning the Safe, Secure and Ecologically Sound Destruction of Chemical Weapons, under which US chemical-weapon destruction assistance was to be provided to Russia at no cost. (This agreement was subject to the provisions of the US–Russian Agreement on the Safe and Secure Transportation, Storage and Destruction of Weapons and the Prevention of Weapons Proliferation, the so-called Weapons Destruction and Non-Proliferation Agreement, concluded on 17 June 1992.) Several other states also provide such assistance to Russia.

Before the entry into force of the CW Convention, there were only three known possessors of chemical weapons – Russia, the United States (the two together holding the bulk of the declared stocks) and Iraq. In the declarations submitted after the Convention had become effective, two more states informed the OPCW that they stored chemical weapons and possessed facilities for their production.

Protection against Chemical Weapons

The parties have the right to conduct research into and to develop, produce, acquire, transfer or use means of protection against chemical weapons for purposes not prohibited under the Convention. They have also the right to participate in the exchange of equipment, material, and scientific and technological information concerning means of protection.

Subject to special procedures, each party is entitled to receive assistance and protection against the use or threat of use of chemical weapons if it considers that: (a) chemical weapons have been used against it; (b) riot control agents have been used against it as a method of warfare; or (c) it is threatened by actions or activities of any state that are prohibited for parties. Assistance is defined as the coordination and delivery of protection against chemical weapons, including detection equipment and alarm systems; protective equipment; decontamination equipment and decontaminants; medical antidotes and treatments; and advice on any of the protective measures. The establishment of a voluntary fund for assistance is provided for. Parties may also conclude agreements with the OPCW concerning the procurement of assistance upon demand, or declare the kind of assistance they might provide in response to an appeal by the OPCW. In 1999 Iran offered to set up an international centre for the treatment of chemical warfare casualties.

Economic and Technological Development

Provisions of the CW Convention must be implemented in such a way as to avoid hampering the economic or technological development of the parties as well as international cooperation in the field of chemical activities for purposes not prohibited by the Convention.

In particular, parties may not maintain among themselves any restrictions incompatible with obligations undertaken under the Convention which would restrict or impede trade and the development and promotion of scientific and technological knowledge in the field of chemistry for industrial, agricultural, research, medical, pharmaceutical or other peaceful purposes. This was specifically understood to mean that the Australia Group of exporters would lift restrictions on trade in commercial chemicals and related technology between the parties complying with the Convention.

Amendments

An amendment proposed by a party may be considered only by an Amendment Conference. Such a conference must be convened if one-third or more of the parties notify the Director-General of the OPCW that they support consideration of the proposal.

To enter into force for all parties, amendments must be adopted by a positive vote of a majority of all parties, with no party casting a negative vote, and ratified or accepted by all those parties casting a positive vote at the Amendment Conference.

Provisions of the annexes to the convention are subject to changes only if the proposed changes relate to matters of an administrative or technical nature. If the Executive Council of the OPCW recommends that a proposal of such nature be adopted, it shall be considered approved if no party objects to it within 90 days after receipt of the recommendation. If the Executive Council recommends that the proposal be rejected, it shall be considered rejected in the absence of an objection to the rejection also within 90 days. In case a recommendation of the Executive Council is not accepted, a decision on the proposal for a change is to be taken up as a matter of substance by the Conference of the States Parties. Any changes adopted under this procedure enter into force for all parties 180 days after the date of notification by the Director-General of their approval, unless another time period is recommended by the Executive Council or decided by the Conference of States Parties.

Final Clauses

Relation to Other Agreements. The CW Convention stipulates that its provisions should not be interpreted as in any way limiting or detracting from the obligations assumed by states under the 1925 Geneva Protocol and the 1972 BW Convention. This means that parties to the latter two agreements remain bound by them whether or not they have become parties to the CW Convention. For the parties to the CW Convention, the reservations they may have made to the 1925 Geneva Protocol, those concerning the right to employ the banned weapons under certain circumstances, must be considered as invalid.

Duration and Withdrawal. The CW Convention is of unlimited duration. Each party has the right to withdraw from the Convention if it decides that extraordinary events related to the subject matter of the Convention have jeopardized its supreme interests. Notice of withdrawal must be given 90 days in advance to all parties, the Executive Council of the OPCW, the depositary and the UN Security Council.

Entry into Force. As stipulated in the CW Convention, it entered into force 180 days after the date of the deposit of the 65th instrument of ratification with the UN Secretary-General, designated as the depositary. For states depositing their instruments of ratification or accession after entry into force of the Convention, it enters into force on the 30th day following the deposit of the relevant instruments.

Assessment

Despite certain shortcomings, which are difficult to avoid in any document adopted by consensus, the CW Convention constituted a great achievement. In establishing an international legal norm against the possession of chemical weapons, it complemented and reinforced the ban on the use of these weapons, which is embodied in the 1925 Geneva Protocol.

Unlike the 1968 Non-Proliferation Treaty, which admitted, for an indefinite period of time, the continued existence of two categories of states – nuclear-weapon and non-nuclear-weapon states – and which accorded different rights and obligations to each category, the CW Convention treats all nations alike. All are prohibited from producing or retaining chemical weapons and all are subject to the same monitoring procedures.

The elaborate verification envisaged by the Convention has the potential of ensuring that militarily significant amounts of chemical weapons are not being produced and that militarily significant stockpiles have been accounted for and destroyed. In any event, because of the complexity of the destruction operations and their costs, and also because of the environmental hazards, the process of elimination of chemical weapons poses more problems than verification. The ten-year destruction period seems too ambitious.

The relative attractiveness of the CW Convention is due, among other reasons, to the arrangements among the parties for assistance in the event of chemical-weapon attack or threat of such attack. Indeed, in many cases chemical weapons were directed at countries having no such weapons or means of protection against them. The danger of becoming a target for restrictions on transfers of chemicals from parties to non-parties, as well as expectations that chemical export controls would loosen up among parties, may have also played a role in attracting adherents. However, not all nations have set aside the idea that chemical weapons are a poor country's 'nuclear deterrent'.

Implementation

The United States, India, South Korea and Russia declared their chemical weapon stockpiles amounting to some 70 thousand tons of chemical agents and nearly 8.4 million munitions and containers. During the first four years of the implementation of the CW Convention the first three countries began destroying their chemical weapons, and by April 2001 approximately 7–10% of the world's chemical agents and 15–20% of its chemical munitions had been eliminated under the supervision of the OPCW. The goal of total abolition of chemical weapons was still far from being reached. Russia, the possessor of the largest stockpile of chemical weapons in the world, had not even met the first deadline for destruction (see above), mainly because of financial difficulties. The United States, the second largest possessor of chemical weapons, may also fail to comply with the relevant Convention provisions within the prescribed time limits, mainly because of stringent environmental regulations. Moreover, the OPCW was compelled to significantly reduce its verification activities owing to a budgetary crisis. In April 2002, on the insistence of the US government, the Director-General of the OPCW was removed from office by the special session of the Conference of States Parties, allegedly because of the financial mismanagement of the Organisation and 'ill-considered initiatives'.

9

Environmental and Radiological Weapons

9.1 The 1977 Environmental Modification Convention

In the early 1970s much attention was devoted to the possibility of using environmental forces for military ends. Interest in such new means of warfare arose, in part, because of the rainmaking operations and large-scale destruction of vegetation during the war which was then going on in Indo-China. Concern about the consequences of environmental manipulation for the world ecological system led to suggestions for reaching an international agreement to prevent this danger. The issue became the subject of US–Soviet as well as multilateral negotiations.

As a result of these negotiations, the Convention on the Prohibition of Military or Any Other Hostile Use of Environmental Modification Techniques (the so-called Enmod Convention) was signed on 18 May 1977 and entered into force in 1978. Four of the ten Convention articles are clarified and amplified in Understandings. These Understandings have not been written into the text of the Convention, but they form part of the *travaux préparatoires* and are important for the comprehension of the drafters' intentions.

Subject of the Prohibition

The Enmod Convention deals with changes in the environment brought about by deliberate human manipulation of natural processes, as distinct from conventional acts of warfare, which might result in adverse effects on the environment. The Convention covers those changes which affect the dynamics, composition or structure of the earth, including its biota, lithosphere, hydrosphere and atmosphere, or of outer space. The employment of techniques producing such modifications as the means of destruction, damage or injury to another state party is prohibited. In the opinion of the United States, expressed in the course of the negotiations, the targets alluded to include the enemy's military forces and civilian population as well as its cities, industries, agriculture, transportation systems, communication systems, and natural resources and assets. Nor is a state allowed to assist, encourage or induce other states to engage in these activities. However, the threat to use the techniques in question has not been expressly forbidden.

Scope of the Prohibition

The ban under the Enmod Convention applies to the conduct of military operations during armed conflicts, as well as to hostile use (whether by military or non-military personnel) when no other weapon is being employed. It is applicable both to offence and defence, regardless of geographical boundaries. In the light of these explanations, which were given by the Soviet and US sponsors of the text, the term 'hostile' alone would have sufficed as a purpose criterion upon which to base the Convention.

Threshold of Damage. Not all hostile uses causing harm to others are prohibited by the Convention – only those having 'widespread, long-lasting or severe effects'. The meaning of these terms, according to the Understanding relating to Article I and describing the main obligations of the parties, is as follows: 'widespread' means encompassing an area on the scale of several hundred square kilometres; 'long-lasting' means lasting for a period of months or approximately a season; and 'severe' means involving serious or significant disruption or harm to human life, natural and economic resources or other assets.

It is noted in the Understanding that the above interpretation is intended exclusively for this Convention and should not prejudice the interpretation of the same or similar terms used in connection with any other international agreement. This proviso was found necessary in order to prevent giving an identical interpretation to the terms 'widespread, long-term and severe', used in the 1977 Protocol I Additional to the Geneva Conventions of 1949 and relating to the protection of victims of international armed conflicts then under negotiation. Indeed, the two documents pursue different aims. The 1977 Protocol I is meant to ban the employment in armed conflict of methods or means of warfare which are intended, or may be expected, to cause serious damage to the environment, regardless of which weapons are used; to make this ban applicable, the presence of all three criteria – widespread, long-term and severe – is required. On the other hand, the Enmod Convention forbids the use (or manipulation) of the forces of the environment as 'weapons', both during hostilities and when there is no overt conflict; in this case, the presence of only one of the three criteria is enough for the environmental modification technique to be deemed outlawed. Thus, the use of environmental modification techniques is prohibited if two requirements are met simultaneously: (a) that the use is hostile; and (b) that it causes destruction, damage or injury at, or in excess of, the threshold described above.

Exemptions. Exempted from the prohibition are non-hostile uses of modification techniques, even if they produce destructive effects exceeding the threshold. Equally permissible are hostile uses that produce destructive effects below the threshold. Assuming, therefore, that hostile intent has been proved (which may not be an easy task), it would still not be illegal, according to the Understanding, to devastate an area smaller than several hundred square kilometres; or to cause adverse effects lasting for a period of weeks instead of months, or less than a season; or to bring about disruption or harm to human life, natural and economic resources or other assets if the disruption is not 'severe', 'serious' or 'significant' – whatever these subjective terms might mean to countries of different sizes, of different population densities or at different stages of economic development. The perpetrator's perception of the gravity of such acts may not agree with that of the victim.

Applicability. The Enmod Convention clearly prohibits causing such phenomena as earthquakes, tsunamis (seismic sea waves), an upset in the ecological balance of a region, changes in weather patterns (clouds, precipitation, cyclones of various types or tornadic storms), changes in climate patterns, changes in ocean currents, changes in the state of the stratospheric ozone layer and changes in the state of the ionosphere, when produced by hostile use of environmental modification techniques. It is understood that all these phenomena would result, or could reasonably be expected to result, in widespread, long-lasting or severe destruction, damage or injury. It is further recognized, in the Understanding relating to Article II that contains the definition of the term 'environmental modification techniques', that the above list is only illustrative and that the use of techniques producing other phenomena would also be illegal, insofar as the criteria of hostility and destructiveness were met. Nevertheless, only the most fanciful events are enumerated in this Understanding – those unlikely to be caused through deliberate action for warlike purposes, that is, in such a way that the effects would be felt only (or primarily) by the enemy. Attempts were made to extend the 'illustrative' list by including, for example, an upset in the hydrological balance of a region through the diversion of rivers, but these attempts failed. As regards hostile use of herbicides, the 1992 Enmod Review Conference confirmed the interpretation, given in 1976 by the US negotiator, that such use is prohibited only if it upsets the ecological balance of a region, thus causing widespread, long-lasting or severe effects.

As a consequence of the threshold approach, the techniques that can produce more limited effects (such as precipitation modification short of changing the 'weather pattern') and which are therefore more likely to be used in a selected area to affect the environment with hostile intent have escaped proscription. Even the deliberate setting on fire of the Kuwaiti oil wells by Iraq (a signatory to the Convention) during the 1991 Gulf War, which must have been expected to produce at least one of the effects covered by the Convention, was not generally recognized as a prohibited act. Research into as well as development of the environmental modification techniques for hostile purposes are not prohibited; nor does the Convention ban the use of environmental modification techniques aimed at increasing the effectiveness of other weapons in producing destruction, damage or injury.

Comprehensive versus Partial Approach. The narrow scope of prohibition under the Enmod Convention stands in contrast to the Soviet draft, which was submitted in 1974. Under this proposal, the parties would have agreed not to use 'any' means of influencing the environment for military or any other purpose incompatible with the 'maintenance of international security, human well-being and health'. It is also worth noting that a study of possible international restraints on environmental warfare, prepared by the US National Security Council and submitted to the US President in 1974, envisaged, as one option, a 'comprehensive' prohibition on hostile use of environmental modification techniques. In departing from this all-inclusive approach the United States argued that a comprehensive ban would give rise to disputes over 'trivial' issues and could create a risk of unprovable claims of violation. However, what is deemed trivial by the party carrying out modification activities may not seem so to the victim. The imprecise definition of the terms 'widespread, long-lasting or severe' may generate greater controversies than an unqualified ban. There is no reason why any hostile modification of the environment or any amount of damage caused by such modification should be tolerated. Even the right to use modification techniques on a state's own territory to forestall foreign invasion – for example, by opening dams to cause catastrophic floods or by producing massive landslides – is challenged by some.

A partial approach may have some justification in an agreement which restricts the possession of a certain category of weapon but leaves other categories unaffected. However, in an agreement such as the Enmod Convention, which prohibits the use of certain methods of warfare and thereby establishes a new international norm of behaviour in armed conflict, it is incongruous to speak of a threshold of damage or injury below which the parties retain freedom of action. This is certainly out of place with regard to non-conventional means of warfare capable of causing mass destruction, nor is it in harmony with the humanitarian principles underlying the law of armed conflict. The 1925 Geneva Protocol, prohibiting the use of chemical and bacteriological methods of warfare, makes no distinction between quantitatively more or less severe effects caused by these methods. It has never been suggested that allowance should be made for some degree of harm to human life with the use of weapons indisputably covered by the 1925 Protocol.

The protection from hostile uses of environmental modification techniques extends only to parties, that is, to states that have ratified or acceded to the Convention. The negotiators were of the view that, if non-parties were also to be covered by such protection, there would be no incentive for them to assume contractual obligations.

A number of environmental modification techniques may have peaceful applications. For example, fog or cloud dispersion could be applied at civilian airports, seaports or other major civilian enterprises. Suppression of conditions that lead to hailstone precipitation could help reduce damage to crops. Manipulation of storms could be used to moderate the intensity of hurricanes or to disperse or redirect them. Rainmaking could be employed for the relief of drought or for extinguishing forest fires. Stimulation of weak earthquakes could be applied to relieve stress conditions that otherwise might lead to destructive natural earthquakes. Precipitating a snow avalanche is used for controlled avalanche release, and river diversion is commonly used for irrigation, navigation or power-generating purposes

The parties to the Convention have undertaken to facilitate and participate in the 'fullest possible' exchange of scientific and technological information on the use of environmental modification techniques for peaceful purposes. They are to contribute, as far as they are in a position to do so, to international economic and scientific cooperation in the preservation, improvement and peaceful utilization of the environment, with due consideration for the needs of the developing areas of the world. These pledges have proved to be of no consequence.

Assessment

Preventing environmental forces from being used as weapons of war could be of value as an arms control measure and as a rule of the law of armed conflict, even though very few environmental modification techniques with significant military utility have as yet been identified. However, to be effective, the constraints would have to be unambiguous and as nearly all-inclusive as possible, that is, covering all modification techniques for hostile purposes, regardless of their sophistication. The Enmod Convention does not meet these requirements, because it is not clear what it actually bans. The Convention prohibits the use of techniques that are the subject of scientific speculation or which, if proved feasible, could hardly be used as rational weapons of war. It thus appears to condone hostile manipulation of the environment with some unspecified 'benign' means – those producing effects below the set threshold. It is therefore not surprising that the Convention has attracted considerably fewer parties than most other multilateral arms control agreements.

To become a meaningful contribution to the cause of halting the arms race – one of the main purposes proclaimed in its preamble – as well as a useful addition to international humanitarian law, the Enmod Convention would have to be substantially amended. In the first place, the list of phenomena that the parties are not allowed to cause under any circumstances should be expanded by removing the threshold limiting the ban to uses having only 'widespread, long-lasting or severe effects'. In other words, the Convention should be made applicable to *any* hostile use of environmental modification techniques.

Second, the parties should undertake to abstain not only from the hostile use of environmental modification techniques but also from preparations for such use. This implies constraints on militarily oriented research and development of the techniques in question. States should assume an obligation to give advance notification of all major experiments in environmental modification and subject them to international observation in order to demonstrate that their purposes are genuinely peaceful. Large-scale internationalization of research and development in the field of peaceful uses of environmental modification techniques could, apart from the obvious scientific, economic and technological advantages, provide additional assurance that substantial resources were not being diverted to hostile military ends.

Third, it would be desirable to prohibit hostile uses of modification techniques against any state or people instead of confining the ban – as the Enmod Convention does – to uses against parties. An environmental weapon would indiscriminately strike both combatants and non-combatants in contravention of the rule of international law requiring protection of the civilian population. Another justification for such an absolute prohibition is the difficulty, if not the impossibility, of circumscribing the effects of the use of an environmental modification technique within geographic boundaries so as to injure a non-party without injuring a party. Threats to use environmental modification techniques for hostile purposes should also be clearly prohibited.

According to the final clauses of the Enmod Convention, any party may propose amendments by submitting the proposed text to the UN Secretary-General, the depositary of the Convention. The amendments would enter into force for all parties which had accepted them, upon the deposit of the instruments of acceptance by a majority of the parties. Proposals to amend the Convention may also be considered at review conferences. A crucial point is the removal of the threshold limitation on the scope of the environmental modification ban.

Without such amendments and new understandings, the Enmod Convention will remain inapplicable and therefore irrelevant to the security concerns of states.

9.2   Consideration of a Ban on Radiological Weapons

The 1948 UN definition of weapons of mass destruction (see Chapter 8.1) included 'radioactive material weapons'. Referring to this, in 1979 the United States and the Soviet Union proposed the conclusion of a convention prohibiting radiological weapons. The declared aim was to prevent the misuse of radioactive material, which, as a result of the development of nuclear energy, was becoming available in large amounts to many countries. In the course of negotiations at the Conference on

Disarmament (CD), divergent views emerged regarding the definition of the weapons in question as well as the scope of the proposed ban.

Definition

The United States and the Soviet Union defined a radiological weapon as any device (including any weapon or equipment), other than a nuclear explosive device, specifically designed to employ radioactive material by disseminating it to cause destruction, damage or injury by means of the radiation produced by the decay of such material, as well as any radioactive material (other than that produced by a nuclear explosive device) specifically designed for such use. Thus, a clear distinction was drawn between a weapon relying for its destructive effect on radiation emitted by radioactive material contained in it and a weapon relying for its destructive effect on heat and blast as well as radiation caused by the processes occurring at the time of a nuclear explosion. The former would be prohibited, the latter would not. Several nations objected to the definition of radiological weapons, which contained a clause excluding nuclear explosives; it could, in their opinion, 'legitimize' the use of nuclear weapons. An alternative formulation suggested at the CD to overcome the definitional hurdle was to consider as a radiological weapon any device containing radioactive material or waste as its principal harmful element and specifically designed or used to cause injury, death, environmental damage or destruction through the direct or indirect effects of ionizing radiation, without involving the critical assembly of any fissile material.

A radiological weapon should not be confused with the enhanced radiation/reduced blast weapon, commonly referred to as a 'neutron' weapon. The latter is a nuclear explosive device that kills mainly (but not exclusively) by radiation. The prohibition of the production, stockpiling, deployment and use of neutron weapons was proposed in 1978 by the Soviet Union as a separate measure. The Soviet Union then contended that the introduction of neutron weapons would lower the nuclear threshold, increasing the possibility that an armed conflict would escalate to the level of an all-out nuclear war. However, the Soviet proposal, reiterated in 1981 when the United States decided to start the production of neutron weapons (intended to repel tank attacks in Europe by incapacitating the crews manning the tanks), was rejected by the Western powers. They argued that there was no reason to single out for special arms control treatment this particular nuclear weapon, which was less destructive than other nuclear weapons.

Scope of the Intended Prohibition

The envisaged convention would prohibit the development, production, stockpiling, possession, transfer and use of radiological weapons. However, so far, no nation is known to have manufactured a radiological weapon. In view of the enormous practical difficulties connected with the use of such a weapon in war, it is even doubtful whether any serious thought has been given to developing one. A very high radiation dose would be required to kill or injure people on the battlefield. One would need radioactive isotopes having a very short half-life, but these cannot be stored; they would decay before being used. Alternatively, one would need such large amounts of isotopes with long half-lives that the whole proposition would be impractical. Transport of significant quantities of radioactive material to the battlefield, or to areas destined to be denied to the enemy, would be a very cumbersome task, mainly because of the heavy protective shielding which would be needed. It is also hard to conceive of the delivery of this material to intercontinental targets for strategic purposes.

On the other hand, it is technically possible to use radioactive material of lower activity so as to cause long-term effects harmful to life or health after months or years, or even to future generations. For this purpose one might use materials having a relatively long half-life, for instance strontium-90, which has a half-life of about 28 years. These materials can be obtained from the radioactive waste of reactors. There would be little military rationale for producing long-term harmful effects with radioactive materials, but the danger of their use for terrorist purposes is not negligible.

Banning Attacks on Nuclear Facilities

To make the envisaged ban more meaningful, a proposal was put forward to prohibit deliberate damage to nuclear reactors or other nuclear facilities, which could cause release of radioactive material and contamination of the environment. The Chernobyl reactor accident in 1986 demonstrated the disastrous consequences of such a release. In fact, attacking nuclear facilities would seem to be at present the only conceivable way of waging radiological warfare.

According to the 1977 Protocol I relating to the protection of victims of international armed conflicts, 'nuclear electrical generating stations' are not to be made the object of attack if such attack may cause the release of dangerous forces and consequent severe losses among the civilian population. However, the protection may cease if the station provides electric power 'in regular, significant and direct support of military operations and if such attack is the only feasible way to terminate such support'. This reservation is vague enough to bring to naught the ban to which it is attached. Moreover, the Protocol prohibition does not cover facilities committed to military use, while in the field of civilian use it leaves out installations with large quantities of radioactive materials. The latter include research reactors, cooling ponds which contain fuel elements removed from the reactor before they are shipped to reprocessing plants, reprocessing plants where the spent fuel elements are chemically treated to separate uranium and plutonium from the waste products, and storage tanks containing high-level radioactive wastes. A more adequate and much stricter legal norm would be needed than the existing rule of international humanitarian law.

So far, only India and Pakistan – countries with a comparable level of nuclear development – have succeeded in reaching, in 1988, an agreement banning the destruction of or damage to nuclear installations or facilities. (In the spring of 1998, during the period of heightened tension between the two countries, Pakistan accused India of preparing to attack the Pakistani nuclear facilities, but India dismissed the allegation.) The agreement remained in force in spite of the armed clashes along the border between the two countries, and the parties continued to exchange information concerning the emplacement of their nuclear installations and facilities. However, the CD, which had been trying for years to work out elements of a global ban on attacks against such nuclear objectives, encountered many obstacles. Indeed, establishing a relevant international rule, with a degree of certainty that it will be universally observed under all circumstances, is extremely complex. Whereas wanton destruction of nuclear power stations and of other peaceful nuclear facilities could be inhibited by the projected ban, a country suspecting that its actual or potential non-nuclear-weapon enemy is engaged in clandestine production of nuclear-weapon material might not hesitate, in a situation of acute international crisis or war, to attack the relevant, ostensibly civilian, installations, invoking the imperative of ultimate defence. This is what happened in 1981, when Israeli aircraft attacked the Iraqi nuclear centre, and in 1991, when US aircraft attacked Iraqi nuclear facilities during the Gulf War. There was no dangerous release of contaminants, but there could have been if the reactors had been loaded with large quantities of nuclear fuel and if they had been in operation at the time of the bombing.

The question of prohibiting radiological weapons was considered along with proposals for the prohibition of new types of weapons and new systems of weapons of mass destruction. Certain countries favoured a general ban on such weapons, possibly through a single treaty. However, an agreement encompassing all imaginable weapons based on new scientific or technological principles could not be sufficiently clear as regards its object or sufficiently precise as regards its scope to produce real arms control effects. In addition, verification of an omnibus treaty would encounter enormous difficulties, as it would involve monitoring a wide gamut of activities, the military implications of which are often not obvious.

It would seem more practical to tackle each specific and clearly identified new weapon of mass destruction separately, taking account of its peculiarities. On the other hand, it is generally considered easier to ban arms at the research and experimentation stage than to eliminate those already developed, manufactured and stockpiled. In order to detect signs of the development of a new weapon with the potential to cause mass destruction, pertinent scientific discoveries would need to be internationally reviewed on a current basis and their possible military impact examined.

10

Outer Space and Celestial Bodies

In the 1960s international attention shifted to outer space, which was becoming a new arena of military competition between the superpowers. The first attempt to control this competition was made on 27 January 1967, when the Treaty on Principles Governing the Activities of States in the Exploration and Use of Outer Space, Including the Moon and Other Celestial Bodies (the Outer Space Treaty) was opened for signature.

10.1 The 1967 Outer Space Treaty

Although primarily concerned with the peaceful uses of outer space, the Outer Space Treaty contains an article directly related to arms control.

Arms Control Provisions

Elaborating on a UN General Assembly resolution unanimously adopted in 1963, the Treaty prohibits the placing in earth orbit of any objects carrying nuclear weapons or any other kinds of weapon of mass destruction, as well as the installation of such weapons on celestial bodies, or the stationing of them in outer space in any other manner. 'Weapons of mass destruction' are not defined here, but the understanding of the negotiators was that, in addition to nuclear weapons, weapons of mass destruction include chemical and biological weapons. It was also understood that the principle of peaceful use could accommodate passive military use, such as the orbiting of military satellites for reconnaissance, surveillance, early warning or communications.

Also banned are the establishment of military bases, installations and fortifications, the testing of any type of weapon and the conduct of military manoeuvres on celestial bodies. However, the use of military personnel for scientific research or for any other peaceful purpose is allowed.

Any state party may give notice of its withdrawal from the Treaty by written notification to the depositaries – the governments of Russia, the United Kingdom and the United States. The withdrawal would take effect one year from the date of receipt of such notification.

Assessment

From the technical point of view, weapons of mass destruction in orbit around the earth would have serious drawbacks. Hitting a predetermined target on the earth's surface lying on the path defined by the orbit would be feasible only at certain hours or on certain days. Malfunction of the orbiting weapon could cause unintended large-scale damage to the enemy, to a third state or even to the launching state itself. There would also be problems of maintenance and of command and control. The

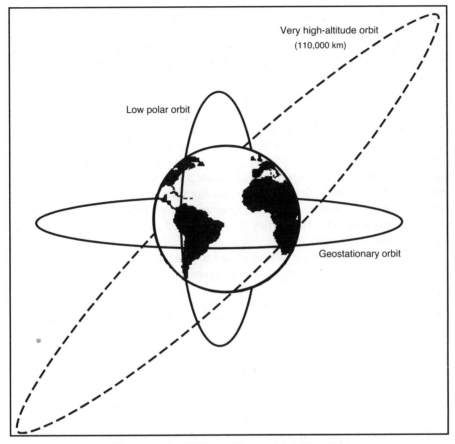

Figure 10.1 *Different Types of Satellite Orbit*

weapon could be intercepted or rendered inoperative. Placing weapons on manned orbiting stations would remove only some of these operational inconveniences.

On balance, the disadvantages of placing nuclear or other weapons of mass destruction in outer space outweigh their military usefulness. In banning them, the great powers have not sacrificed much, especially since outer space has not been fully denuclearized: it is not forbidden to launch ballistic missiles carrying nuclear weapons into outer space. Nor is deployment in outer space of weapons not capable of mass destruction subject to any restriction; only the moon and other celestial bodies are to be used 'exclusively' for peaceful purposes.

Since October 1967, when the Outer Space Treaty entered into force, repeated proposals have been made to amend its arms control clause so as to render it applicable to all kinds of weapon. However, by 2001 no such amendment had been formally submitted.

Despite all its drawbacks, the Outer Space Treaty became a legal obstacle to placing nuclear explosive charges in space to power X-ray lasers, as had been envisaged by the US Strategic Defense Initiative in the 1980s (see Chapter 5.2).

10.2 The 1979 Moon Agreement

Following a Soviet initiative, the Agreement Governing the Activities of States on
the Moon and Other Celestial Bodies (the Moon Agreement) was opened for signature on 18 December 1979.

Arms Control Provisions

The 1979 Moon Agreement, which amplifies the relevant articles of the Outer Space
Treaty, entered into force in 1984. It confirms the demilitarization of the moon – as
provided for in the 1967 Outer Space Treaty – and prohibits any threat or use of
force or any other hostile act or threat of hostile act on the moon. Similarly, it prohibits the use of the moon in order to commit any such act or to engage in any such
threat in relation to the earth, the moon, spacecraft, or the personnel of spacecraft or
man-made space objects. The parties are not allowed to place in orbit around the
moon, or in other trajectory to or around it, objects carrying nuclear weapons or any
other kinds of weapon of mass destruction, or to place or use such weapons on or in
the moon. The placing of conventional weapons in orbit around the moon is not
prohibited.

Unlike the Outer Space Treaty, the Moon Agreement did not require acceptance
by the great powers to enter into force. It has only one depositary – the UN
Secretary-General.

Assessment

The Moon Agreement has attracted few parties, probably because the danger of war
conducted from another planet against a state on earth seems an unrealistic prospect.
The arms control effect of the undertaking to use the moon or any other celestial
body exclusively for peaceful purposes is thus even scantier than that of banning the
orbiting of weapons of mass destruction around the earth.

In emphasizing the freedom of scientific investigation, the Moon Agreement reiterates an Outer Space Treaty provision to the effect that the use of any equipment or
facility necessary for peaceful exploration of the moon is not prohibited. However,
the terms 'equipment' and 'facility' have not been defined. They may lend themselves to interpretations undermining the purpose of the Agreement.

10.3 The 1975 Registration Convention

The Convention on Registration of Objects Launched into Outer Space (the Registration Convention), complementing the Outer Space Treaty, was opened for signature on 14 January 1975. The Registration Convention also complemented the 1972
Convention on International Liability for Damage Caused by Space Objects, which,
like the 1968 Agreement on the Rescue of Astronauts, the Return of Astronauts and
the Return of Objects Launched into Outer Space, deals with technical and legal
aspects of international cooperation in the exploration and use of outer space for
peaceful purposes.

Main Provisions

Under the Registration Convention, in force since 1976, any space object launched into earth orbit or beyond is to be recorded in an appropriate national registry.

The launching state must furnish to the UN Secretary-General – the depositary of the Convention – as soon as practicable, information on the designator of the space object or its registration number, the date and territory or location of the launch, basic orbital parameters and general function of the object. Each state of registry also has the duty to notify the depositary, to the greatest extent feasible, of space objects concerning which it has previously transmitted information and which have been but no longer are in earth orbit.

Assessment

Arms control measures regarding outer space – other than the 1967 Outer Space Treaty and the 1979 Moon Agreement – include the prohibition, under the 1963 Partial Test Ban Treaty, on testing nuclear weapons in this environment; the ban, under the 1972 ABM Treaty, on the deployment of space-based anti-ballistic missile systems; an undertaking, under the 1977 Enmod Convention, not to engage in military or any other hostile use of environmental modification techniques, defined as techniques for changing the dynamics, composition or structure of the earth or of outer space; and the banning, under the 1979 SALT II Treaty, of fractional orbital bombardment systems (FOBS). Whereas all these measures appear to have been observed, the Registration Convention provisions have often been disregarded. There have been inordinate delays in announcing launches of objects into outer space and, as corroborated by information from non-governmental sources, some launches have never been announced. Nor do the space powers provide a description of any military functions of the objects they launch.

To strengthen the Registration Convention, the following important supplementary information would be needed: precise description of the space object, including its mass, size and energy sources available on board; mission of the object; presence or absence of weapons; and possible changes in the stated orbit. The term 'as soon as practicable' should be made more specific. There seems to be no reason why launch forecasts could not be provided, nor why notifications of actual launches could not be given prior to launch and/or immediately afterwards. An agency for overseeing the Convention and checking compliance might also be useful. Any state party to the Registration Convention has the right to propose amendments. These would enter into force for each accepting party upon their acceptance by the majority.

10.4 Anti-Satellite Weapons

Since the 1960s the superpowers have employed satellites for military purposes: communication, navigation, intelligence gathering, early warning of missile attack, weather forecasting and verification of arms control agreements. The military utility of satellites has made them an attractive target – and possibly also an easy one. This circumstance prompted the development of anti-satellite (ASAT) weapons.

States could take various measures to enhance the survivability of their satellites –
for example, by hardening them or by equipping them with manoeuvring capabilities. Such measures, however, would be costly and difficult to carry out; they may
not even be adequate to guard against all possible threats. Hence the interest in
agreed arms control.

Comprehensive ASAT Ban

A comprehensive approach to the ASAT problem would imply outlawing all systems capable of attacking and seriously impairing satellites in their assigned function by kinetic energy, as well as explosive, electronic or thermodynamic means.
States accepting such an approach would forgo the possession of dedicated ASAT
systems; the testing – on earth or in space – of specialized ASAT capabilities; the
testing in an 'ASAT mode' (including testing against targets in space or against
points in space) of non-dedicated systems with inherent ASAT capabilities, such as
ICBMs or ABMs; and the deployment in space of any weapon. The existing ASAT
interceptors would have to be destroyed. However, an absolute ban on ASAT
weapons would be difficult to verify.

ASAT Weapon Test Ban

In a partial approach to the ASAT problem, the parties would renounce all testing in
an 'ASAT mode', as well as the deployment of any weapon in space. However, such
a regime would not offer lasting protection for satellites because possession of
ASAT and space-based weapons would not be banned; they could be developed and
held in a state of readiness on earth.

Limitation of ASAT Weapons

In a still less restrictive regime, states would only forgo ASAT systems capable of
hitting satellites in high orbits. Although most military satellites are launched into
low orbits, high-orbit satellites, which include early-warning satellites, are particularly sensitive; their preservation is considered essential for the maintenance of
strategic stability. Such a measure, however, is prone to circumvention; a licence to
build low-orbit ASAT weapons could make limitations on high-orbit ASAT
weapons difficult to check.

Prospects

Under any of these regimes, various types of weapon system could still be used to
damage satellites. Nevertheless, if all testing were prohibited, the first use of ASAT
weapons at the outset of an international crisis would be less likely to occur because
of the uncertainty as to whether and how they would perform.

Talks on control of ASAT weapons were held in 1978–79 between the Soviet
Union and the United States, but they were suspended indefinitely without solving
any of the problems discussed. Progress in ASAT arms control efforts will to a great
extent depend on the fate of strategic ballistic missile defence because of the similar-
ity of the two technologies.

10.5 Confidence-Building Measures in Space

Since 1982 an item called 'prevention of an arms race in outer space' (PAROS) figured on the agenda of the Disarmament Conference (CD). Under this item considerable attention was devoted to confidence building, which could promote transparency and predictability in the space activities of states. A code of conduct (also
called 'rules of the road') in space was proposed, containing a formal renunciation
of all action that might interfere with the operation of space objects, whether civilian
or military. States would undertake to observe minimum distances between space
objects placed in the same orbit, as well as speed limits for objects approaching one
another. Restrictions on overflights, as well as on trailing foreign satellites, would
also have to be accepted. The risks of accidental collision that could be mistaken for
aggression would be thus reduced.

It was suggested by some that 'keep-out zones' around satellites could provide a
measure of security against space mines capable of shadowing satellites and detonat-
ing on command. However, setting up such zones could present problems because
of the great number of satellites orbiting the earth, especially in geostationary orbit
(36,000 kilometres above the equator). Others suggested that a multilateral agreement be negotiated on *prior* notification of all launches of ballistic missiles. Still
others saw the need for international inspection at launch sites.

The possibility of establishing various international institutions to deal with outer
space matters was also discussed. Those proposed included: a world space organization, an international satellite monitoring agency, a satellite image processing
agency, an international space inspectorate, an international trajectography centre,
an international centre for the collection and dissemination of data regarding space
launches and regional agencies to pool information gathered by satellites.

10.6 Further Initiatives

In 2000 China proposed that a new international legal instrument (or instruments) be
negotiated regarding the militarization of outer space as the instruments in existence
were not – in its view – sufficiently effective. The agreement sought by China
would prohibit testing, deployment and use of weapons, weapon systems or components of weapon systems in outer space, without affecting the right to use outer
space for peaceful purposes. Organizations would be designated or specially established to ensure that the parties were implementing their obligations. International
cooperation was to promote exchanges and technical assistance for peaceful purposes, so that all countries could share the economic and technological benefits of
the scientific advances in outer space. Inspections or alternative means would be
resorted to in order to prevent violations. Confidence-building measures might
reduce suspicions about compliance and a mechanism for consultations would be set
up to address possible disputes.

Russia favoured the elaboration of a comprehensive agreement on the non-
deployment of weapons in outer space. In June 2002 it produced, together with
China, a working paper containing possible elements of a treaty which would prohibit: the placing in orbit around the earth of any objects carrying *any* kinds of
weapon; the installation of such weapons on celestial bodies or their stationing in

outer space in any other manner; resorting to the threat or use of force against outer space objects; and assisting or encouraging other states, groups of states or international organizations to participate in activities prohibited by this treaty. Canada preferred that a relevant protocol to the 1967 Outer Space Treaty be worked out. However, the United States said that the current international regime regulating the use of space was meeting all its purposes and that, consequently, it saw no need for new agreements. The danger of a space-weapon race is nevertheless real, and its impact on world security is incalculable.

11

The Sea Environment

Over 70% of the surface of the globe is sea, and more than two-thirds of the world's human population live within 300 kilometres of a sea coast. Hence the continuous interest with which nations have followed the build-up of naval forces and the effects of militarization of the seas on international security.

Several attempts were made in the 19th century to control naval armaments through international agreements. One example is the 1817 US–British agreement called the Rush–Bagot Convention (after the US Acting Secretary of State and the British Envoy to the United States). This agreement – motivated by the desire to reduce defence spending and improve mutual relations – demilitarized the border between the United States and Canada (at that time a British colony) and reduced, limited and equalized the two sides' naval forces in the Great Lakes. Formally, it remains in force, but none of its original provisions is currently observed. A later example is the Pactos de Mayo of 1902, under which Argentina and Chile cancelled their orders for war vessels under construction and undertook to give advance notice of any new warship construction; this pact held for six years. Several conventions dealing with the law of naval warfare were adopted at The Hague in 1907.

11.1 Post-World War I Naval Treaties

Of the proposals aimed at stabilizing world conditions after World War I, the one that appeared to hold the greatest promise of success concerned the limitation of naval armaments. The initial step in this direction was made in 1921–22, when representatives of France, Great Britain, Italy, Japan and the United States met in Washington, DC, to conclude a treaty limiting the size of their navies.

The 1922 Washington Treaty

The 1922 Washington Treaty established limits on the size of individual capital ships (battleships and cruisers) and aircraft carriers, and set a definite ratio of tonnages of such types of ship among the five signatory countries. Other essential obligations included: (a) scrapping an agreed number of warships by sinking or breaking them up, or converting them to target use; (b) prohibiting the construction and acquisition of warships, other than capital ships or aircraft carriers, exceeding 10,000 tons standard displacement, and limiting the calibre of guns carried by all warships; and (c) undertaking not to establish new fortifications or naval bases in the territories and possessions of Great Britain, Japan and the United States, as specified in the Treaty; not to take any measures to increase the existing naval facilities for the repair and maintenance of naval forces; and not to make any increases in the coastal defences of the mentioned territories and possessions.

The Treaty was to remain in force until 1936, subject to withdrawal of any signatory on two years' notice. It was generally considered a success even though no agreement could be reached on the total tonnage of destroyers and submarines. However, as a consequence of this omission, the signatories began extensive programmes of warship construction in those categories that were not restricted.

It is worth noting that in the early 1920s there was a strong opinion, especially in Great Britain, favouring the complete abolition of submarines. Opponents of such a disarmament measure, mainly France, saw the submarine as a legitimate weapon of defence. They argued that it was the only means available to a small naval power against a nation with an overwhelming superiority in capital ships.

The 1930 London Naval Treaty

At the second naval conference, held in London in 1930, attempts were made to extend the provisions of the 1922 Washington Treaty to include additional naval craft. However, France refused to follow the Washington Treaty ratio of tonnages rule and rejected Italy's claim to parity with France in cruisers, destroyers and submarines, while Italy would accept nothing less than parity. As a result, the most important part of the London Treaty (Part III) – which established limits on these types of warship – applied only to Japan, Great Britain and the United States. Nevertheless, the pact was significant in that, for the first time in history, the three major sea powers accepted, by international arrangement, quantitative and qualitative limitations on all categories of warship.

Other parts of the 1930 London Naval Treaty reaffirmed the general provision of the Washington Treaty governing capital ships and aircraft carriers and included an agreement to postpone for six years replacement construction of capital ships. In addition, the number of capital ships of Japan, Great Britain and the United States was to be reduced.

In 1934, Japan, which considered the capital ship ratio to be unjust, abrogated the 1922 Washington Treaty. At a conference subsequently called in London, it became apparent that Japan's proposal for a common upper limit and its claim to parity would be rejected. Japan then decided to withdraw from the conference.

The 1936 London Naval Treaty

The 1936 naval conference was held under highly unfavourable international conditions. Italy had already embarked on the conquest of Ethiopia; Japan had taken control of Manchuria and expanded into northern China; Germany was preparing the occupation of the Rhineland; and the failure of the League of Nations Disarmament Conference prompted the seapowers to contemplate naval rearmament. For these reasons the 1936 London Naval Treaty – as distinct from the 1922 and 1930 treaties – provided for neither quantitative limitations nor reductions in existing fleets. It did, however, regulate certain aspects of naval competition through qualitative restrictions that were to govern ship construction until 1942 – namely, restrictions on ship displacement and gun calibres by class of ship. The Treaty also introduced an interesting innovation: the parties were required to regularly exchange detailed information regarding the construction and acquisition of vessels.

The main drawback of the Treaty was that it did not bind such important powers as Germany, Italy, Japan or the Soviet Union. Any of these countries could bring about a collapse of the Treaty by constructing vessels which did not conform to its provisions, because this would release the signatories from the limitations on displacement and armaments. Denmark, Finland, Norway and Sweden acceded to the Treaty in 1938. Separate agreements incorporating the principal features of the 1936 London Naval Treaty were signed by Great Britain with Germany and the Soviet Union in 1937, and with Italy in 1938. However, the outbreak of World War II in 1939 put an end to all these agreements.

11.2 The 1936 Montreux Convention

The post-World War I Peace Treaty with Turkey, signed at Sèvres, France, in 1920, contained provisions for the demilitarization of a zone surrounding the Straits of the Dardanelles and the Bosphorus as well as a zone in the Aegean Sea comprising a few large islands opposite the mouth of the Dardanelles. In imposing these measures, the Allies wanted to ensure control over Turkey and especially to prevent Turkey from dominating the Straits. However, the Treaty was never ratified. Fearing that the nationalists who had by then come to power in Turkey could be driven to an alliance with Bolshevik Russia, the Allies decided not to enforce it.

In late 1922, following a war between Greece and Turkey, in which the former suffered a series of defeats while the latter recovered large portions of lost territory, a conference was convened at Lausanne, Switzerland, to renegotiate the 1920 peace settlement. This resulted in a new treaty, the Treaty of Lausanne, which was signed on 24 July 1923 and entered into force a year later. This Treaty also contained demilitarization clauses, but the geographical extent and the severity of restrictions were considerably less onerous for Turkey than under the Treaty of Sèvres. The Straits Convention – which formed a part of the Treaty of Lausanne – placed some limitations on the number and size of warships that could pass through the Dardanelles.

The Treaty of Lausanne remained in effect until 1936, when at the demand of Turkey – which invoked changes in the international situation – a conference was convened at Montreux, Switzerland, to revise its terms. On 20 July 1936 the states participating in the conference signed the Convention Regarding the Regime of the Straits, which replaced the Lausanne Straits Convention. Ratified in November of the same year, the Montreux Convention restored Turkish sovereignty over the Straits.

Main Provisions

According to the Montreux Convention, 'light surface vessels, minor war vessels and auxiliary vessels', whether belonging to Black Sea or non-Black Sea powers, are to enjoy, in time of peace, freedom of transit through the Straits, defined as comprising the Dardanelles, the Sea of Marmora and the Bosphorus. In time of a war in which Turkey is engaged, the passage of warships is to be left entirely to the discretion of the Turkish government.

The maximum aggregate tonnage of all non-Turkish naval forces permitted to transit through the Straits at any one time must not exceed 15,000 tons; these forces are not to comprise more than nine vessels. Vessels paying visits to a port in the

Straits are not included in the tonnage. The aggregate tonnage which non-Black Sea powers may have in the Black Sea in time of peace is not to exceed 30,000 tons. If at any time the tonnage of the strongest fleet in the Black Sea exceeds by at least 10,000 tons the tonnage of the strongest fleet in that sea at the date of the signature of the Convention, the aggregate tonnage of 30,000 may be increased by the same amount, up to a maximum of 45,000 tons. The tonnage which any one non-Black Sea power may have in the Black Sea is limited to two-thirds of the aggregate tonnage provided for above. Any non-Black Sea power may supplement its naval forces already present in the Black Sea by up to 8000 tons, for humanitarian purposes, provided that permission is obtained from Turkey and all other Black Sea powers. Vessels of war belonging to non-Black Sea powers may not remain in the Black Sea for more than 21 days, whatever the purpose of their presence there.

Unlike non-Black Sea powers, the Black Sea powers may send through the Straits capital ships of a tonnage exceeding 15,000, provided that such ships pass singly, escorted by no more than two destroyers. (The definition of capital ships, as set out in an annex to the Convention, expressly excludes aircraft carriers.) Black Sea powers also have the right to send through the Straits, for the purpose of rejoining their base, submarines constructed or purchased outside the Black Sea, provided that adequate notice is given to Turkey. Submarines of these powers are also entitled to pass through the Straits in order to be repaired in dockyards outside the Black Sea. In either case, the submarines must travel by day and on the surface and must pass through the Straits singly.

Restrictions on warships do not apply to merchant ships; the latter retain the guaranteed right of passage through the Straits, with the exception of ships belonging to states at war with Turkey. The passage of civil aircraft between the Mediterranean and the Black Seas is also assured, but only along air routes indicated by the Turkish government.

The Montreux Convention was to remain in force for 20 years from the date of its entry into force. Two years prior to its expiry, each party could have given notice of denunciation to the French government, the depositary of the Convention. Since this did not happen, the Convention remains in force today. At the expiry of each period of five years from the date of the Convention's entry into force, each party is entitled to initiate a proposal for amending its provisions.

Assessment

The Montreux Convention reflected the international situation of the 1930s, characterized by US isolationism (the United States did not join the Convention) as well as Anglo-Soviet rivalry in the Black Sea region. The Convention served primarily Turkish and Soviet interests: Turkey recovered military control of the Straits, while the Soviet Union was assured naval dominance in the Black Sea. Serious restrictions were imposed on the rights of non-Black Sea powers to send warships through the Straits, but the right of the Black Sea powers to send warships into the Mediterranean was also restricted – although less severely so. The Convention helped to prevent the Black Sea from becoming an area of competition between the Black Sea powers and outsiders.

The Montreux Convention became the subject of sharp controversy after World War II, when the Soviet Union made claims on Turkish territory along the Black

Sea coast and demanded bases in the Straits. Insisting on a revision of the Straits regime, the Soviet Union proposed, *inter alia*, free passage at all times for the warships of Black Sea states; no passage for warships of non-Black Sea states, except in certain special (non-specified) cases; that this new regime be established by the Black Sea states alone; and that the defence of the Straits be assured jointly by the Soviet Union and Turkey. The government of Turkey appeared ready to consider a revision of transit rights, but it was opposed to discussing any arrangement that would affect Turkish sovereignty. Diplomatic exchanges on this matter produced no agreed action.

The demilitarized status of the Eastern Aegean islands under Greek sovereignty, as established by the 1923 Treaty of Lausanne, is a matter of dispute. Greece argues that since the Montreux Convention, adopted after the Lausanne Convention, was silent on this issue and since the conditions prevailing when the Treaty of Lausanne was concluded have changed fundamentally, the demilitarization rules have become null. Turkey, however, contends that the demilitarization rules remain valid and that, given the geographical proximity of the islands to the Turkish coast, their demilitarization is essential for Turkish security. Turkey also complained that the Dodecanese Islands, ceded to Greece by Italy on the basis of the 1947 Peace Treaty with Italy, had been remilitarized in contravention of that Treaty.

11.3 The 1971 Seabed Treaty

In 1968, when arms control measures concerning the seabed had begun to receive international attention, the Soviet Union proposed that the Eighteen-Nation Committee on Disarmament (ENDC) should consider prohibiting the use for military purposes of the seabed beyond the limits of the territorial waters. The United States, for its part, proposed that the question of arms limitation on the seabed should be taken up with a view to preventing the use of this environment for the emplacement of weapons of mass destruction. Negotiations began in the spring of 1969, with the two powers presenting draft treaties reflecting their respective positions.

It soon became apparent that no comprehensive ban on the military use of the seabed could be achieved in the foreseeable future. Following the concessions made by the Soviet Union, the US and Soviet co-chairmen of the Conference of the Committee on Disarmament (CCD), which succeeded the ENDC, tabled a joint draft treaty under which the parties would undertake not to emplace nuclear weapons or other weapons of mass destruction on the seabed beyond a zone defined in the draft. Since many nations found this joint draft inadequate, the United States and the Soviet Union submitted revised versions which took into account some of the criticisms. The resulting text of the Treaty on the Prohibition of the Emplacement of Nuclear Weapons and Other Weapons of Mass Destruction on the Seabed and the Ocean Floor and in the Subsoil thereof – known as the Seabed Treaty – was judged acceptable and was commended by the UN General Assembly in 1970. It was opened for signature in February 1971 and entered into force on 18 May 1972.

Scope of the Prohibitions

UN resolutions calling for the use of the seabed and ocean floor and the subsoil thereof exclusively for peaceful purposes formed a framework for possible arms control measures. However, controversy arose over the meaning of the phrase 'exclusively for peaceful purposes'.

The non-aligned countries contended that the United Nations had invariably understood the use of a given environment for exclusively peaceful purposes to mean the prohibition of all military activities, whatever their purpose, and that there should be no departure from this approach in the case of the seabed. Some of them reasoned that, since the seabed must be used for the benefit of all states (as stated in the above resolutions), any military use of the seabed represented an unjustified usurpation hampering peaceful exploitation of the environment. The Soviet Union also equated 'peaceful purposes' with non-military purposes. Accordingly, the first Soviet draft treaty aimed at completely demilitarizing the seabed and the ocean floor as well as its subsoil.

The United States, however, interpreted the phrase 'for peaceful purposes' as not barring military activities generally. It argued that specific limitations of certain military activities would require detailed agreements and that activities not pre-cluded by such agreements would continue to be conducted in accordance with the principle of the freedom of the seas. It saw an analogy with the 1967 Outer Space Treaty, which does not provide for the use of outer space exclusively for peaceful purposes but specifically prohibits the placing in earth orbit of objects carrying nuclear weapons or other weapons of mass destruction in time of peace. Accord-ingly, the United States proposed that states undertake not to emplant or emplace fixed nuclear weapons or other weapons of mass destruction or associated fixed launching platforms on, within or beneath the seabed and ocean floor. In advocating these measures, the United States asserted that only weapons of mass destruction could have enough military significance to justify the expense of stationing them on the seabed. It expressed the belief that realistic possibilities did not and would not soon exist for such conventional military uses of the seabed as would be threatening to the territories of states. Some non-nuclear but clearly military uses of the seabed (e.g., placing devices for detection and surveillance of submarines) were seen as essential to the security of states and therefore indispensable. In the opinion of the United States, complete demilitarization would moreover raise verification problems since it would impose the task of deciding whether each object or installation emplaced on the seabed was of a military nature. In any event, the United States was not prepared to accept a ban on all military activities on the seabed.

The text which was eventually agreed provided for an undertaking by states par-ties to the Seabed Treaty not to emplant or emplace on the seabed and the ocean floor or in the subsoil thereof any nuclear weapons or any other types of weapon of mass destruction as well as structures, launching installations or any other facilities specifically designed for the storage, testing or use of such weapons. The parties also undertook not to assist, encourage or induce any state to carry out activities prohibited by the Treaty and not to participate in any other way in such actions. The term 'other types' of weapon of mass destruction was understood as including bio-logical, chemical and radiological weapons.

When asked for greater precision, the sponsors of the Treaty explained that it prohibits, *inter alia*, nuclear mines anchored to or emplaced on the seabed. It does not apply to facilities for research or for commercial exploitation not specifically designed for storing, testing or using weapons of mass destruction; however, facilities specifically designed for the use of such weapons could not be exempted from the prohibitions of the Treaty on the grounds that they could also use conventional weapons. The prohibitions are not intended to affect the use of nuclear reactors or other non-weapon applications of nuclear energy consistent with the Treaty obligations. It was also explained that, while submersible vehicles able to navigate in the water above the seabed would be viewed as any other ships and would not violate the Treaty when anchored to or resting on the bottom of the sea, the ban did apply to bottom-crawling vehicles which could navigate only when in contact with the seabed and which were specifically designed to use nuclear weapons. Thus, the prohibition embraces not only fixed facilities (as originally provided for in the US draft) but also certain mobile facilities.

Geographical Coverage

There was a general understanding that, as indicated in several UN resolutions, seabed disarmament measures were to include the area underlying the high seas beyond the limits of national jurisdiction. This somewhat vague language reflected the disagreement then existing as to where the limits of national jurisdiction actually lay. However, the view prevailed that agreement should be reached on a precise boundary, devised specifically for arms control purposes and expressed in terms of distance from the coast.

Under the Soviet draft treaty the prohibition was to cover an area beyond a 12-mile maritime zone of coastal states. The US draft provided for a prohibition beyond a three-mile band adjacent to the coast. The area of prohibition was eventually defined as lying beyond the outer limit of a seabed zone coterminous with the 12-mile outer limit of the zone referred to in Part II of the Convention on the Territorial Sea and the Contiguous Zone, signed at Geneva on 29 April 1958. Only a few nations have ratified the 1958 Geneva Convention, which was described by several delegations negotiating the Seabed Treaty as both highly controversial and antiquated. They considered it inappropriate to invite non-parties to an agreement to accept its formulations in defining new obligations. Indeed, there was no need to refer to this Convention: a simple formula, without such reference, would have served the same purpose.

The undertakings by states parties to the Treaty are applicable also to the 12-mile seabed zone, except that within such a zone they shall not apply either to the coastal state (which is free to place any object there) or to the seabed beneath its territorial waters. In other words, since the Treaty does not contain an absolute prohibition on the placement of weapons of mass destruction beyond the parties' own seabed zone, and since an exception has been made regarding territorial waters, states have the right, according to the language of the Treaty, to install weapons of mass destruction on the seabed beneath the territorial waters within the 12-mile seabed zone of other states, presumably with the consent and authorization of the states concerned ('allied option'). This, however, would not be permitted in the band between the outer limit of the territorial sea and the 12-mile limit of the seabed zone if the breadth of the territorial waters were narrower than 12 miles. The United States and the Soviet Union pointed out that the exception regarding the seabed beneath the territorial waters within the seabed zone left unaffected the sovereign authority and control of the coastal state within its territorial sea.

The 1979 SALT II Treaty prohibited the development, testing or deployment of fixed ballistic or cruise missile launchers in any area of the ocean floor and the seabed, or on the beds of internal waters and inland waters, or in the subsoil thereof, as well as mobile launchers of such missiles which move only in contact with the ocean floor, the seabed, or the beds of internal waters and inland waters, and missiles for such launchers. A similar clause was included in the 1991 START I Treaty. However, there was resistance to proposals for extending the geographical scope of the Seabed Treaty through a formal amendment of the Treaty, so as to make it applicable from 'shore to shore', mainly because such an amendment would entail international verification in the territorial waters of the parties. Rather, in a declaration adopted in 1989 by the Third Seabed Treaty Review Conference, the parties stated that they had not emplaced any nuclear weapons or other weapons of mass destruction on the seabed outside the zone of application of the Treaty and had 'no intention to do so'.

Implementation

Upon signing or ratifying the Seabed Treaty, certain states made reservations to ensure that their rights under the existing law of the sea were not adversely affected, or to reiterate the points of view not taken into account during the negotiating process. Some states reserved the right to verify, inspect, remove or destroy any weapon, structure, installation, facility or device placed by other countries on or beneath their continental shelves beyond the outer limit of the seabed zone. Italy pointed out that, in the case of agreement on further measures in the field of disarmament relating to the seabed, the question of the delimitation of the area within which these measures would find application would have to be examined and solved in each specific instance.

Periodic conferences convened to review the operation of the Seabed Treaty are meant to examine the effects of developments in underwater and weapon technology on military uses of the seabed and the implications of such developments for efforts to control arms on the seabed. However, countries which possess sophisticated underwater technologies and military resources, and which might be in a position to identify the developments that could affect the purposes and provisions of the Treaty, are reluctant to make relevant information available, for reasons of military security or commercial confidentiality. Nevertheless, it seems that any advantage that new deep-water technologies would confer on a state wishing to emplace nuclear weapons or other weapons of mass destruction on the seabed would probably be outweighed by parallel advances in technologies for detecting such weapons.

Assessment

Because of its limited scope and geographic coverage, the Seabed Treaty has low arms-control value. Nuclear installations on the seabed, once considered a possibility, have proved to be unattractive to the military. They would be extremely costly, very difficult to maintain and control and, above all, vulnerable and redundant. Placing other weapons of mass destruction on the seabed is even less probable. The Seabed Treaty banned something which did not exist and which was not likely to be developed. However, the parties undertook to continue negotiations 'in good faith' concerning further measures in the field of disarmament for the prevention of an arms race on the seabed, the ocean floor and the subsoil thereof.

11.4 Prevention of Incidents at Sea

Incidents at sea relate mainly to ship manoeuvres which create the danger of collision. They also include close air surveillance ('buzzing') of foreign vessels, simulated attacks on such vessels or their harassment with flares or searchlights, as well as accidental firing during naval exercises. Such incidents may increase international tension and thereby the risk of war. Awareness of this risk grew considerably following several serious US–Soviet confrontations at sea in 1967 and 1968. This led the two powers to conclude an Agreement on the Prevention of Incidents on and over the High Seas, commonly called the Incidents at Sea Agreement. The Agreement was signed and entered into force on 25 May 1972.

Main Provisions

The US–Soviet Incidents at Sea Agreement regulates dangerous manoeuvres, prohibits certain forms of harassment and requires increased communication at sea as well as regular exchanges of information and consultation.

Regulations. The parties undertook to instruct the commanding officers of their respective naval ships to observe strictly the International Regulations for Preventing Collisions at Sea, known as the Rules of the Road. Ships operating near each other must remain well clear to avoid the risk of collision, and ships operating near a formation of the other party must avoid manoeuvring in a manner which would hinder the evolutions of the formation. Formations may not conduct manoeuvres through heavily trafficked areas where internationally recognized traffic separation schemes are in effect. Ships engaged in surveillance of other ships must stay at a distance and may not conduct manoeuvres which endanger the ships under surveillance.

Restrictions. Ships of the parties shall not simulate attacks by aiming guns, missile launchers, torpedo tubes or other weapons in the direction of a passing ship of the other party, nor launch any object in the direction of such a ship, nor use searchlights or other powerful devices to illuminate its navigation bridges. Commanders of aircraft may not permit simulated attacks by the simulated use of weapons against aircraft and ships, or performance of various aerobatics over ships, or dropping objects near them in such a manner as to constitute a hazard to navigation.

Communication. When ships of both parties manoeuvre in sight of one another, their operations and intentions are to be signalled. In particular, proper signals must be given concerning the intent of the ships to begin launching or landing aircraft. Aircraft flying over the high seas in darkness are expected to display navigation lights. When conducting exercises with submerged submarines, exercising ships must show the signals prescribed by the International Code of Signals to warn ships of the presence of submarines in the area. Three to five days' advance notification must be given of actions on the high seas, such as missile launches, which represent a danger to navigation or to aircraft in flight.

Exchange of Information and Consultation. The parties are committed to exchanging information concerning incidents at sea between their ships and aircraft. The US Navy is to provide such information through the Soviet naval attaché in Washington, DC, whereas the Soviet Navy is to do so through the US naval attaché in Moscow. Consultations regarding the implementation of the Agreement are to be held at least once a year. A special committee is to consider the 'practical workability' of concrete fixed distances to be observed in encounters between ships, aircraft, and ships and aircraft.

Duration. The Agreement was concluded for a period of three years, after which it may be renewed without further action by the parties for successive three-year periods. It can be terminated by either party upon six months' written notice.

Protocol. On 22 May 1973, the United States and the Soviet Union signed a Protocol extending certain applicable provisions of the Incidents at Sea Agreement to non-military ships. The Protocol is considered as an integral part of the Agreement.

Assessment

The US–Soviet Incidents at Sea Agreement is generally regarded as a successful international instrument. It has contributed to a decline in the number and severity of incidents despite the increased maritime activities of both sides. Moreover, the channels of communication between the parties, which the Agreement established, have proved useful in resolving questions about the incidents that have occurred.

The Agreement is not applicable to submerged submarines. They have been excluded because of the inherently covert nature of submarine operations. However, there have been several collisions between US and Soviet submarines, including those carrying nuclear weapons. Some such incidents have even taken place in territorial waters. Notwithstanding this drawback, the US–Soviet Incidents at Sea Agreement has served as a model for other bilateral agreements for the prevention of incidents at sea. These have been concluded between the Soviet Union and a dozen other states, including the United Kingdom, France and Japan. The first such agreement not involving the United States or the Soviet Union was signed in 1990 between Germany and Poland.

The 1985 UN study on the naval arms race suggested that consideration should be given to transforming the US–Soviet Incidents at Sea Agreement into a multilateral agreement. Such a global regime would render illegal all dangerous manoeuvres and harassment at sea, but this suggestion was not followed up.

11.5 The 1982 Law of the Sea Convention

On 10 December 1982, as a result of a nine-year-long conference, an overwhelming majority of states signed the UN Convention on the Law of the Sea (UNCLOS). The Convention is not directly concerned with arms control, but several of its provisions deal with military matters. It entered into force on 16 November 1994, pursuant to the approval by the UN General Assembly of the Agreement Relating to the Implementation of Part XI of the UNCLOS, which deals with the exploration for, and exploitation of, the resources of the seabed and ocean floor and subsoil thereof beyond the limits of national jurisdiction, in the area recognized as the 'common heritage of mankind'. (The Agreement, which met the concerns of several industrialized states, entered into force in July 1996.)

While reaffirming the principle of the freedom of the seas, the Convention also confirms the existing restrictions on this freedom and introduces some new ones.

Innocent Passage

All ships of all states, including warships and other ships in government non-commercial service, may exercise the right of passage through the territorial sea of other states. However, such passage must be continuous and expeditious. It must also be innocent – that is, not prejudicial to the peace, good order or security of the coastal state.

Passage is not considered innocent if, among its other activities, a foreign ship engages in a threat or use of force against the sovereignty, territorial integrity or political independence of the coastal state; in an exercise or practice with weapons of any kind; in collecting information to the prejudice of the defence or security of the coastal state; or in the launching, landing or taking on board of an aircraft or any military device. Submarines must navigate on the surface and show their flag, whereas nuclear-powered ships and ships carrying nuclear or other inherently dangerous or noxious substances are required to observe special precautionary measures established for such ships by international agreements. Aircraft have no right of overflight and may enter the airspace above the territorial sea only under arrangements that include the consent of the coastal state.

UNCLOS empowers the coastal state to ask foreign ships exercising the right of innocent passage to use lanes and follow traffic separation schemes prescribed by it; to take the necessary steps in its territorial sea to prevent passage which is not innocent; and to suspend innocent passage temporarily in specified areas of the territorial sea, whenever this is essential for the protection of its security.

A controversy which has remained unresolved is whether, before exercising the right of innocent passage, warships should obtain the authorization of the coastal state or at least notify it. Several states maintain that warships must, by their very nature, be presumed to be on passage that is not innocent unless explicitly recognized as such by the coastal state. The weakness of UNCLOS resides in the fact that warships are exempted from the enforcement jurisdiction of the coastal state. If a warship does not comply with the laws and regulations of the coastal state and disregards calls for compliance, all the coastal state can do is to require that the warship leave the territorial sea immediately.

Transit Passage

UNCLOS adopted the concept of unimpeded transit through and over straits used for international navigation. In exercising the right of this so-called transit passage, ships and aircraft must refrain from any activities other than those incident to their normal modes of continuous and expeditious transit, unless rendered necessary by *force majeure* or by distress.

The question of 'innocence' does not arise with regard to transit passage as it does with regard to passage through the territorial sea; submarines are not required to navigate on the surface. The right of transit passage enjoyed by all ships, as well as its exercise, may not be suspended. Having agreed not to exercise enforcement jurisdiction against ships in transit passage, including warships and ships in government non-commercial service, border states may not even require such ships to leave the strait immediately. Claims for losses or damages resulting from acts contrary to their laws and regulations may be made only through diplomatic channels.

UNCLOS does not affect the legal regime in straits in which passage is regulated by long-standing international conventions.

Assessment

Most of the UNCLOS provisions reflect the general understanding of the existing law of the sea. Certain rules have already acquired the status of customary law and have been incorporated into military manuals.

The Convention has provided an additional proof that the principle of *mare liberum* is not incompatible with restrictions on the uses of the sea. The restrictions in the military field have been adapted to the interests of the great powers, but this does not preclude multilateral arms control agreements in the marine environment that might be negotiated separately from UNCLOS.

11.6 Confidence-Building Measures at Sea

In the late 1980s and early 1990s, remarkable progress was made in diminishing the threat posed by nuclear-armed navies. As a result of bilateral agreements between Russia and the United States, the numbers of strategic ballistic missiles deployed on the submarines of these two powers have been limited and are to be drastically reduced. The number of nuclear warheads which each of these missiles carries will also be significantly cut. Moreover, by virtue of unilateral undertakings, US, Russian and British tactical nuclear weapons deployed on all kinds of warship have been withdrawn, to be stored on land or destroyed. France has scaled down the nuclear component of its navy as well. Even the movements of nuclear-armed ships may be somewhat restricted by multilateral treaties setting up nuclear-weapon-free zones.

Whereas important negotiated cuts have been made in several categories of non-nuclear land-based armament, in particular in Europe, prospects for significantly restricting non-nuclear naval armaments have remained uncertain. It is difficult to see why, in the search for improved world security, conventional naval forces and activities should be treated differently from conventional ground or air forces. However, in addition to geostrategic asymmetries among the potential parties, several obstacles stand in the way of negotiated naval arms control that would limit naval forces substantially, both quantitatively and qualitatively. Warships will continue to navigate in distant waters in support of national political and economic interests, taking advantage of the exceptional mobility and flexibility of maritime power. The establishment of 200-nautical-mile exclusive economic zones and the growing exploitation of the seas, as well as the awareness of the vast unused resource potential of the seas, have increased the need for surveillance and for enforcement of international rules of conduct at sea. Other missions of naval ships which states are unlikely to renounce include the defence of their coastal waters, training exercises and protection of fishing fleets, as well as power projection or simply showing the flag. Such activities may lead to dangerous situations and conflicts – hence the need for increased confidence at sea. Maritime confidence building may encourage attitudes of cooperation with political, economic and security consequences beyond the maritime field.

Unlike the CBMs related to conventional ground and air forces, those related to conventional naval forces do not form a distinct class of international instruments. (CBMs adopted for Europe cover naval activities in the sea area adjoining Europe only if they are functionally linked with notifiable military activities on land.) Some naval CBMs are incorporated in arms limitation or other treaties and intermingled with norms regulating various other activities.

Proposals for Maritime CBMs

In recent years, a wide range of measures have been proposed by both governmental and non-governmental bodies to lower the risk of incidents at sea, improve the security of coastal states and render non-military maritime activities safer.

Constraints. Although UNCLOS does not prohibit innocent passage of foreign warships or ships in government non-commercial service through the territorial sea, flag states should ensure that, barring exceptional situations, such ships do not normally pass within 12 nautical miles of the baselines of the coastal states. If the passage is necessary for the conduct of peacetime naval activities, the coastal state should be notified in advance. A similar constraint has been suggested regarding all nuclear-powered ships as well as ships carrying nuclear or other dangerous or noxious substances.

The nuclear-weapon states should abandon the policy of neither confirming nor denying the presence or absence of nuclear weapons on board their ships.

The passage of ships carrying nuclear weapons through the territorial waters of foreign countries should not be considered 'innocent' in the meaning of UNCLOS.

A limit to the frequency and size of naval exercises, as well as to their duration, should be agreed.

States should not conduct naval exercises in international straits or in the exclusive economic zones of foreign states.

States should refrain from constructing military installations and emplacing weapons or other devices on the continental shelves of other states without the express consent of the latter.

Separate areas could be established for the submarine operations of different nations, to reduce the dangers of close-quarter situations between submarines in peacetime; submarines should be required to avoid simulated attacks on ships or submarines of other nations and to minimize submerged operations in coastal areas.

The law of naval warfare should be modernized by making restrictions on the use of mines at sea applicable to all types of mine, not only to automatic contact mines as covered by the 1907 Hague Convention VIII. The laying of mines in international straits for offensive purposes should be prohibited. States should also refrain from laying mines in areas of intense shipping or fishing. Mines should be equipped with a neutralizing mechanism which renders them harmless once they are no longer of military use. Each party to a conflict should keep detailed records of the location of its minefields and of the technical characteristics of the mines. Upon the cessation of hostilities, all such information should be made available to the other party, to third countries, or to appropriate international organizations, and the belligerent states should be responsible for removing or rendering safe the mines they have laid.

Openness and Communication. Information on naval force structure, deployment and capabilities, as well as on other naval matters of general interest should be regularly exchanged, especially among countries within the same region; communication links among coastal states should be improved.

Naval manoeuvres of agreed categories and above a certain size should be notified in advance – with an indication of the numbers and classes of vessels involved – and be open to observers from other states.

In addition to warships, the UN Register of Conventional Arms should include naval construction plans.

Exchange of ship visits, as well as contacts among the naval personnel of different countries, should be intensified, including high-level meetings to discuss maritime doctrines.

Assessment

Certain naval CBMs, for example those regarding naval manoeuvres, could be relatively easily verified. Others, for example those regarding movements of submarines, present obstacles to both national and international verification which may be difficult to surmount. The fear of excessive intrusiveness is a limiting factor. However, unlike in arms control measures which directly affect military forces or hardware, verifiability – although desirable – need not be a *sine qua non* for CBMs.

Some of the proposals listed above, if accepted, may require agreed understandings of the existing agreements. Others may call for new agreements.

12

Demilitarized Areas

Arms restrictions regarding certain geographical areas rather than entire countries have been negotiated several times in the past. For example, in 1905 Sweden and Norway established on both sides of their common border a permanently neutral zone, in which there were to be no fortifications, no armed units stationed, nor storage of military material, and all war operations were prohibited. (With the passage of time, this agreement has lost its political and military significance; it formally ceased to be valid some 90 years after its conclusion.)

In the period immediately following World War I, demilitarization provisions were included in the territorial settlements regarding two groups of islands, the sovereign status of which had been the subject of international dispute: the Archipelago of Spitsbergen and the Aaland Islands. However, in both cases arms control was not the main issue. It was chiefly a means to achieve a compromise solution by providing a *quid pro quo* to those countries whose territorial claims had not been accepted.

After World War II, the islands ceded by Italy to Greece by virtue of the 1947 Treaty of Peace were to be and 'remain' demilitarized. In the 1959 Antarctic Treaty, the continent of Antarctica was declared free of military activity.

Certain countries, such as Costa Rica and Iceland, decided not to possess armed forces. However, the demilitarized status of these states is based exclusively on their national policies and legislation. There exists no international instrument preventing them from changing this status and reinstituting armed forces if they so wish.

12.1 The 1920 Spitsbergen Treaty

Situated in the Arctic Sea, several hundred kilometres north of Norway, and devoid of indigenous population, the Archipelago of Spitsbergen (also referred to as Svalbard) was considered for a long time as a 'no man's land'. It used to be only occasionally visited by fishermen and hunters, mainly from the Nordic countries. At the turn of the 20th century, when large deposits of coal and iron ore were discovered there, the archipelago attracted the attention of entrepreneurs from several countries. In addition, the islands began to be viewed as a possible location for a naval base. A conflict over the economic and military assets was avoided thanks to a treaty signed on 9 February 1920.

Main Provisions

The Treaty Concerning the Archipelago of Spitsbergen (the Spitsbergen Treaty), in force since 1925, recognized the 'full and absolute' sovereignty of Norway over the archipelago.

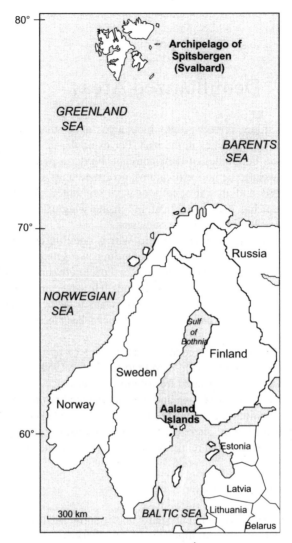

Figure 12.1   *Geographical Location of the Archipelago of Spitsbergen
and the Aaland Islands*

Ships and nationals of all states parties may enjoy equally the rights of fishing and
hunting in the territories covered by the Treaty. The parties also have free access to
the waters, fjords and ports of these territories; their nationals are allowed to conduct, without impediment, all maritime, industrial, mining and commercial operations on a footing of equality, but subject to the observance of local laws and regulations. No monopoly may be established on any account or for any enterprise.

Norway undertook not to create nor to allow the establishment of any naval base,
nor to construct any fortification in the archipelago of Spitsbergen, which may never
be used for warlike purposes. This is the only clause in the Treaty dealing with arms
control.

Assessment

Whereas the political significance of the Spitsbergen Treaty is undisputed, its arms control content is less impressive.

The Treaty provides for 'defortification', but does not expressly forbid such military activities as the stationing of troops or storing of weapons. The fact that only naval bases are prohibited could be interpreted as legitimizing army and airbases. There are no restrictions on the entry of warships into the territorial waters of the islands or on naval operations in the area. On the other hand, one can argue that the ban on using Spitsbergen 'for warlike purposes' implies a ban on militarization. No mechanism for supervision and control is provided for, but respect for the provisions of the Spitsbergen Treaty has not been disputed.

In 1944 the Soviet Union put forward a demand for revision of the Spitsbergen Treaty. It proposed establishing a system of joint Soviet–Norwegian rule, a kind of condominium, with an equal right for both countries to have troops and fortifications on the islands. This demand, reiterated in 1946, was clearly aimed at using Norwegian territory to promote Soviet strategic interests. Norway declared it unacceptable.

12.2   The 1921 Aaland Islands Convention

The Finnish islands of Aaland are situated in the northern part of the Baltic Sea, at the entrance of the Gulf of Bothnia. An overwhelming majority of the inhabitants are Swedish in terms of their culture and language. In 1917, when, in the aftermath of the Russian Revolution, Finland declared its independence from Russia, a strong popular movement developed on the Aaland Islands demanding their separation from Finland and unification with Sweden. (In fact, it would have been 'reunification', because for hundreds of years – until 1809 – Finland had formed part of Sweden.) The separatists were actively supported by the Swedish government – not least because of the strategic importance of the area. This dispute over sovereignty seriously strained relations between the two countries.

At the initiative of Great Britain, the question was referred to the League of Nations, whose Council declared that the sovereignty over the Aaland Islands belonged to Finland. The Council also decided that an international agreement should be concluded to guarantee to all countries concerned that the Islands 'never become a source of danger from the military point of view'. On 20 October 1921, at a conference convened in Geneva at the invitation of the Secretary-General of the League of Nations, the ten participating states signed the Convention Relating to the Non-Fortification and Neutralisation of the Aaland Islands (the Aaland Islands Convention).

Main Provisions

In its preamble, the Aaland Islands Convention refers to the 1856 Convention on the Demilitarization of the Aaland Islands, which had been annexed to the Paris Peace Treaty ending the Crimean War. Under the 1856 Convention, the Russian Empire (of which Finland at that time formed a part as an autonomous Grand Duchy) undertook not to fortify the Aaland Islands and not to maintain or erect any military or naval establishment there. The parties to the 1921 Convention assumed the same obligation and, in addition, committed themselves not to maintain or set up in the area in question any military aircraft establishment or base of operations, nor any other installation used for war purposes.

In addition to the demilitarization provisions, the Aaland Islands Convention contains neutralization provisions prohibiting war operations from taking place in the strictly delimited zone. No land, naval or air force of any state may enter this zone or stay there. The production, import, transport and re-export of arms and implements of war are also strictly forbidden. In time of war, the Aaland Islands, as a neutral area, may not be used for any purpose connected with military operations.

There are, however, exceptions applicable only to Finland. In peacetime Finland may, in exceptional circumstances, bring to the zone such armed forces as are strictly necessary to maintain order. Furthermore, from time to time, one or two Finnish light surface warships may visit the Islands. Finland may also, in special circumstances, bring to the zone other warships, and keep them there temporarily, but the total displacement of these ships may not exceed 6,000 tons. Bringing in submarines is not permissible. The right to enter the archipelago and anchor there temporarily cannot be granted by the Finnish government to more than one warship of any other power at a time. Finnish aircraft are allowed to fly over the zone but may not land there except in cases of emergency. In wartime, if the Baltic Sea becomes involved, Finland has the right, in order to assure respect for the neutrality of the Aaland Islands, temporarily to lay mines in the territorial waters of the islands. Should the neutrality of the zone be endangered by a sudden attack, Finland has the right to take the necessary measures to stop and repel the aggressor until the other parties can intervene.

Assessment

The main purpose of the demilitarization and neutralization measures taken with regard to the Aaland Islands since 1856 has been to prevent the islands from falling into the hands of a state which could threaten the operations of fleets in the Baltic Sea and to prevent the use of the islands as a base for an armed attack. This purpose has been largely achieved.

Although the Aaland Islands have been fortified several times in the past, and although the demilitarization provisions of the conventions in force have been suspended in times of war, all these measures were of a defensive character. After each war, the demilitarization and neutralization obligations have been renewed; upon the termination of World War II, they were reiterated in the 1947 Peace Treaty regarding Finland. It was chiefly due to their status that the islands have been spared war destruction. The provisions of the 1921 Aaland Islands Convention are considered to be binding irrespective of any changes that may occur in the status quo of the Baltic Sea.

12.3 The 1959 Antarctic Treaty

The 1957/58 International Geophysical Year provided an opportunity for scientists from 12 countries to cooperate and to establish and expand scientific bases in Antarctica. The success of that undertaking gave rise to a search for an international regime for the region. In 1958, the United States invited 11 other nations involved in scientific activities in Antarctica to negotiate a treaty for such a regime. As a result of a conference convened in Washington, DC, the sought-for Antarctic Treaty was opened for signature on 1 December 1959; it entered into force in 1961.

Scope of the Obligations

According to the 1959 Antarctic Treaty, Antarctica shall be used exclusively for peaceful purposes. The Treaty prohibits any measures of a military nature, such as the establishment of military bases or fortifications, the carrying out of military manœuvres or the testing of any type of weapon. However, it does not prohibit the use of military personnel or equipment for scientific research or for any other peaceful purpose.

The Treaty bans nuclear explosions in Antarctica, whatever their nature, as well as the disposal of radioactive waste material. It stipulates, however, that, should international agreements be concluded concerning the use of nuclear energy, including nuclear explosions and the disposal of radioactive waste material, the rules established under such agreements will apply in Antarctica. It is unlikely that any new agreement would invalidate the above bans.

The arms control purpose of the Antarctic Treaty derives from its other three main objectives: to establish a foundation for international cooperation in scientific investigation in Antarctica; to protect the unique Antarctic environment; and to avert discord over territorial claims. Cooperative exploration of the Antarctic continent has been ensured by the undertaking of the parties to exchange scientific personnel and information. Protection of the Antarctic environment, including the preservation and conservation of living resources, is included in the list of topics to be reviewed regularly by the parties and has figured prominently on the agenda of their meetings. What is most sensitive is the question of territorial claims.

Seven states – Argentina, Australia, Chile, France, New Zealand, Norway and the United Kingdom – have claimed sovereignty over areas of Antarctica on the basis of discovery, exploration, geographic proximity or territorial continuity. In the case of Argentina, Chile and the United Kingdom, the claims overlap. Only some 15% of the Antarctic landmass remains unclaimed. The United States and the Soviet Union have made no claims of their own, nor have they recognized the claims made by others, but they have established a de facto presence throughout Antarctica by setting up scientific stations in different parts of the continent. The Antarctic Treaty introduced a moratorium implying neither renunciation nor recognition of previously asserted rights of or claims to territorial sovereignty in Antarctica and prohibiting the making of new claims or the extension of existing ones. This moratorium could be terminated 30 years from the date of entry into force of the Treaty, that is, after 1991, at which time a conference would review the operation of the Treaty. This, however, has not happened.

Area of Application

The Treaty applies to the area south of 60 degrees South latitude, including the ice shelves, but the rights of states under international law with regard to the high seas in that area are not to be affected. This proviso has given rise to an argument over what should be considered as 'high seas' in Antarctica. If the territorial claims

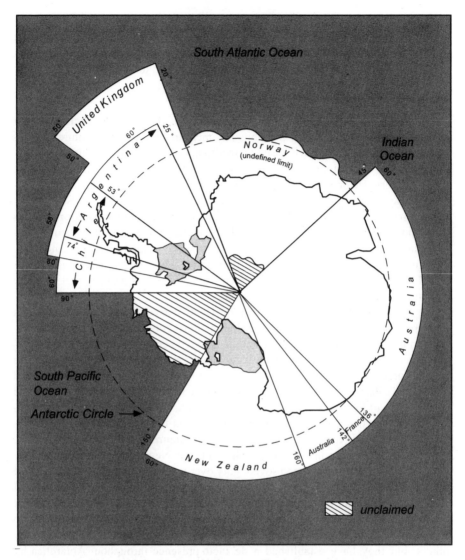

Figure 12.2 *National Claims to Territory in Antarctica*

The 1959 Antarctic Treaty neither renounces nor recognizes these claims, prohibits new claims and extension of existing claims, and declares that Antarctica shall be used exclusively for peaceful purposes.

Claimant	Year of Claim	Claimant	Year of Claim
UK	1908	Norway	1939
New Zealand	1923	Chile	1940
France	1924	Argentina	1943
Australia	1933		

advanced by certain states were valid, there would exist a territorial sea contiguous to the coast, and the high seas, as everywhere else, would begin where the territorial sea ended. If it were generally admitted that no state exercised sovereignty in Antarctica, there could be no territorial sea there, and the high seas would begin at the coast. The latter interpretation would make it permissible for states to deploy naval vessels, whether nuclear or conventional, close to the shores of the Antarctic continent. However, from the military point of view, deployment in such remote places would not make much sense.

Parties

Although the Antarctic Treaty is an international agreement of universal interest, it has fewer parties than many other multilateral arms control agreements. One reason why many states have not joined may be the structure of the Treaty, which provides for different categories of parties.

The signatory states – former participants in the scientific investigation in the Antarctic during the International Geophysical Year – enjoy full rights under the Treaty as the so-called consultative parties. They are entitled to participate in consultative meetings; they have the right to carry out inspections; they may modify or amend the Treaty at any time through agreement among themselves; they are empowered to decide whether or not non-UN members should be allowed to accede; and only they may call a conference to review the operation of the Treaty. States which have acceded to the Treaty acquire the rights of the consultative parties only if they conduct 'substantial scientific research activity' in the Antarctic, such as the establishment of a scientific station or the despatch of a scientific expedition. They may be deprived of these rights if they cease to be actively engaged in the Antarctic research, whereas the original consultative parties maintain their status irrespective of their actual research activity. All other parties may only attend consultative meetings as observers.

Mineral Exploitation

It has been known for some time that there are deposits of precious minerals on the Antarctic continent and that the waters in that part of the world are rich in living resources. The special interest of recent years in Antarctica is related to indications that its continental shelf may contain oil and gas, although estimates of these resources are speculative.

Economic activity in Antarctica is neither expressly permitted nor prohibited by the Antarctic Treaty, but it is not considered contrary to its principles or purposes. In fact, exploitation of the marine resources of the area is being conducted. In 1980, the parties signed a Convention on the Conservation of Antarctic Marine Living Resources – complementing the provisions of the 1972 Convention for the Conservation of Antarctic Seals – which requires that both the population levels of the exploited species and the balance of the ecosystem be conserved.

It was feared, however, that if exploitation of the Antarctic mineral resources became a practical proposition, a struggle could erupt over national rights to territorial possessions containing these non-renewable resources. This could be a struggle among the original claimants, especially where claims overlapped, or between them and non-claimants active in Antarctica, or also with new claimants demanding a share, whether party or non-party to the Antarctic Treaty. To assert their declared rights over other contenders, or to guard against infringements on their economic activities, nations might resort to the use of force. This would bring about a collapse of the order prevailing under the Treaty. Antarctica would cease to be a non-militarized zone and would instead become a zone of rivalry and conflict. Conscious of this danger as well as of possible serious adverse consequences of unregulated exploitation of minerals, the Antarctic Treaty consultative parties launched negotiations on a minerals regime for Antarctica.

CRAMRA. The Convention on the Regulation of Antarctic Mineral Resource Activities (CRAMRA), adopted in Wellington in 1988 (sometimes referred to as the Wellington Convention), was to apply to all such activities taking place on the continent of Antarctica and all Antarctic islands, including all ice shelves, south of 60 degrees South latitude, and in the seabed and subsoil of the adjacent offshore areas. CRAMRA would have provided an institutional mechanism for assessing the possible impact on the environment of Antarctic mineral resource activities and for determining their acceptability. It stipulated a set of environmental conditions to be met by prospective operators. These conditions would be enforced through a system of regulations and powers vested in a commission and regulatory committees. An operator would be allowed to commence exploratory work only after an application for a specific area had been approved by the commission by consensus. Regulatory committees would perform the functions of issuing exploratory and development permits on the basis of an approved management scheme, and of monitoring exploration and development activities. The Convention also provided for inspection of installations and stations associated with mineral resource activities.

Instead of dissuading mineral exploration and mining, as several states desired, CRAMRA would have actually promoted such activities by creating a legal and political framework within which mining rights could be obtained and by attenuating the uncertainties which usually deter large investments. Moreover, despite very strict environmental requirements, many considered a 'mining convention' to be incompatible with the protection of the fragile Antarctic environment. Since several countries, including Australia and France, decided not to ratify CRAMRA, the Convention did not enter into force. Joint Franco-Australian efforts led to the drafting of a new agreement.

The Madrid Protocol. On 4 October 1991, the Antarctic consultative parties, meeting in Madrid, adopted the Protocol on Environmental Protection to the Antarctic Treaty. The signatories of the Madrid Protocol committed themselves to comprehensive protection of the Antarctic environment as well as dependent and associated ecosystems, and designated Antarctica as a 'natural reserve, devoted to peace and science'. The comprehensiveness agreed upon consists in setting a uniform standard to assess all human activity on the continent. The Committee for Environmental Protection, established under the Madrid Protocol, is to provide advice and formulate recommendations to the parties in connection with the implementation of the Protocol.

The most striking aspect of the Madrid Protocol is its prohibition on 'any activity relating to mineral resources, other than scientific research'. Many consider the relevant clause as a moratorium because, 50 years from the date of entry into force of the Protocol, any of the Antarctic Treaty consultative parties may request a conference to review the operation of the Protocol and to amend it. A modification or amendment could enter into force after its ratification by three-fourths of the consultative parties, including all those states that were consultative parties when the Protocol was adopted. However, as regards the clause dealing with Antarctic mineral resource activities, the prohibition on such activities would continue even after the clause had been modified or amended, unless there were in force a legally binding regime specifying agreed means for determining whether mining activities were acceptable and the conditions under which they would be permitted. The proposed modification or amendment of the clause in question should include such a regime. In view of all these requirements, the ban on mining is, for all practical purposes, of indefinite duration. Nevertheless, a provision has been included that if a proposed modification or amendment has not entered into force three years after its adoption, any party may withdraw from the Protocol with two years' notice.

Assessment

The Antarctic Treaty has established an important arms control regime. In particular, its non-nuclearization clause has helped to prevent the use of the vast expanses of the Antarctic continent as a nuclear testing ground, a nuclear-weapon base or a nuclear waste storage.

To reinforce and perpetuate the demilitarized status of the Antarctic, to forestall developments dangerous to the environment and ecology, and to preserve Antarctica as a zone of peace and of international scientific cooperation, several states have insisted that the concept of the common heritage of mankind should be applied to the region. They point out that Antarctica, which has never been controlled by any state, has, from the legal point of view, the same characteristics as outer space and celestial bodies, now generally recognized as the province of all mankind, or as the seabed beyond the limits of national jurisdiction, and must be used in the interest of all nations. Demands have also been made for equal rights for all parties, irrespective of the degree of their involvement in scientific research in the Antarctic. In the late 1980s, the UN General Assembly called upon the Antarctic Treaty consultative parties to deposit with the UN Secretary-General information and documents covering all aspects of Antarctica.

13

Denuclearized Zones

The idea of establishing nuclear-weapon-free zones was conceived with a view to preventing the emergence of new nuclear-weapon states. As early as in 1958, ten years before the signing of the Non-Proliferation Treaty (NPT), the Polish government, which feared the nuclearization of West Germany and wanted to prevent the deployment of Soviet nuclear weapons on its territory, put forward a proposal, called the Rapacki Plan (after the Polish Foreign Minister), for a nuclear-weapon-free zone in Central Europe. The zone was to comprise Poland, Czechoslovakia, the German Democratic Republic and the Federal Republic of Germany, but other European countries would have the possibility to accede. In the area in question, the stationing, manufacture and stockpiling of nuclear weapons and of nuclear delivery vehicles would be prohibited and strict control of compliance exercised. The nuclear powers would undertake to respect the nuclear-weapon-free status of the zone and not to use nuclear weapons against the territory of the zone. In the political climate of the 1950s, the Rapacki Plan had no chance of becoming the subject of an international transaction. Nonetheless, several of its elements were later adopted as guidelines for the establishment of denuclearized zones.

Efforts to ensure the absence of nuclear weapons in other populated parts of the world have been more successful. By 2001 four regional denuclearization agreements – namely, the 1967 Treaty of Tlatelolco regarding Latin America, the 1985 Treaty of Rarotonga regarding the South Pacific, the 1992 Declaration on the Denuclearization of the Korean Peninsula and the 1995 Treaty of Bangkok regarding South-East Asia – had entered into force, whereas the 1996 Treaty of Pelindaba regarding Africa had been signed but was not yet in force. The denuclearization of Central Asia was then under negotiation.

Certain uninhabited areas of the globe have also been formally denuclearized. They include Antarctica under the 1959 Antarctic Treaty (see Chapter 12.3); outer space, the moon and other celestial bodies under the 1967 Outer Space Treaty and the 1979 Moon Agreement (see Chapter 10.1 and 10.2); and the seabed, the ocean floor and the subsoil thereof under the 1971 Seabed Treaty (see Chapter 11.3).

Article VII of the NPT, affirmed the right of states to establish nuclear-weapon-free zones in their respective territories. The United Nations, in numerous resolutions, went further by encouraging the creation of such zones, and the 1995 NPT Review and Extension Conference expressed the conviction that regional denuclearization measures enhance global and regional peace and security. Nuclear-weapon-free zones have become part and parcel of the nuclear non-proliferation regime.

13.1 Guidelines for Denuclearized Zones

In 1975 the UN General Assembly formulated a set of principles which should guide states in setting up nuclear-weapon-free zones. These principles were later expanded and included in a consensus report of the UN Disarmament Commission issued in 1999. The main recommendations are as follows:

Nuclear-weapon-free zones should be established on the basis of arrangements freely arrived at among the states of the region concerned.

The initiative to establish such a zone should emanate exclusively from states within the region and be pursued by all the states of that region.

Assistance should be provided, including through the United Nations, to the states concerned in their efforts to establish a zone.

All the states of the region concerned should participate in the negotiations on and the establishment of a zone.

The status of a nuclear-weapon-free zone should be respected by all states parties to the treaty establishing the zone as well as by states outside the region, including the nuclear-weapon states and, if there are any, states with territory or that are internationally responsible for territories situated within the zone.

The nuclear-weapon states should be consulted during the negotiations of each treaty and its relevant protocol(s) in order to facilitate their signature and ratification of the protocol(s) through which they undertake legally binding commitments to the status of the zone and not to use or threaten to use nuclear weapons against states parties to the treaty.

If there are states with territory or that are internationally responsible for territories within the zone, these states should be consulted during the negotiations of each treaty and its relevant protocol(s) with a view to facilitating their signature and ratification of the protocol(s).

The process of establishing the zone should take into account all the relevant characteristics of the region concerned.

The obligations of the parties should be clearly defined and be legally binding.

The arrangements should be in conformity with the principles and rules of international law, including the UN Convention on the Law of the Sea.

States parties to a nuclear-weapon-free zone exercising their sovereign rights and without prejudice to the purposes and objectives of such a zone remain free to decide for themselves whether to allow visits by foreign ships and aircraft to their ports and airfields, transit of their airspace by foreign aircraft and navigation by foreign ships in or over their territorial sea, archipelagic waters or straits that are used for international navigation, while fully honouring the rights of innocent passage, archipelagic sea lane passage or transit passage in straits that are used for international navigation.

States parties to the current nuclear-weapon-free zones should ensure that their adherence to other international and regional agreements does not entail any obligation contrary to their obligations under the zone treaties.

A nuclear-weapon-free zone should provide for the effective prohibition of the development, manufacturing, control, possession, testing, stationing or transporting by the states parties to the treaty of any type of nuclear explosive device for any purpose, and should stipulate that states parties to the treaty do not permit the stationing of any nuclear explosive devices by any other state within the zone.

A nuclear-weapon-free zone should provide for effective verification of compliance with the commitments made by the parties to the treaty.

A zone should constitute a geographical entity whose boundaries are to be clearly defined by prospective states parties to the treaty through consultations with other states concerned, especially in cases where territories in dispute are involved.

Nuclear-weapon states should, for their part, assume in full their obligations with regard to nuclear-weapon-free zones upon signing and ratifying relevant protocols.

A nuclear-weapon-free zone should not prevent the use of nuclear science and technology for peaceful purposes and could also promote international cooperation for the peaceful use of nuclear energy in the zone.

However, given the dissimilar geographical circumstances as well as different political, cultural, economic and strategic considerations of the states concerned, there can be no uniform pattern of denuclearized zones. The differences may relate to the scope of the obligations assumed by the parties; the responsibilities of extra-zonal states; the geographical area subject to denuclearization; the verification arrangements; and the conditions for the entry into force of the zonal agreement as well as for its denunciation.

13.2 The 1967 Treaty of Tlatelolco

During the 1962 Cuban Missile Crisis, a draft resolution calling for a nuclear-weapon-free zone in Latin America was submitted at the UN General Assembly by Brazil but was not put to a vote. In April 1963, at the initiative of the President of Mexico, the presidents of five Latin American countries announced that they were prepared to sign a multilateral agreement that would make Latin America a nuclear-weapon-free zone. This announcement received the support of the UN General Assembly, and the Latin American nations started negotiations among themselves. On 14 February 1967, at Tlatelolco, a district of Mexico City, the Treaty for the Prohibition of Nuclear Weapons in Latin America was signed by a number of Latin American states. Two Additional Protocols annexed to the Treaty of Tlatelolco were intended for signature by extra-zonal states.

Scope of the Obligations

The Treaty of Tlatelolco prohibits the testing, use, manufacture, production or acquisition by any means as well as the receipt, storage, installation, deployment and any form of possession of nuclear weapons in Latin America. Encouraging or authorizing or in any way participating in the testing, use, manufacture, production, possession or control of any nuclear weapon is equally prohibited. Research and development directed towards acquiring a nuclear-weapon capability is not expressly forbidden. Each party must conclude an agreement with the International Atomic Energy Agency (IAEA) for the application of safeguards to its nuclear activities.

Explosions of nuclear devices for peaceful purposes are allowed under the Treaty, and procedures for carrying them out are specified in Article 18. However, a proviso is made that such activities must be conducted in conformity with Article 1, which bans nuclear weapons, as well as with Article 5, which defines a nuclear weapon as any device capable of releasing nuclear energy in an uncontrolled manner and having characteristics appropriate for use for warlike purposes. An instrument that may be used for the transport or propulsion of the device is not included in this definition if it is separable from the device and not an indivisible part thereof. Most countries interpret all these requirements as prohibiting the manufacture of all nuclear explosive devices, unless or until nuclear devices are developed which cannot be used as weapons. This interpretation had for a long time been contested by Argentina and Brazil. Subsequently, however, both countries undertook to prohibit in their respective territories the testing, use, manufacture, production or acquisition by other means of any nuclear explosive device, as long as no technical distinction can be made between nuclear explosive devices for peaceful purposes and those for military purposes. Thus, the controversy over whether indigenous development of nuclear explosive devices for peaceful purposes is compatible with the participation in the Treaty of Tlatelolco has been set aside. It is obvious that allowance for any kind of nuclear explosion would defeat the purpose of a nuclear-weapon-free zone.

One of the purposes of the treaties establishing zones free of nuclear weapons is to make a nuclear attack against states parties militarily unjustifiable and, consequently, less likely. To achieve this goal, all potential targets of a nuclear strike would have to be removed from the denuclearized areas. These targets include nuclear-weapon-related support facilities, such as communication, surveillance and intelligence-gathering facilities, as well as navigation installations, serving the nuclear strategic systems of the great powers. The Treaty of Tlatelolco does not, however, specifically ban such facilities.

Area Subject to Denuclearization

The zone of application of the Treaty of Tlatelolco embraces the territory, territorial sea, airspace and any other space over which the zonal state exercises sovereignty in accordance with its own legislation. It will also include vast areas in the Atlantic and Pacific Oceans, hundreds of kilometres off the coasts of Latin America (Article 4), upon fulfilment of several requirements specified in Article 28. These requirements are: adherence to the Treaty by all states of the region; signature and ratification of the Additional Protocols to the Treaty by all the states concerned; and conclusion of agreements with the IAEA for the application of safeguards to the nuclear activities of the parties. The extra-continental or continental states which are internationally responsible, de jure or de facto, for territories lying within the limits of the geographical zone established by the Treaty – France, the Netherlands, the United Kingdom and the United States – have undertaken to apply the statute of military denuclearization to these territories by adhering to Additional Protocol I of the Treaty. All nuclear-weapon powers have unreservedly assumed an obligation under Additional Protocol II to respect the denuclearization of Latin America as 'defined, delimited and set forth' in the Treaty, that is, as covering the designated portions of the high seas as well. However, in statements contradicting this obligation, the signatories of Additional Protocol II pointed out that they would not accept any restrictions on their freedom at sea.

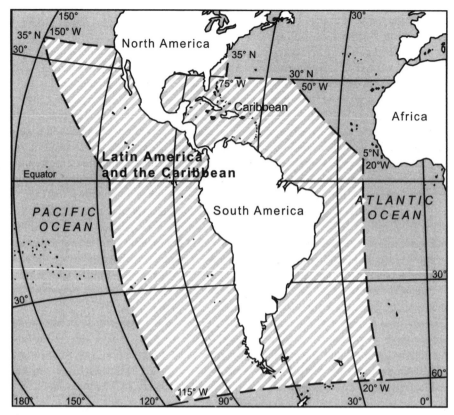

Figure 13.1   *Zone of Application of the Treaty of Tlatelolco*

The continental part of the territory of the United States of America and its territorial waters are excluded from the zone of application of the Treaty.

Furthermore, since the Treaty has not explicitly prohibited transit of nuclear weapons, the question arose whether such activity is actually permitted. According to the interpretation given in 1967 by the Preparatory Commission for the Denuclearization of Latin America (COPREDAL), it is the prerogative of the territorial state, in the exercise of its sovereignty, to grant or deny permission for transit. In joining the Additional Protocols of the Treaty, the United States and France made a declaration of understanding to the same effect, while the Soviet Union expressed the opinion that authorizing transit of nuclear weapons in any form would be contrary to the objectives of the Treaty. China considers that the passage of means of transport or delivery carrying nuclear weapons through Latin American territory, territorial sea or airspace is prohibited by the Treaty. Indeed, once nuclear weapons are allowed in transit, even if such transit is limited to port visits or overflights, it will be difficult to maintain that the zone has been denuclearized. In any event, since the great powers refuse, as a matter of policy, to disclose the whereabouts of their nuclear weapons, they are unlikely to request permission of transit for specific nuclear-weapon-carrying ships or aircraft. The right of zonal states to deny permission for transit of nuclear weapons is thus purely hypothetical.

Security Assurances of Extra-Zonal States

Under Additional Protocol II to the Treaty of Tlatelolco, the 'powers possessing nuclear weapons' must fully respect the statute of denuclearization of Latin America in respect of warlike purposes, not to contribute to the performance of acts involving a violation of the Treaty, and not to use or threaten to use nuclear weapons against the parties to the Treaty. However, the obligations which the nuclear-weapon powers have actually assumed under this Protocol are conditional. The United States and the United Kingdom made interpretative statements at the time of signing and ratifying Protocol II, which reflected their current military doctrines. They reserved the right to reconsider their non-use obligations with regard to any state in the nuclear-weapon-free zone in the event of an armed attack by that state carried out with the support or assistance of a nuclear-weapon power. The Soviet Union formulated a similar qualification with regard to a party to the Treaty committing an act of aggression with the support of, or together with, a nuclear-weapon state. For France, its non-use undertaking would present no obstacle to the full exercise of the right of self-defence enshrined in the UN Charter.

Entry into Force and Denunciation

The Treaty of Tlatelolco enters into force among states that have ratified it only when certain conditions have been met – the same conditions as are required under Article 28 for the extension of the geographical area of the Treaty's application. These conditions may be waived, and most parties have in fact done so. The Treaty became operative in April 1968, when El Salvador joined Mexico in ratifying it and in waiving the requirements for its entry into force.

The Treaty is of a permanent nature and is not subject to reservations. However, any party may denounce it with three months' notice if, in its opinion, there have arisen or 'may arise' circumstances connected with the content of the Treaty or of the Additional Protocols to the Treaty which affect its supreme interests or the peace and security of one or more parties.

After the entry into force of the Treaty for all countries of the zone, the rise of a new power possessing nuclear weapons could have the effect of suspending the execution of the Treaty for those countries which had ratified it without waiving the requirement that Additional Protocol II be signed and ratified by all powers possessing nuclear weapons, and which would request such suspension. The Treaty would then remain suspended until the new power ratified the Protocol.

Amendments

In 1992, at the initiative of Argentina, Brazil and Chile, several articles of the Treaty of Tlatelolco were amended. The most important amendments concerned the so-called special inspections which, according to a new treaty paragraph, would be carried out exclusively by the IAEA.

Another amendment, adopted in 1990, added to the official title of the Treaty of Tlatelolco the words 'and the Caribbean' in order to incorporate the English-speaking states of the Caribbean area into the zone of application of the Treaty. By yet another amendment, adopted in 1991, all the independent states of the region became eligible to join the regime of denuclearization, whereas, according to the original version, a 'political entity', part or all of whose territory was the subject of a dispute or claim between an extra-continental country and one or more Latin American states, could not be admitted. Owing to this amendment, Belize and Guyana could join the Treaty.

Conditions for the entry into force of the amendments are not clearly stated in the Treaty. The government of Mexico, the depositary of the Treaty, considers the amendments to be in force for those states that have ratified them and waived the requirements specified in Article 28.

13.3 The 1985 Treaty of Rarotonga

In 1983, in the context of growing concern over the activities of the nuclear-weapon powers in the South Pacific, and especially over nuclear test explosions, Australia proposed the establishment of a nuclear-free zone in the region. The proposal was officially submitted at the annual South Pacific Forum, the high-level meeting of independent or self-governing South Pacific countries. It was endorsed the following year. Subsequently, as a result of negotiations among Australia, Cook Islands, Fiji, Kiribati, Nauru, New Zealand, Niue, Papua New Guinea, Solomon Islands, Tonga, Tuvalu, Vanuatu and Western Samoa – all member-states of the South Pacific Forum – a treaty establishing the proposed zone was signed on 6 August 1985, at Rarotonga in the Cook Islands. (The Republic of Marshall Islands and the Federated States of Micronesia became eligible to sign only upon joining the Forum in 1987. Three protocols annexed to the Treaty were intended for signature by extra-zonal states.)

Scope of the Obligations

The South Pacific Nuclear Free Zone Treaty, called the Treaty of Rarotonga, in force since 1986, prohibits the manufacture or acquisition by other means, as well as the possession or control, of any nuclear explosive device by the countries of the zone. It also bans seeking or receiving assistance in the manufacture or acquisition of nuclear explosive devices. Protocol 3, prohibiting tests of any nuclear explosive device anywhere within the zone, was opened for signature by all five declared nuclear-weapon powers, but it was clearly addressed to France, the only state which at the time of signing was engaged in such tests in the region.

By 'nuclear explosive device' the Treaty means any nuclear weapon or other explosive device capable of releasing nuclear energy, irrespective of the purpose for which it could be used. The term includes such a weapon or device in unassembled and partly assembled forms, but does not include the means of transport or delivery of such a weapon or device if separable from and not an indivisible part of it. As in the Treaty of Tlatelolco, research and development directed towards acquiring a nuclear-weapon capability are not expressly forbidden.

In addition to banning nuclear explosive devices, the Treaty of Rarotonga contains a ban on dumping radioactive matter at sea anywhere within the South Pacific Zone. Hence the zone is called 'nuclear-free', which conveys a wider notion than 'nuclear-weapon-free'. The relevant provision reflects the concern, often voiced in the United

Nations and other international organizations, over the inability of the nuclear industry to dispose safely of its wastes.

As regards weapon-related prohibitions, the Treaty of Rarotonga appears to be stricter than the Treaty of Tlatelolco, because it prohibits the possession or testing of nuclear explosive devices for peaceful purposes. Nevertheless, as in the Treaty of Tlatelolco, the denuclearization measures taken in the South Pacific region have not removed all the potential targets for nuclear attack, because the Treaty of Rarotonga does not prohibit the facilities serving nuclear strategic systems.

Full-scope IAEA safeguards must be applied to nuclear activities of the parties, and no nuclear exports to any non-nuclear-weapon state may take place without the application of such safeguards.

Area Subject to Denuclearization

Although it is claimed that the Treaty of Rarotonga set up a nuclear-free zone stretching to the border of the Latin American nuclear-weapon-free zone in the east and to the border of the Antarctic demilitarized zone in the south, it bans the presence of nuclear weapons only within the territories of the South Pacific states, up to the 12-mile territorial sea limit. It does not seek – as the Treaty of Tlatelolco does – to have nuclear-weapon prohibitions applied to a larger ocean area. This omission seems to be justified by a specific reference to international law with regard to freedom of the seas, although no law, including the law of the sea, can exclude constraints on any activity, if the constraints are internationally agreed. Establishment of extensive nuclear-weapon-free maritime areas adjacent to nuclear-weapon-free territories would reinforce the sense of security of zonal states.

Each party may allow visits by any foreign ships and aircraft to its ports and airfields, transit of its airspace by foreign aircraft, and navigation by any foreign ships in its territorial sea or archipelagic waters in a manner not covered by the rights of innocent passage, archipelagic sea lane passage or transit passage of straits. The frequency and duration of such permitted visits and transits are not limited. It is therefore not clear to what extent they differ from the 'stationing' (defined in the Treaty as 'emplantation, emplacement, transportation on land or inland waters, stockpiling, storage, installation and deployment') of nuclear weapons, which is prohibited. Under Protocol 1 to the Treaty of Rarotonga, open for signature by France, the United Kingdom and the United States, the signatories are to apply the prohibitions contained in the Treaty in respect of the territories in the zone for which they are internationally responsible.

Security Assurances of Extra-Zonal States

Protocol 2 to the Treaty of Rarotonga provides for assurances to be given by the nuclear-weapon powers not to use or threaten to use nuclear explosive devices against the parties to the Treaty or any territory within the zone for which a state that has become a party to Protocol 1 is internationally responsible. In signing this Protocol, the Soviet Union stated that in case of action taken by a party or parties violating their major commitments concerning the status of the zone, it would consider itself free from its non-use commitments. The same would apply in case of aggression committed by one or several parties to the Treaty, supported by a nuclear-weapon state, or together with it, with the use by such a state of the territory,

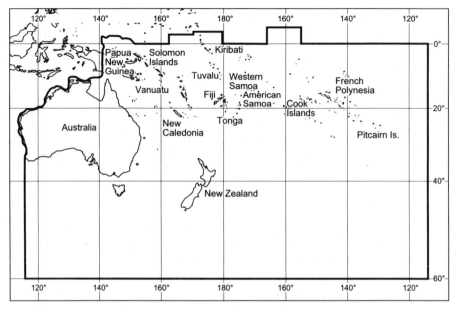

Figure 13.2 *Zone of Application of the Treaty of Rarotonga*

airspace, territorial sea or archipelagic waters of the parties for visits by nuclear-weapon-carrying ships and aircraft or for transit of nuclear weapons. Eventually, Protocols 2 and 3 were ratified by the Soviet Union without reference to the above statement.

China signed the same Protocols with an understanding that it might reconsider its obligations if other nuclear-weapon states or parties to the Treaty took action in gross violation of the Treaty and its Protocols, thus changing the status of the zone and endangering the security interests of China. This understanding was not referred to at the time of ratification.

In 1996, after the termination of the last series of French nuclear tests in the Pacific, France and the United Kingdom became parties to the Protocols. In its statement of reservation and interpretation, the French government made it clear that it did not consider France's inherent right to self-defence to be restricted by the signed documents and that the assurances provided for in Protocol 2 were the same as those given by France to non-nuclear-weapon states parties to the NPT. The British government stated that it would not be bound by its undertaking under Protocol 2 in the case of an invasion or any other attack carried out or sustained by a party to the Treaty in association or alliance with a nuclear-weapon state, or if a material breach of the non-proliferation obligations under the Treaty were committed. The US government signed the Protocols without a formal reservation but in 2002 the United States was still not a party to them.

Entry into Force and Denunciation

The Treaty of Rarotonga entered into force upon the deposit of the eighth instrument of ratification. This procedure was much simpler than that provided for in the Treaty of Tlatelolco. The denunciation formula of the Treaty of Rarotonga is also different. It is more restrictive than that of the Treaty of Tlatelolco, because it concedes the right of withdrawal only in the event of violation of a provision essential to the achievement of the objectives of the Treaty, and it requires 12 months' notice. Reservations are not allowed.

13.4 The 1992 Declaration on Korea

Whereas the Republic of Korea (South Korea) – which joined the NPT in 1975 – has all along been subject to full-scope safeguards, as provided for in that Treaty, the Democratic People's Republic of Korea (North Korea) – party to the NPT since 1985 – refused to sign a safeguards agreement with the IAEA within the time limit prescribed by the Treaty. It put forward several political conditions for signing which were not directly related to the NPT.

Following the decision by the United States to withdraw tactical nuclear weapons deployed outside its borders and the statement by the South Korean President that there were no such weapons in his country, the government of North Korea finally accepted the NPT safeguards. On 20 January 1992, both Korean states signed a Joint Declaration on the Denuclearization of the Korean Peninsula. The stated aim of the Declaration was to 'eliminate the danger of nuclear war' and, in particular, to 'create an environment and conditions favourable for peace and peaceful unification of our country'.

The parties agreed not to test, manufacture, produce, receive, possess, store, deploy or use nuclear weapons. They further undertook to use nuclear energy solely for peaceful purposes, and not to possess nuclear reprocessing or uranium enrichment facilities. To verify compliance, each side may conduct inspections of the objects agreed upon by both sides. (South Korea's proposal for a system of challenge inspections to be conducted upon the initiative of the requesting party was not accepted by North Korea.) A South–North Joint Nuclear Control Commission is to be in charge of implementing the obligations of the parties.

The Joint Declaration entered into force – together with the Agreement on Reconciliation, Non-aggression and Exchanges and Co-operation between the South and the North – upon the exchange of appropriate instruments, which took place on 19 February 1992. However, the 1993 decision by North Korea to withdraw from the NPT – although subsequently suspended – placed in jeopardy and, in any event, delayed the realization of the nuclear-weapon-free zone agreement.

If brought fully into effect, the Korean Declaration would significantly complement the global non-proliferation regime. Its ban on reprocessing and enrichment activities – which goes beyond the obligations assumed by the parties to other nuclear-weapon-free zone treaties – is particularly noteworthy. However, since these activities, which have legitimate civilian applications, are not prohibited by the NPT, they may not be banned in other zonal denuclearization agreements.

13.5 The 1995 Treaty of Bangkok

In South-East Asia, the idea of setting up a nuclear-weapon-free zone was developed as part of the Declaration on the Zone of Peace, Freedom and Neutrality (ZOPAN), issued in 1971 by the Association of South-East Asian Nations (ASEAN). In the 1990s the states of the region revitalized the denuclearization proposal, and a working group was established by the Association to implement the initiative. This work gathered momentum after the United States.had closed its military bases in the Philippines and appeared to support the ASEAN project. On 16 December 1995 the Treaty on the Southeast Asia Nuclear-Weapon-Free Zone was signed in Bangkok. It is referred to as the Treaty of Bangkok.

Scope of the Obligations

Parties to the Treaty of Bangkok may use nuclear energy for their economic development and social progress, but are prohibited from developing, testing, manufacturing or otherwise acquiring, possessing or having control over nuclear weapons, both inside and outside the zone. (Research on nuclear explosive devices is not expressly banned.) The parties will not allow other states to engage in such activities on their territories, including the use of nuclear weapons. 'Nuclear weapon' is defined simply as any explosive device that is capable of releasing nuclear energy in an uncontrolled manner. The means of transport or delivery of such a device are not included in this definition if they are separable from and not an indivisible part thereof. Nuclear explosive devices in unassembled or partly assembled forms are not explicitly covered. Dumping at sea or discharge into the atmosphere within the zone of any radioactive material or wastes is not allowed. Nor is it allowed to dispose of radioactive material or wastes on land, unless the disposal is carried out in accordance with IAEA standards and procedures. Seeking or receiving assistance in the commission of acts which would violate the above provisions, as well as assisting in or encouraging the commission of such acts, is equally prohibited.

Parties which have not yet done so must conclude an agreement with the IAEA for the application of full-scope safeguards to their peaceful nuclear activities. Prior to embarking on a peaceful nuclear energy programme, each party must subject the programme to rigorous nuclear safety assessment conforming to the guidelines and standards recommended by the IAEA for the protection of health and minimization of danger to life and property.

Stationing – defined as deploying, emplacing, emplanting, installing, stockpiling or storing nuclear weapons – in the South-East Asia zone is prohibited. However, each party, on 'being notified', may decide for itself whether to allow visits by foreign ships and aircraft to its ports and airfields, transit of its airspace by foreign aircraft, navigation by foreign ships through its territorial sea or archipelagic waters and overflight of foreign aircraft above those waters in a manner not governed by the rights of innocent passage, archipelagic sea lanes passage or transit passage. As elsewhere, it is doubtful whether the presence of nuclear weapons on foreign ships or aircraft would ever be notified.

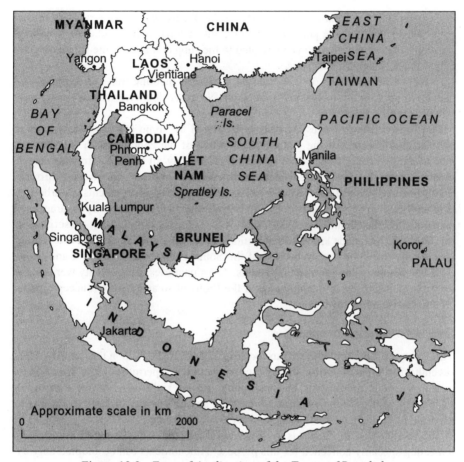

Figure 13.3   *Zone of Application of the Treaty of Bangkok*

The zone of application of the 1995 Treaty of Bangkok comprises, in addition to the territories, the continental shelves and exclusive economic zones of the states parties within the
zone.

Area Subject to Denuclearization

The South-East Asia nuclear weapon-free zone comprises the territories of Brunei
Darussalam, Cambodia, Indonesia, Laos, Malaysia, Myanmar, the Philippines, Singapore, Thailand and Viet Nam, as well as their respective continental shelves and
exclusive economic zones (EEZs). The inclusion of continental shelves and EEZs is
a novelty, but, according to the language of the Treaty, the right of states with regard
to freedom of the high seas is not to be prejudiced.

Security Assurances of Extra-Zonal States

Under the Protocol annexed to the Treaty of Bangkok and open for signature by
China, France, Russia, the United Kingdom and the United States, the signatories
would assume the following obligations: to respect the Treaty and not to contribute
to any act which would constitute its violation, and not to use or threaten to use

nuclear weapons against any state party to the Treaty and, in general, within the zone. The Protocol is of a permanent nature, but each party may withdraw from it if it decides that extraordinary events related to the subject matter of the Protocol have jeopardized its supreme interests.

In the event of breach of the Protocol, a special meeting may be convened to decide on appropriate measures to be taken. No other denuclearization treaty provides for such action.

The United States expressed concerns (shared by some other nuclear-weapon powers) that because of the geographical extent of the zone – which it considers , inconsistent with the 1982 UN Convention on the Law of the Sea – regular movement of nuclear-powered and nuclear-armed naval vessels and aircraft through South-East Asia would be restricted and regional security arrangements disturbed. It is unwilling to provide what it deems to be sweeping negative security assurances – demanded by the South-East Asian states – to a zone as large as that prescribed in the Treaty. China made known its objection to the geographical scope of the Treaty, specifically to the inclusion of parts of the South China Sea to which it and some ASEAN members have conflicting claims. The signatories of the Treaty were asked by some states to revise the language of the Protocol so as to make it acceptable to all nuclear-weapon powers.

Entry into Force and Denunciation

The Treaty of Bangkok entered into force on 27 March 1997, upon the deposit of the seventh instrument of ratification. Reservations are not permitted. The Treaty is to remain in force indefinitely, but each party has the right to withdraw from it, at 12 months' notice, in the event of a breach by any other party of a provision that is essential to the achievement of the objectives of the Treaty.

The operation of the Treaty is to be reviewed ten years after its entry into force at a meeting specially convened for this purpose. Amendments can be adopted only by a consensus decision.

13.6 The 1996 Treaty of Pelindaba

On 24 November 1961, in the aftermath of the first French nuclear-weapon tests in the Sahara desert, the UN General Assembly called on member-states to refrain from carrying out such tests in Africa and from using the African continent for storing or transporting nuclear weapons. Nearly three years later, the African heads of state and government, participating in a summit conference of the Organization of African Unity (OAU), solemnly declared that they were ready to undertake, through an international agreement to be concluded under United Nations auspices, not to manufacture or control atomic weapons. The declaration was endorsed in resolutions of the United Nations, but no concrete action was taken to carry it into effect. Only in 1991, after South Africa – the only country on the African continent that possessed the technical capability to produce nuclear weapons – had acceded to the NPT, real prospects opened up for the establishment of an African nuclear-weapon-free zone.

In 1995, as a result of several years' work, OAU and UN experts succeeded in elaborating a draft treaty which, after a few amendments, was approved by the OAU Assembly. As in the previously concluded denuclearization treaties, the Protocols annexed to the Treaty on the African Nuclear-Weapon-Free Zone, called the Treaty of Pelindaba (after the former seat of the South African nuclear-weapon-related activities), are to be signed by extra-zonal states. Also in many other respects the Treaty of Pelindaba followed the pattern of the nuclear-weapon-free-zone arrangements in force in other parts of the world. On 11 April 1996 the Treaty of Pelindaba was opened for signature.

Scope of the Obligations

The Treaty of Pelindaba prohibits the manufacture, testing, stockpiling or acquisition by other means, as well as possession and control, of any nuclear explosive device (in assembled, unassembled, or partly assembled forms) by the parties. In addition – and this is an important novelty – research on, and development of, such a device is banned. The Treaty also bans seeking, receiving or encouraging assistance in these activities. Under Protocol II, open for signature by the five declared nuclear-weapon states, the signatories should undertake not to test or assist in or encourage the testing of any nuclear explosive device within the African zone. Nuclear explosive device is defined in the same way as in the Treaty of Rarotonga.

In a clear allusion to the past South African nuclear-weapon programme, the Treaty of Pelindaba requires the dismantlement and destruction of any nuclear device that was manufactured prior to the entry into force of the Treaty, as well as the destruction of the relevant facilities or their conversion to peaceful uses. All such operations must take place under the supervision of the IAEA. These provisions aim at dispelling any lingering suspicion that some nuclear items have been hidden away in South Africa or that certain prohibited activities are still taking place there. They have set a precedent for possible future nuclear-weapon-free-zone treaties concluded with the participation of nuclear-capable states.

The Treaty of Pelindaba prohibits armed attacks against nuclear installations (reactors and other relevant facilities enumerated in the Treaty) and the dumping of radioactive matter anywhere within the African zone. It also contains an undertaking by the parties to implement or to use as guidelines the measures contained in the 1991 Bamako Convention on the Ban of the Import into Africa and the Control of Transboundary Movement and Management of Hazardous Wastes within Africa, in so far as it is relevant to radioactive waste. The parties undertake to strengthen the mechanisms for cooperation, at the bilateral, sub-regional and regional levels, with a view to promoting the use of nuclear science and technology for economic and social developments.

Whereas stationing of nuclear explosive devices in the territory of the zonal states is prohibited, visits and transit by foreign ships and aircraft – in a manner not covered by the rights of innocent passage, archipelagic sea lane passage or transit passage of straits – may be allowed by the parties on the (rather unclear) condition that no prejudice should be caused to the purposes and objectives of the Treaty. Nor is it prohibited in the African zone to establish facilities serving the nuclear strategic systems of the nuclear-weapon powers.

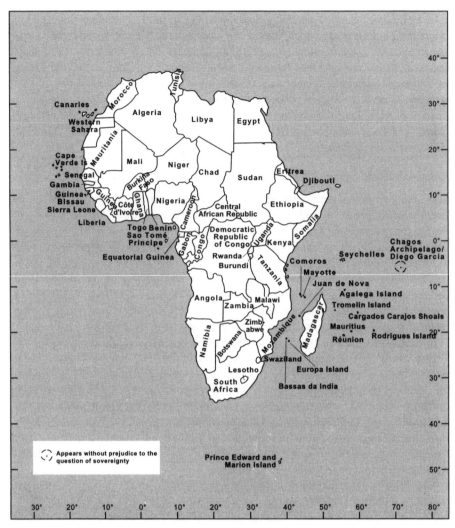

Figure 13.4   *Zone of Application of the Treaty of Pelindaba*

Verification of the uses of nuclear energy is to be performed by the IAEA, which must apply full-scope safeguards to prevent the diversion of nuclear material to nuclear explosive devices. Furthermore, the Treaty obliges the parties to observe international rules regarding the security and physical protection of nuclear materials, facilities and equipment in order to prevent their theft or unauthorized use.

Area Subject to Denuclearization

The Treaty of Pelindaba bans nuclear weapons in the territory of the continent of Africa, island states members of the OAU and all islands considered in OAU resolutions (presumably also resolutions which may be adopted in the future) to be part of

Africa. For the purpose of the Treaty, 'territory' means land territory, internal waters, territorial seas and archipelagic waters and the airspace above them, as well as the seabed and subsoil beneath. A reference made to the freedom of the seas is clearly intended to preclude restrictions on the presence of nuclear weapons beyond the territorial sea limits of the zonal states. Under Protocol III of the Treaty of Pelindaba, open for signature by France and Spain, the signatories should undertake to apply, in respect of the territories for which each of them is de jure or de facto internationally responsible, and which are situated in the African zone, the denuclearization provisions contained in the Treaty, and to ensure the application of IAEA safeguards there.

The geographic extent of the application of the Treaty of Pelindaba and of its Protocols is illustrated in a map annexed to the Treaty. The main difficulty in drawing up this map was the status of the Chagos Archipelago, which comprises the island of Diego Garcia harbouring a US military base. The Archipelago is covered by the map with a proviso that this is 'without prejudice to the question of sovereignty' claimed by both the United Kingdom and Mauritius. It was thus made clear that the resolution of the sovereignty issue would have to take place outside the framework of the Treaty. However, the United Kingdom stated that it did not accept the inclusion, without its consent, of the British Indian Ocean Territory, of which Diego Garcia is a part, within the African nuclear-weapon-free zone, and that it did not accept any legal obligations in respect of that Territory. In a related statement, the United States noted that neither the Treaty nor Protocol III apply to the activities of the United Kingdom, the United States or any other state not party to the Treaty on the Island of Diego Garcia or elsewhere in the British Indian Ocean Territories, and that, accordingly, no change was required in US armed forces operations there. Russia, however, pointed out that, as long as a military base of a nuclear-weapon power is situated on the Chagos Archipelago islands, and as long as certain nuclear powers consider themselves free from the obligations under the Protocols to the Treaty of Pelindaba with regard to these islands, Russia could not regard them as meeting the requirements of nuclear-weapon-free territories.

Security Assurances of Extra-Zonal States

Under Protocol I, open for signature by China, France, Russia, the United Kingdom and the United States, the signatories should undertake not to use or threaten to use a nuclear explosive device against any party to the Treaty, or any territory within the African zone for which a state that has become party to Protocol III is internationally responsible. However, in signing this Protocol, the United States, the United Kingdom and France declared that they would not be bound by it in case of an invasion, or any other attack upon them, carried out or sustained by a party to the Treaty in association or alliance with a nuclear-weapon state. Russia made a similar statement, but added that it did not consider itself bound by the obligations under Protocol I in respect of the Chagos Archipelago islands.

Parties to the Protocols would undertake not to contribute to any act constituting a violation of the Treaty or the relevant Protocol. This undertaking is unverifiable without the transparency of the nuclear powers' naval and air deployments in the nuclear-weapon-free zone as well as in the areas adjacent to the zone.

Entry into Force and Denunciation

The Treaty of Pelindaba is not subject to reservations. It will enter into force on the date of the deposit with the Secretary-General of the OAU of the 28th instrument of ratification. (In 2001 the OAU member-states decided that the Organization would be replaced by the African Union.) The Treaty is of unlimited duration, but any party may withdraw from it at 12 months' notice if some extraordinary events have jeopardized its supreme interests. The denunciation clause is thus less rigorous here than in the Treaty of Rarotonga which permits withdrawal only in the event of a material breach of the Treaty.

13.7 Negotiations for the Denuclearization of Central Asia

On 27 February 1997, following the proposal made by the President of Uzbekistan, the leaders of Central Asian states – Kazakhstan, Kyrgyzstan, Tajikistan, Turk-menistan and Uzbekistan – meeting in Almaty (Kazakhstan), declared their intention to establish a Central Asian Nuclear-Weapon-Free Zone (CANWFZ). The initiative was endorsed by the UN General Assembly, and in 1998 governmental experts from the five republics, assisted by the United Nations, started negotiating a denucleariza-tion treaty.

The negotiations proved difficult. The problems that arose were directly or indi-rectly related to the special characteristics of the region, which are as follows. As distinct from other nuclear-weapon-free zones, the CANWFZ is to border on two nuclear-weapon states, China and Russia, whereas India and Pakistan, countries which have demonstrated their nuclear-weapon capability and potential, are situated in relative proximity to the envisaged zone. The territory of Central Asia had served, until the late 1980s, as the main site of Soviet nuclear explosions, and nuclear weapons were withdrawn from the area only in the 1990s. Moreover, some Central Asian states are bound by the collective security arrangements (under the 1992 Tashkent Treaty) within the framework of the Commonwealth of Independent States (CIS), dominated by Russia, and would not assume commitments which they (and/or Russia) consider inconsistent with these arrangements. In addition, political rivalry among the states of the region impeded progress.

Area Subject to Denuclearization

Whereas land territories, waters within these territories and the air space above them, belonging to the five Central Asian states, are to be included in the nuclear-weapon-free zone, the Caspian Sea, although lying in the region, cannot be included, because only two (Kazakhstan and Turkmenistan) of the five littoral states will be covered by the CANWFZ. Nor can the territorial waters of the parties become part of the CANWFZ – as is the case in other zones – because the Caspian Sea is not subject to the Law of the Sea regime; there is no legally recognized division there between territorial and international waters. It appeared, therefore, necessary to leave the Caspian Sea, in its entirety, outside the geographic scope of the CANWFZ for all time, or until an agreement is reached on the legal status of this sea, in partic-ular, on the maritime boundaries.

Figure 13.5 *Proposed Nuclear-Weapon-Free Zone in Central Asia*

Transit of Nuclear Weapons

Most treaties establishing nuclear-weapon-free zones contain a clause (imposed by the nuclear-weapon powers) according to which parties may allow or deny visits or transit through their land territories, waters or air space by foreign ships or aircraft. Those opposed to inserting a similar clause in the CANWFZ treaty – which could be understood as sanctioning temporary presence of nuclear weapons in the zone – referred to UN guidelines which require that states forming a nuclear-weapon-free zone should ensure the 'total absence' of nuclear weapons in the zone.

Parties to the Treaty

The signers of the 1997 Almaty Declaration expressed the wish that the CANWFZ be open to other states in the region. After extensive negotiations it was agreed that only neighbouring states, those contiguous to the CANWFZ, would be allowed to accede to the CANWFZ treaty after its entry into force. A precedent for such an enlargement can be found in the Treaty of Rarotonga, which envisages the possibility that a member of the South Pacific Forum whose territory is outside the South Pacific nuclear-free zone could become a party to this Treaty. However, even if no clause expressly permitting accession were incorporated in the CANWFZ treaty, the original parties would always be free to amend it to allow accession. It is equally obvious that the parties to the protocol to the treaty would be free to refuse to accept alterations to their obligations under the protocol brought about by the enlargement of the CANWFZ.

Parties to the Protocol

The nuclear-weapon-free-zone agreements (with the exception of the Declaration on the Denuclearization of the Korean Peninsula) are accompanied by a protocol containing assurances of no use and of no threat of use of nuclear weapons against the parties to the treaty. In this connection, the Treaty of Tlatelolco refers, in general terms, to all powers 'possessing nuclear weapons', whereas the protocols to other treaties specify that the assurances are to be given by China, France, the Soviet Union/Russia, the United Kingdom and the United States. After India and Pakistan had carried out a series of nuclear test explosions and asserted themselves as nuclear-weapon states, a question arose as to whether these countries, too, should be invited to sign the projected protocol to the CANWFZ treaty. However, the parties to the NPT do not consider that India and Pakistan have acquired the status of a nuclear-weapon state because, according to the NPT, only countries which have exploded a nuclear device before 1 January 1967 may enjoy this status. An amendment to the NPT to accommodate the aspirations of the new de facto nuclear powers is unlikely to be adopted or even proposed. Another question was whether the security assurances to be provided by the nuclear-weapon powers should be unconditional, that is, valid under any circumstances, or conditional, that is, permitting the use of nuclear weapons against the parties to the CANWFZ treaty under certain circumstances – as postulated by most nuclear-weapon states.

Status of Other Security Arrangements

The sharpest controversy arose over the validity of the security arrangements already in force. Some negotiators insisted on making a proviso in the CANWFZ treaty to the effect that the rights and the obligations under other treaties signed by the parties must not be affected. Those opposed to this proviso argued that, in entering a nuclear-weapon-free-zone treaty prohibiting the deployment of nuclear weapons on its territory, a state renders invalid any previous agreement (open or secret) which may allow such deployment. In this context, they referred to the *lex posterior derogat legi priori* principle enshrined in the 1969 Vienna Convention on the Law of Treaties.

13.8 Proposals for Further Denuclearized Zones

Middle East

In the Middle East – one of the most explosive regions in the world – the concept of a nuclear-weapon-free zone was advanced by Iran and Egypt in 1974. Since then, the UN General Assembly has adopted several resolutions supporting this concept, often by consensus. UN Security Council ceasefire Resolution 687, passed after the 1991 Gulf War, also emphasized the need for a denuclearized Middle East.

The zone in the Middle East, as envisaged, to a large extent overlaps with the nuclear-weapon-free zone in Africa, but it is aimed in the first place at Israel, which refuses to join the NPT. Israel acknowledges having nuclear-weapon capabilities but has neither confirmed nor denied the possession of nuclear weapons. It has repeatedly made ambiguous statements to the effect that it will not be the first country to introduce nuclear weapons into the Middle East. A UN study on 'effective and verifiable measures which would facilitate the establishment of a nuclear-weapon-free zone in the Middle East', published in 1990, suggested that the process of creating such a zone should be preceded by confidence-building measures.

In April 1990 Egyptian President Mubarak proposed the establishment in the Middle East of a zone free of all types of weapons of mass destruction. Thus, not only nuclear weapons would be banned in the area in question, but chemical and biological weapons as well, and probably also certain categories of ballistic missiles. In any event, complete and verified denuclearization of the Middle East is not likely to take place before the conclusion of the peace process, which would end threats of the use of force in the region.

South Asia

The proposal for the establishment of a nuclear-weapon-free zone in South Asia has been on the agenda of the UN General Assembly since the early 1970s, and a number of resolutions have been adopted recommending such a measure. The resolutions were aimed at India and Pakistan, the major powers in the region not bound by the NPT. India has rejected the nuclear-weapon-free zone proposal, arguing that without a proper definition of the geographic extent and the security needs of the region (an allusion to neighbouring China) endorsement of the concept of regional denuclearization would be inappropriate. It also considers that nuclear disarmament is a matter requiring a global rather than regional approach. After the 1998 nuclear test explosions by both India and Pakistan, further efforts to create a denuclearized zone covering these two countries seem futile.

Europe

All the European states have joined the NPT. Moreover, the elimination by the United States and Russia of ground-launched missiles with a range of 500–5,500 kilometres, in compliance with the 1987 INF Treaty, as well as the two powers' unilateral withdrawals of their short-range missiles and most other tactical weapons, have transformed much of the European continent into a zone of considerably thinned-out nuclear armaments (see Chapter 5). In addition, according to the 1990 Treaty on the Final Settlement with respect to Germany, after the withdrawal of Russian forces from the former German Democratic Republic no nuclear weapons may be stationed in that part of Germany. Nevertheless, over the years, proposals have been made for a formal denuclearization of different parts of Europe.

In the mid-1990s, the government of Belarus suggested creating a nuclear-weapon-free zone which would comprise countries situated between the Baltic Sea and the Black Sea. The suggestion grew out of concern that the eastward expansion of the North Atlantic Treaty Organization (NATO) could lead to the deployment of Western tactical nuclear weapons on the territories of the former members of the Warsaw Treaty Organization (WTO). There was, however, no agreement among the countries of Central and Eastern Europe about the need for denuclearization undertakings. In 1996 NATO declared that it had 'no intention, no plan and no reason' to deploy nuclear weapons on the territory of new member-states. Proposals for a Nordic nuclear-weapon-free zone and a nuclear-weapon-free Baltic Sea, repeatedly made during the Cold War, were never the subject of actual interstate negotiation.

Other Regions

The suspected acquisition of nuclear weapons by North Korea gave rise to a debate, at the non-governmental level, about the advisability of setting up a nuclear-weapon-free zone in North-East Asia. The zone would cover not only the two Korean states and Japan but also Mongolia, Taiwan and, possibly, parts of the territories of some nuclear-weapon states. Moreover, the idea of establishing a 'nuclear-weapon-free southern hemisphere and adjacent areas' was launched at the 1996 UN General Assembly. Proposals for 'zones of peace' in the Indian Ocean, the Mediterranean, Central America and the South Atlantic implied measures of denuclearization as well.

13.9 Nuclear-Weapon-Free Countries

New Zealand

In 1987, the Parliament of New Zealand decided to establish the New Zealand Nuclear Free Zone. The Zone comprises all of the land, territory and inland waters within the territorial limits of New Zealand; the internal waters and the territorial sea of New Zealand; as well as the airspace above all these areas. In addition to prohibitions on the acquisition, stationing and testing of nuclear explosive devices in the Zone, the Prime Minister may grant approval for the entry of foreign warships into the internal waters of New Zealand only if he is satisfied that the warships will not be carrying any nuclear explosive device upon their entry into these waters. Similarly, approval for the landing in New Zealand of foreign military aircraft may be granted by the Prime Minister only if he is satisfied that the aircraft will not be carrying any nuclear explosive device when it lands. Entry into the internal waters of New Zealand by any ship whose propulsion is wholly or partly dependent on nuclear power is prohibited. In this respect, the Parliamentary Act establishing the New Zealand Zone went beyond the restrictions set by other nuclear-weapon-free zones; none of the zone treaties prohibits the presence of nuclear-powered engines. The Act reflected public opinion in New Zealand, where – according to a poll – an overwhelming majority of people desired that their defence be arranged in a way which ensured that the country remained nuclear-free.

New Zealand's anti-nuclear posture proved unacceptable to the United States, which cancelled its naval exercises with New Zealand, stopped its long-established intelligence relationship with that country, and suspended its security obligations to it. The argument put forward by the United States was that, by barring US warships, New Zealand had placed in jeopardy the collective capacity of the 1951 Australia–New Zealand–United States (ANZUS) alliance to resist armed attack. (ANZUS continued to govern security relations only between Australia and the United States and between Australia and New Zealand.) In 2000 proposals were submitted to the New Zealand Parliament to extend the existing restrictions by prohibiting the passage of nuclear-armed vessels, nuclear-powered ships and radioactive materials through the country's maritime exclusive economic zone (EEZ).

Nordic Countries

In 1988, the Parliament of Denmark passed a resolution requesting the government to notify (or rather remind) all visiting warships that they must not carry nuclear arms into Danish ports. In a sense, the resolution merely elaborated on the official Danish policy proclaimed more than three decades earlier, namely, that in time of peace introduction of nuclear weapons to the country was prohibited. In fact, however, the resolution appeared to reject the practice of 'neither confirming nor denying' the presence of nuclear weapons, which is followed by the navies of all the nuclear-weapon powers. Eventually, under pressure exercised within NATO, mainly by the United States and the United Kingdom, Denmark (a member of NATO) agreed, as a compromise, to proceed on the assumption that its decision to keep its territory free of nuclear weapons in peacetime is respected by visiting foreign ships or aircraft and not to seek specific assurances. Norway and Iceland (two other members of NATO) also stated that their governments presume that nuclear weapons are not carried on board foreign warships visiting their ports.

In non-aligned Sweden, where visiting warships are not permitted to carry nuclear weapons, the ruling Social Democratic Party decided, in 1987, that efforts should be made to convince the nuclear-weapon powers to forgo the practice of not giving information regarding the presence of nuclear weapons on their warships.

In Finland (also non-aligned), visiting naval vessels are subject to the provision that no nuclear weapons or other nuclear explosive devices may be introduced, even temporarily, into Finnish territory. The Finnish government expects that this provision will be strictly observed.

Mongolia

In 2000, following the 1992 declaration of its territory a nuclear-weapon-free zone, Mongolia adopted a law to preserve the country 'in its entirety' free from nuclear weapons. In addition to non-proliferation obligations, the law specifies that 'transportation through the territory of Mongolia of nuclear weapons, parts and components thereof, as well as of nuclear waste or any other nuclear material designed or produced for weapon purposes shall be prohibited'.

In response to this initiative, China, France, Russia, the United Kingdom and the United States jointly issued a statement on security assurances, in which they reaffirmed their commitment to seek immediate UN Security Council action to provide assistance to Mongolia, as a non-nuclear-weapon state party to the NPT, in accordance with the provisions of the 1995 UN Security Council Resolution 984, if Mongolia should become a victim of an act of aggression or an object of a threat of aggression in which nuclear weapons are used. They also reaffirmed, in the case of Mongolia, their respective unilateral negative security assurances, as stated in their declarations issued in 1995 and referred to in the above-mentioned resolution. (See Chapter 6.2.)

In addition, China and Russia confirmed the legally binding commitments undertaken by them with respect to Mongolia through bilateral treaties.

In fact, Mongolia has received the same security assurances from the nuclear-weapon powers as have all the other non-nuclear-weapon parties to the NPT, whether or not they have formally declared themselves to be nuclear-weapon-free.

Other Countries

In 1999 the Austrian Federal Parliament passed a constitutional law prohibiting the production of nuclear weapons as well as the storage, transportation, testing and use of such weapons. Installations intended for the stationing of nuclear weapons may not be built. Transportation through Austria of fissile material is also banned, unless it takes place in conformity with the international obligations.

In January 2001 the Latvian government adopted regulations stipulating that warships with nuclear-powered engines and on-board nuclear weapons would be banned from entering Latvia's territorial waters and ports. However, in October of that year the Latvian government decided that foreign warships with nuclear reactors or weapons on-board would be allowed to enter Latvia's territorial waters, internal waters and ports in order to ensure that Latvia could participate in 'NATO joint defence measures'.

A few countries, including members of military alliances (for example, Japan and Spain), have formally prohibited foreign ships or aircraft from entering their territories with nuclear weapons aboard. However, to avoid antagonizing the great powers, the governments of some of the denuclearized states chose to pretend not to be aware of the presence of nuclear weapons on board the visiting foreign craft. Certain cities have also declared themselves nuclear-weapon-free, but such declarations are not binding on the governments of the countries concerned.

13.10   Assessment

To the extent that the incentive to acquire nuclear weapons may emerge from regional considerations, the establishment of areas free of nuclear weapons is an important asset for the cause of nuclear non-proliferation. Countries confident that their enemies in the region do not possess nuclear weapons may not be inclined to acquire such weapons themselves. The zones which have been established so far meet other postulates as well. Besides prohibiting the acquisition of nuclear weapons by zonal states, they proscribe (unlike the NPT) the stationing of these weapons in the territories of non-nuclear-weapon states. Zonal procedures to verify compliance with non-proliferation obligations are even stricter than the procedures prescribed by the NPT. Moreover, zonal states benefit from some legally binding security assurances of the great powers.

Nevertheless, as pointed out in the preceding sections, the present nuclear-weapon-free zone treaties are deficient in several respects. In particular:

1. None of the treaties specifies that the denuclearization provisions are valid both in time of peace and in time of war.

2. Research on nuclear explosive devices is explicitly prohibited only in the Treaty of Pelindaba.

3. Only the Treaty of Rarotonga and the Treaty of Pelindaba make it clear that the bans cover nuclear explosive devices also in unassembled or partly assembled forms.

4. So-called peaceful nuclear explosions may be allowed by the Treaty of Tlatelolco (although only under certain specified conditions).

5. Nuclear-weapon-related support facilities serving the strategic systems of the nuclear-weapon powers are not banned by any nuclear-weapon-free-zone treaty.

6. Only the Treaty of Pelindaba prohibits attacks on nuclear installations.

7. Only the Treaty of Tlatelolco and the Treaty of Bangkok provide for the denuclearization of maritime areas adjacent to the territorial waters of zonal states.

8. The possibility of nuclear weapons transiting the territories of zonal states, including visits by foreign ships and aircraft with nuclear weapons aboard, is not excluded under any of the treaties.

9. The withdrawal clauses of the Treaty of Tlatelolco and the Treaty of Pelindaba, which refer to the 'supreme interests' of the parties, are too permissive as compared to the Treaty of Rarotonga and the Treaty of Bangkok, which concede the right of withdrawal only in the event of a material breach of the parties' obligations.

10. The nuclear-weapon powers' undertaking to respect the status of the denuclearized zones is unverifiable.

11. Assurances not to use nuclear weapons against zonal states, as given by the nuclear powers, are conditional.

12. Only the Treaty of Bangkok calls for action in the event of violation of the obligations assumed by the nuclear-weapon powers.

The above deficiencies may be removed through amendments of the existing nuclear-weapon-free zone treaties and avoided in the drafting of new such treaties, provided that due account is taken of the particularities of each region. Unilateral formal declarations on the denuclearization of individual countries are also important; they may contain broader and stricter undertakings than treaties. They ought, therefore, to be encouraged to further strengthen the nuclear non-proliferation regime.

14

Conventional Arms Control

For many years after World War II, Europe had the largest concentration of armed forces and weapons ever known in peacetime. In the early 1970s, when relations between the opposed military blocs – the North Atlantic Treaty Organization (NATO) and the Warsaw Treaty Organization (WTO) – improved, it became possible to consider limitations not only on nuclear armaments (see Chapter 5) but also on the conventional military potential of European states.

14.1 The MBFR Talks

Negotiations on force reductions in Europe opened in Vienna in October 1973. Officially called the talks on Mutual Reduction of Forces and Armaments and Associated Measures in Central Europe, they were referred to by the West as the Mutual and Balanced Force Reduction (MBFR) Talks, to emphasize NATO's aim of achieving a balance of forces through lower but equal force ceilings.

For the US Administration, an important motivation for entering the MBFR Talks, held in Vienna, was to open the prospect of troop cuts negotiated with the Soviet Union and thereby thwart the passage by the US Congress of Senator Mansfield's proposal for a substantial unilateral reduction of the US military presence in Europe. The Soviet Union agreed to engage in these talks as a concession to NATO, in order to secure its consent to convening a conference on European security, which would ratify the post-war political and territorial status quo in Europe.

Eleven states with indigenous or stationed forces in Central Europe were full participants in the Vienna talks – the United States, Canada, the United Kingdom, the Federal Republic of Germany (FRG), Belgium, the Netherlands and Luxembourg, on the NATO side; and the Soviet Union, the German Democratic Republic (GDR), Czechoslovakia and Poland, on the WTO side. Eight additional countries had special status of 'indirect' participants – Denmark, Greece, Italy, Norway, Turkey, Bulgaria, Romania and Hungary. The envisaged reductions were to take place in the MBFR area comprising the territories of Belgium, the FRG, Luxembourg and the Netherlands, as well as of Czechoslovakia, the GDR and Poland. France refused to participate in the talks because it was opposed, as a matter of principle, to bloc-to-bloc transactions and because it questioned the value of a regional approach to arms control in Europe.

Initial Positions

In November 1973 the four WTO participants submitted a draft agreement for a three-phase reduction of forces and their weapons: 20,000 men on each side in two years' time; 5% the following year; and 10% in 1977. These successive reductions were to affect foreign forces as well as national forces and their equipment. The equipment of the withdrawn forces was to be returned to the country of origin, whereas the equipment of demobilized national forces was to be eliminated.

Two weeks later, the seven NATO participants countered by proposing ground force reductions in two phases: first, a 15% reduction to affect only US and Soviet forces; second, a reduction of all forces, so as to arrive at common ceilings of 700,000 men on either side. Associated measures were to be taken to prevent the possibility of a surprise attack.

Further Developments

The initial positions having proved irreconcilable, in December 1975 NATO offered to withdraw 1,000 nuclear warheads and 90 nuclear delivery vehicles (54 Phantom aircraft and 36 Pershing missiles) in exchange for the withdrawal of a Soviet tank army (68,000 men and 1,700 tanks). In response, the WTO countries proposed an equal reduction of nuclear delivery vehicles (combat aircraft, ballistic missiles and ground-to-air missiles) and a freeze on nuclear systems remaining in the MBFR area. The NATO offer was subsequently abandoned, and the talks centred on data regarding forces stationed in the zone of envisaged reductions, including air forces. However, the two sides disagreed on the criteria to be used in counting the effectives. NATO preferred a collective, alliance-wide ceiling, whereas WTO wanted national sub-ceilings imposed on individual MBFR participants in order to restrict the possibilities of force restructuring within NATO and, in particular, to constrain the West German Bundeswehr – a major component of the NATO forces.

In June 1978, WTO amended its initial proposal by accepting equal ceilings: 700,000 men for ground forces and 200,000 for air forces, as well as a reduction, in the first phase, of Soviet and US troops – by 30,000 and 14,000, respectively. Nevertheless, WTO insisted on a balanced withdrawal of nuclear warheads and delivery vehicles, and this was unacceptable to the West. Another major obstacle to progress was the NATO refusal to withdraw units together with their equipment, as demanded by WTO. Later, WTO scaled down its demands, proposing the withdrawal of 20,000 Soviet troops against 13,000 US troops. A Polish proposal, taken up by the Soviet Union, reduced these figures to 11,500 Soviet and 6,500 US troops. There were serious disagreements over verification of compliance.

In January 1986, the West proposed to have 5,000 US and 11,500 Soviet troops withdrawn within one year; to begin, at the end of this period, exchanges of information on forces remaining in the MBFR area; to undertake not to increase the size of the forces present in the MBFR area for a three-year period and to accept verification of the data supplied by the parties; to agree on 30 on-site mutual inspections per year; to provide for observation of notified military activities; and to allow unrestricted use of national monitoring means. The Soviet Union rejected these proposals and the negotiations were deadlocked.

Assessment

Whereas the WTO was seeking reductions that would not affect the existing correlation of forces in Europe, NATO insisted that measures should be taken to remove WTO superiority in the numbers of ground forces and tanks in the agreed area of reductions, and to compensate for the differences in reinforcement capabilities of the two sides. Indeed, while US forces withdrawn from Europe would have to return to the United States, thousands of kilometres away, Soviet forces could be redeployed only a few hundred kilometres from the intra-German border.

The proposals of the two sides converged on a few points: both envisaged reductions by phases and equal ceilings for ground and air forces; both dealt with US and Soviet troop strengths separately from those of the remaining nine states and provided for the return of the withdrawn forces to national territories; and, at one stage, both included tanks and nuclear warheads, as well as aircraft and other nuclear-weapon delivery vehicles, in the categories of weapons to be reduced. However, since different national perspectives generated conflicting perceptions of what was an acceptable military balance, the controversy regarding the scope of reductions and the manner in which they had to be carried out – symmetrically, according to the East, or asymmetrically, according to the West – remained unresolved.

Moreover, from the strategic point of view, the relatively narrow sector of Europe where the reductions were to take place could not be isolated from the remaining European area; it could not be subject to a different regime without affecting the military balance on the European continent as a whole. The artificiality of the exercise was accentuated by the exclusion of Hungary and Italy from the designated zone of possible reductions, as if the forces deployed in these two countries were unrelated to the disposition of forces in the central part of Europe, as well as the absence of France from the negotiating table. The abortive MBFR Talks ended formally in February 1989.

14.2 The 1990 CFE Treaty

In March 1989, new talks on conventional armed forces in Europe (CFE) began in Vienna. As distinct from the MBFR Talks, the CFE negotiations involved all 16 NATO and all seven (six after the unification of Germany) WTO members, as requested by the United States, but they took place under the aegis of the all-European forum, the Conference on Security and Co-operation in Europe (CSCE), as requested by France. In the propitious political climate of the late 1980s, the participants succeeded in producing an agreement in less than two years. Signed on 19 November 1990, the Treaty on Conventional Armed Forces in Europe (CFE Treaty) became the first significant conventional arms control agreement to cover most of Europe.

Main Provisions

In addition to 23 articles, the CFE Treaty incorporated several protocols.

Area of Application. The CFE Treaty applies to the entire land territory of the states parties in Europe, from the Atlantic Ocean to the Ural Mountains. This area, referred to as the ATTU (Atlantic-to-the-Urals) zone, includes all the European island territories of the parties, including the Faroe Islands (Denmark), Svalbard including Bear Island (Norway), the islands of Azores and Madeira (Portugal), the Canary Islands (Spain), and Franz Josef Land and Novaya Zemlya of the former Soviet Union, now Russia.

In the case of Russia (and subsequently of Kazakhstan), the area of application includes all territory lying west of the Ural River and the Caspian Sea. In the case of

Table 14.1 *National CFE Limits under the 1990 CFE Treaty and*
the 1999 Agreement on Adaptation

		Tanks	ACVs	Artillery	Aircraft	Helicopters
NATO	1990	19,142	29,822	18,286	6,662	2,000
	1999[a]	19,096	31,787	19,529	7,273	2,282
WTO/former WTO	1990	20,000	30,000	20,000	6,800	2,000
	1999[b]	16,478	24,783	16,783	5,930	1,712
Total	*1990*	*39,142*	*59,822*	*38,286*	*13,462*	*4,000*
	1999	*35,574*	*56,570*	*36,312*	*13,203*	*3,994*
Difference		− 3,568	− 3,252	− 1,974	− 259	− 6

ACV = armoured combat vehicle.
[a] Enlarged NATO '16 + 3'.
[b] 'Former WTO − 3'.

Turkey, the area of application includes the territory of Turkey north and west of a line extending from the point of intersection of the Turkish border with the 39th parallel to Muradiye, Patnos, Karayazi, Tekman, Kemaliye, Feke, Ceyhan, Dogankent, Gözne and thence to the sea. Such a detailed delimitation and the exclusion of the south-eastern part of the Turkish territory from the area of application proved necessary because some non-member-states of the Organization for Security and Co-operation in Europe (OSCE) are located in the neighbourhood of Turkey and because of the dispute between Greece and Turkey over the port of Mersin. This port had been an outpost for the 1974 invasion of Cyprus by Turkey and – at the insistence of the Turkish government – was to remain free of constraint in order to protect the 'Turkish Republic of Northern Cyprus'.

Scope of the Obligations. The CFE Treaty limitations were to apply to two groups of states: those belonging to NATO and those belonging to the WTO; they were not to apply to the neutral/non-aligned European states. Within the area of application of the Treaty, each party was to limit and, as necessary, reduce five major categories of conventional armaments and equipment – battle tanks, armoured combat vehicles, artillery, combat aircraft and attack helicopters – so that, 40 months after entry into force of the Treaty and thereafter, for the group (that is, the alliance) of states to which the party belonged, the aggregate numbers would not exceed: (a) 20,000 battle tanks, of which no more than 16,500 in active units; (b) 30,000 armoured combat vehicles, of which no more than 27,300 in active units, but of these 30,000 armoured combat vehicles, no more than 18,000 could be armoured infantry fighting vehicles and heavy armament combat vehicles, and of the latter two sub-categories no more than 1,500 could be heavy armament combat vehicles; (c) 20,000 pieces of artillery, of which no more than 17,000 in active units; (d) 6,800 combat aircraft; and (e) 2,000 attack helicopters. To impede destabilizing force concentrations, the ATTU zone was divided into several nested sub-zones subject to varying numerical limitations. The so-called flank limitations for ground armaments were established to address primarily the concerns of Norway and Turkey that the withdrawal of Soviet forces from Central and Eastern Europe might result in a significant, destabilizing build-up of Soviet forces on or near their borders.

Certain armaments and equipment of similar appearance to treaty-limited equipment (TLE) were not limited by the Treaty. These so-called look-alikes – mainly support equipment – were to be notified and possibly checked by inspectors.

To minimize the possibilities of a surprise attack, armoured vehicle-launched bridges (defined as self-propelled armoured transporter–launcher vehicles capable of carrying and of emplacing and retrieving a bridge structure) were to be limited so that for each group of parties their aggregate number in active units within the area of application would not exceed 740. All armoured vehicle-launched bridges in excess of this number were to be placed in designated permanent storage sites. They could be withdrawn from these sites in the event of natural disasters involving flooding or damage to permanent bridges.

Armoured infantry fighting vehicles held by organizations designed and structured to perform internal security functions in peacetime were not limited by the Treaty. However, to avoid possible circumvention, any such armaments in excess of 1,000 were to constitute a portion of the levels set for armoured combat vehicles.

Reductions. Reductions were to be carried out as follows: (a) battle tanks and armoured combat vehicles – by destruction, conversion for non-military purposes, placement on static display, use as ground targets, or, in the case of armoured personnel carriers, modification; (b) artillery – by destruction or placement on static display, or, in the case of self-propelled artillery, by use as ground targets; (c) combat aircraft – by destruction, placement on static display, use for ground instructional purposes, or, in the case of specific models or versions of combat-capable trainer aircraft, reclassification into unarmed trainer aircraft; (d) specialized attack helicopters – by destruction, placement on static display, or use for ground instructional purposes; and (e) multi-purpose attack helicopters – by destruction, placement on static display, use for ground instructional purposes, or recategorization. Destruction was to be carried out by severing, by explosive demolition, by deformation, by smashing or by use as target drones in the case of aircraft.

Reductions in each of the categories of armament and equipment subject to limitations required on-site inspection and were to be effected in three phases after entry into force of the Treaty. However, before the Treaty was signed, thousands of Soviet tanks and other weapons had been moved outside the reduction area. In the reduction area itself, the Soviet Union exempted from limitation many items belonging to the categories limited by the Treaty, by assigning them to coastal defence forces, naval infantry or strategic rocket forces. As a result, fewer Soviet armaments were subject to destruction than the Western countries had originally foreseen.

Final Clauses. The CFE Treaty is of unlimited duration, but any party can withdraw from it if it decides that extraordinary events have jeopardized its supreme interests. Notice must be given at least 150 days prior to the intended withdrawal. In particular, a party may withdraw if another party increases its holdings in such proportions as to pose an obvious threat to the balance of forces within the area of Treaty application. A Joint Consultative Group (JCG) was established to monitor implementation, resolve issues arising from implementation and consider measures to enhance the viability and effectiveness of the Treaty.

Three politically binding declarations were made at the time the CFE Treaty was signed: one, containing the German government's commitment to limit the personnel strength of the armed forces of the united Germany to 370,000, of which no more than 345,000 would belong to ground and air forces; another, containing the original 22 CFE states' commitment not to increase the military manpower during the forthcoming negotiations on the limitation of the personnel strength of conventional armed forces in Europe; and a third, confirming the commitment not to exceed the limit of 430 land-based combat naval aircraft for each group of states, with a single-country limit of 400, and not to hold in the naval forces any permanently land-based attack helicopters.

The Tashkent Document

The breakup of the Soviet Union in December 1991 rendered inapplicable the CFE Treaty ceilings as well as the inspection quotas, because the ceilings had been based on old Soviet military districts that did not always coincide with the old republic boundaries. In some cases sharp antagonisms arose among the newly independent states.

Of the former Soviet republics in the CFE Treaty zone of application, Estonia, Latvia and Lithuania dissociated themselves from the Treaty obligations which had been assumed on their behalf by the Soviet Union; they agreed, however, to keep their territo.ies open to CFE inspectors as long as Russian troops were stationed there. Russia and Ukraine became involved in a dispute over control of the Crimea and the Black Sea Fleet. Moldova was beset by the Trans-Dniestr separatist movement; Armenia and Azerbaijan were at war with each other over control of the Nagorno-Karabakh region; and Georgia was being devastated by civil war. Nevertheless, at a meeting held in Tashkent on 15 May 1992, Armenia, Azerbaijan, Belarus, Georgia, Kazakhstan, Moldova, Russia and Ukraine recognized the CFE Treaty as an essential element of the European security system and agreed, as Soviet successor states, on how to implement it.

According to the Tashkent Document – consisting of the Joint Declaration and the Agreement on the Principles and Procedures for Implementing the Treaty on Conventional Armed Forces in Europe – Russia was to fulfil the obligations under the CFE Treaty and the related documents with regard to the armed forces as well as conventional armaments and equipment which were deployed on the territories of Estonia, Latvia, Lithuania, Germany and Poland, and which were subject to withdrawal to Russia. For each party the Agreement established: (a) maximum levels for holdings of armaments and equipment; and (b) a number of helicopters of certain types equipped for reconnaissance, spotting, or chemical/biological/radiological sampling, not subject to the CFE Treaty limits on attack helicopters. Maximum levels were also set for conventional armaments and equipment belonging to the categories limited by the Treaty but in service with coastal defence forces, naval infantry and strategic rocket forces.

The Oslo Document

The involvement of former Soviet republics required an adjustment of the CFE Treaty to the new situation. The necessary modifications were introduced on 5 June 1992, at an Extraordinary Conference held in Oslo, where the states having their armed forces deployed in the ATTU zone altered the language of the Treaty so as to take account of the enlarged participation. They also approved the reallocation of TLE among the members of the Commonwealth of Independent States (set up in December 1991), as specified in the 1992 Tashkent Document. The adopted Oslo Document included two annexes. One contained the understandings related to the breakup of the Soviet Union and the consequent territorial changes; another contained commitments to the exchanges of information required by the Treaty and a formula for the reallocation of armoured infantry fighting vehicles destined for internal security functions.

The Oslo Document made it possible for the CFE Treaty to enter provisionally into force as early as July 1992. Formally, the Treaty became effective in November 1992, upon ratification by all the states concerned.

The CFE-1A Agreement

The signatories of the CFE Treaty committed themselves to negotiating additional measures. Follow-up negotiations started at the end of November 1990 and led to the signing, on 10 July 1992, of the Concluding Act of the Negotiation on Personnel Strength of Conventional Armed Forces in Europe – the CFE-1A Agreement.

Scope of Limitation. Parties to the CFE-1A Agreement, in force since 17 July 1992, referred to the obligations assumed by all states participating in the CSCE to maintain only such military capabilities as may be necessary to prevent war and provide for effective defence. Each party undertook to limit the personnel of its conventional land-based armed forces within the area of application of the CFE Treaty, namely, all full-time military personnel serving with: land forces (including air defence formations and units); air and air defence aviation forces (including long-range aviation forces and military transport aviation forces); air defence forces other than those specified above; all central headquarters, command and staff elements (excluding naval personnel); all land-based naval formations and units which hold battle tanks, armoured combat vehicles, artillery, armoured vehicle-launched bridges, armoured infantry fighting vehicle look-alikes or armoured personnel carrier look-alikes, or which hold land-based naval combat aircraft; and all other formations, units and other organizations which hold battle tanks, armoured combat vehicles, artillery, combat aircraft or attack helicopters in service with the conventional armed forces. All reserve personnel who have completed their initial military service or training and who are called up or report voluntarily for full-time military service or training in conventional armed forces for a continuous period of more than 90 days are also subject to limitation. Forty months after entry into force of the CFE Treaty and thereafter, the aggregate number of military personnel subject to limitation was not to exceed the ceilings declared by each state. The CFE-1A Agreement met Germany's postulate that it should not be the only European country to have accepted (in connection with its unification) numerical limits on military personnel.

Not covered by the limitation are: personnel serving with organizations designed and structured to perform internal security functions in peacetime; personnel in short-duration (no longer than seven days) transit from a location outside the area of CFE Treaty application to a final destination outside this area; and personnel serving under UN command.

The CFE-1A Agreement calls for annual exchanges of information and advance notification of call-ups of more than 35,000 full-time military personnel from reserves. Any party may revise downward its national personnel limit. A party intending to revise its limit upward would have to provide an explanation. If another party objected to such a revision, an extraordinary conference could be convened to discuss the issue. The relevant provisions of the CFE Treaty can be used to evaluate the observance of the personnel limits under the CFE-1A Agreement. The duration of the Agreement is the same as that of the CFE Treaty. However, as distinct from the latter, the CFE-1A Agreement is only politically (not legally) binding; it does not require parliamentary ratification.

14.3 Adaptation of the CFE Treaty

During the decade of the 1990s the CFE Treaty was becoming inadequate to deal with the rapidly changing post-Cold War situation. To avoid the collapse of the Treaty regime, the parties made some modifications in the flank provisions of the Treaty at the First CFE Review Conference, held in May 1996, and then engaged in negotiations which ended on 19 November 1999 with the conclusion of the Agreement on Adaptation of the Treaty on Conventional Armed Forces in Europe. The Agreement on Adaptation introduced numerous amendments to the original text of the Treaty: certain articles were deleted, completely or in part; in many places they were replaced with new provisions. It incorporated the following documents: the Protocol on Existing Types of Conventional Armaments and Equipment, with an Annex; the Protocol on National Ceilings for Conventional Armaments and Equipment limited by the Treaty; the Protocol on Territorial Ceilings for Conventional Armaments and Equipment limited by the Treaty; the Protocol on Procedures governing the Reclassification of Specific Models or Versions of Combat-capable Trainer Aircraft into Unarmed Trainer Aircraft; the Protocol on Procedures governing the Reduction of Conventional Armaments and Equipment limited by the Treaty; the Protocol on Procedures governing the Categorization of Combat Helicopters and the Recategorization of Multi-purpose Attack Helicopters; the Protocol on Notification and Exchange of Information, with an Annex; the Protocol on Inspection; and the Protocol on the Joint Consultative Group. The Final Act of the Conference of the States Parties, at which the Agreement on Adaptation was signed, contains a package of political commitments made by several states.

The main adaptation consists in ending the military bloc-related, bilateral character of the Treaty. Each state will now have a national ceiling, whereas states with territory in the Treaty area of application will also have a territorial ceiling limiting the total amount of equipment that can be on their soil. The possibility of collective large-scale force concentrations in one state, which are threatening to a neighbouring state, will be considerably diminished.

The Agreement on Adaptation will enter into force after it has been ratified by all 30 signatories. The CFE Treaty will then only exist in its amended version.

Scope of Limitations and Transparency

Parties to the CFE set their initial national limits with an understanding that they would maintain only such military capabilities as are commensurate with their legitimate security interests. The limits cover all the five categories enumerated in the original version of the Treaty, and the ceilings do not exceed the maximum national levels for holdings (MNLH) notified under the Treaty. There are sub-ceilings for active units and for certain sub-categories of armaments. As mentioned above, only states with territory in the ATTU zone have territorial ceilings; conse-quently, the United States and Canada do not have them. The territorial ceilings enable parties to host foreign forces, but the Agreement on Adaptation provides that the TLE of a party may be present on the territory of another party only in conformity with international law, the explicit consent of the host party, or a relevant resolution of the UN Security Council.

The Agreement on Adaptation makes it clear that any upward revision of the national ceiling of a state party must be compensated by a corresponding decrease. Prior notification should be given to all other parties 90 days before the revision becomes effective. Between five-yearly CFE review conferences national ceil-ings/sub-ceilings for active units may be increased by no more than 40 tanks, 60 armoured combat vehicles and 20 artillery pieces or 20% of the established national ceilings, whichever figure is greater, but in no case exceeding 150 tanks, 250 armoured combat vehicles and 100 artillery pieces. For combat aircraft and attack helicopters the upward revision numbers are 30 and 25, respectively. Increases of national ceilings/sub-ceilings for active units in excess of the permitted levels are to be subject to a consensus decision of the parties. The rule and parame-ters for upward revisions of territorial ceilings are similar to those for national ceil-ings. Peace missions mandated by the United Nations or the OSCE are exempted from the territorial ceilings/sub-ceilings of a party on whose territory TLE necessary for the given mission is present. Armaments and equipment in transit within the ATTU zone are exempted from the territorial ceilings/sub-ceilings of the transited parties as well.

Whereas the original text only requires annual reports on the designated peace-time location of tanks, armoured combat vehicles and artillery, the amended text adds annual reporting requirements on the actual location of this TLE. Each state is also required to submit quarterly reports detailing the numbers and actual territorial deployments of its ground TLE.

Military Exercises and Temporary Deployments

Each party has the right to host military exercises on its territory. The number of ground TLE items in excess of the territorial ceiling/sub-ceiling for a military exer-cise must not exceed the number of tanks, armoured combat vehicles and artillery pieces specified for temporary deployment for each state. Temporary deployment in excess of territorial ceilings may be either 'basic', that is, up to the equivalent of one brigade (up to 153 tanks, 241 armoured combat vehicles and 140 artillery pieces) or, for 'exceptional circumstances', up to three brigades (459 tanks, 723 armoured combat vehicles and 420 artillery pieces). Explanatory reports are to be submitted to the JCG.

The duration of temporary deployments is not limited. If a military exercise exceeding a territorial ceiling/sub-ceiling is to last more than 42 days, all relevant information must be provided by the party whose ceiling has been exceeded and by the parties participating in the territorial ceiling/sub-ceiling.

The Flank Issue

One of the chief reasons for the adaptation of the CFE Treaty was the controversy between Russia and NATO over the flank issue. Russia considered that the flank provisions of the original Treaty, which imposed additional restrictions on the deployment of military forces on its territory, were discriminatory. It demanded their suspension or removal to protect its security interests in the southern part of the country, mainly in the Caucasus. NATO argued that the flank regime should remain an integral part of the Treaty and retain its legally binding character, so as not to compromise the security interests of any state.

In January 1999 an agreement was achieved with regard to the southern flank region. Turkey, which had most adamantly opposed relaxation of the flank limits, eventually agreed to allow 2,140 armoured combat vehicles in Russia's revised flank areas, whereas Russia agreed to reduce its holdings in Georgia and Moldova. This 'deal' facilitated the settlement of the overall flank issue. In March 1999 the JCG adopted the 'principles and modalities' to guide the 'maintenance and reconciliation' of the substance of the modified flank provisions. The principles included the prevention of a build-up of forces; the allowance for upward revision of the relevant territorial ceilings and sub-ceilings only through transfers among the flank states; and an enhanced regime of verification and information exchange. The modalities prescribed, among others, the subordination of Russian forces in other countries to general rules regarding national ceilings, territorial ceilings and temporary deployments. Almost all these principles and modalities found their way into the Agreement on Adaptation without direct reference to flank zones.

The Final Act

To emphasize the CFE Treaty's declared aims, Belarus, the Czech Republic, Hungary, Poland, Slovakia and Ukraine stated that their national and territorial ceilings would equal their MNLH. The new NATO members – the Czech Republic, Hungary and Poland – went further, pledging additional future reductions of their territorial ceilings.

Russia reciprocated by pledging that it would show 'due restraint' in tank, armoured combat vehicle and artillery deployments in the region encompassing the Kaliningrad oblast (district), which borders Poland, and in the Pskov oblast, which borders the Baltic states. Echoing the NATO commitment, made in the May 1997 NATO–Russia Founding Act, regarding the non-permanent stationing of substantial ground and air combat forces on the territory of the Czech Republic, Hungary and Poland, Russia said that in the 'present politico-military situation' it had no reasons, plans or intentions to station substantial additional combat forces, whether air or ground forces, in the Kaliningrad and Pskov oblasts on a permanent basis.

In its southern flank, Russia pledged to reduce its TLE holdings in Georgia. Moldova renounced its right to host any temporary deployment. The parties welcomed Russia's commitment to withdraw or destroy by the end of 2001 all of its TLE stationed in Moldova.

Assessment

The declared objective of the CFE Treaty was to eliminate those disparities in conventional forces in Europe which were most prejudicial to strategic stability. This objective has been achieved through a combination of legally and politically binding instruments. Owing to the deep, verified cuts in the non-nuclear arsenals, no state or group of states has today the capability to launch a surprise armed attack in Europe. The proponents of the idea of non-offensive defence (also referred to as non-provocative or non-aggressive defence) see in the Treaty a recognition that the security interests of states can be better met by maximizing defensive and minimizing offensive military options. The constraints and the established mechanisms for regular exchanges of relevant information and for checking compliance with the assumed undertakings strengthen confidence between former enemy states and render their military activities more transparent and predictable. It has also helped to lessen, at least to some degree, Russian concerns about the enlargement of NATO.

Considering the changed political situation, the ceilings for both armaments and military personnel could have been set considerably lower. In fact, the Treaty provisions have to some extent been overtaken by unilateral arms reductions and troop withdrawals. It is, however, regrettable that naval forces are not covered by the Treaty.

The Agreement on Adaptation of the CFE Treaty marked a success for European security because it reinforced the partnership relations among the European states. It has the potential to become a pan-European regime.

Arms control efforts continue in the Forum for Security Co-operation (FSC) of the OSCE. However, the mandate of the FSC does not provide enough room for consideration of further significant quantitative and qualitative arms limitations applicable to all European states.

14.4 The 1996 Florence Agreement

On 21 November 1995, as a result of negotiations initiated by the United States and conducted at the US airbase at Dayton, Ohio, USA, the Republic of Bosnia and Herzegovina, the Republic of Croatia and the Federal Republic of Yugoslavia concluded the General Framework Agreement for Peace in Bosnia and Herzegovina. This Framework Agreement (known as the Dayton Agreement), containing in its Annex 4 a new constitution for Bosnia and Herzegovina, was to end the armed conflicts generated by the disintegration of Yugoslavia. It was signed in Paris on 14 December 1995.

Article IV of Annex 1-B to the Dayton Agreement contains an obligation of the parties to negotiate 'balanced and stable defence force levels at the lowest numbers consistent with their respective security'. On 14 June 1996, in implementation of this obligation, Bosnia and Herzegovina and its two entities, the Federation of Bosnia and Herzegovina and the Republika Srpska, as well as the Republic of Croatia and the Federal Republic of Yugoslavia, signed in Florence, Italy, under the

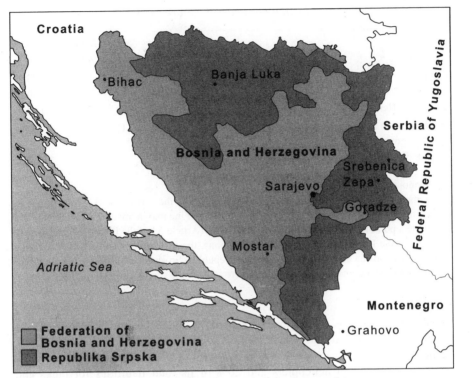

Figure 14.1   *Area of Application of the Dayton Agreement*

auspices of the OSCE, the Agreement on Sub-Regional Arms Control. However, the Florence Agreement does not meet the criteria for what is usually termed an 'arms control agreement' (see Chapter 1), since – like the Dayton Agreement itself – it was imposed from outside on the former belligerents.

Arms Limitations

The Florence Agreement, modelled on the CFE Treaty, set numerical ceilings for major weapon systems, namely, for battle tanks, armoured combat vehicles, artillery pieces of 75 mm and above, combat aircraft and attack helicopters of the parties in a ratio (based on the approximate ratio of populations) of 5 : 2 : 2 for Yugoslavia (Serbia and Montenegro), Bosnia and Herzegovina, and Croatia, respectively, and a ratio of 2 : 1 for Bosnia and Herzegovina's entities – the (Muslim–Croat) Federation of Bosnia and Herzegovina and the Republika Srpska, which maintained their own armies. In separate statements each party declared limitations on the personnel of its armed forces as of 1 September 1996. Within the agreed limits, the parties were free to structure, equip and train their forces as they chose.

Arms Reductions and Other Provisions

The Florence Agreement prescribed the following methods for arms reductions: destruction, conversion to non-military purposes, use for static display or instruc-

Table 14.2 *Limitations under the Florence Agreement*

Party	Tanks	ACVs[a]	AIFVs[b]	Artillery	Air-craft	Heli-copters	Man-power[c]
Yugoslavia	1,025	850	152	3,750	155	53	124,339
Croatia	410	340	76	1,500	62	21	65,000
Bosnia and Herzegovina	410	340	76	1,500	62	21	60,000
Federation of Bosnia and Herzegovina	273	227	38	1,000	41	14	55,000
Republika Srpska	137	113	38	500	21	7	56,000

[a] Armoured combat vehicles.

[b] Armoured infantry fighting vehicles are not limited by the agreement. AIFVs assigned to peacetime internal security forces, however, in excess of the maximum agreed numbers, shall constitute a portion of the permitted levels for ACVs (Article XI of the Florence Agreement).

[c] Unlike the figures in the columns for weapons, the manpower figures declared by the Federation of Bosnia and Herzegovina and the Republika Srpska do not add up to the total for Bosnia and Herzegovina since both entities were reluctant to reintegrate their military forces.

tional purposes, use as ground targets, reclassification and export. Reductions were to be carried out in two phases and completed within 16 months. By 1 January 1997, at the end of phase I, each party was to have reduced 40% of its total liabilities for artillery, aircraft and helicopters and 20% of its total liabilities for tanks and armoured combat vehicles. By the end of phase II each party was to have reduced all agreement-limited equipment in each of the five categories. The limits on holdings are subject to a verification regime similar to that under the CFE Treaty. A Commission was established to act as an implementation review body and provide a mechanism for the parties to work out differences that might arise in the course of implementation.

The following protocols were incorporated in the Agreement as its integral parts: the Protocol on Reduction; the Protocol on Procedures governing the Reclassification of Specific Models or Versions of Combat-capable Trainer Aircraft into Unarmed Trainer Aircraft; the Protocol on Exchange of Information and Notifications; the Protocol on Existing Types of Armaments; the Protocol on Inspection; and the Protocol on the Sub-Regional Consultative Commission.

The Florence Agreement was linked to the so-called 'train-and-equip' programme, under which the United States undertook to provide military support for the armed forces of the Federation of Bosnia and Herzegovina. This programme was criticized by many European politicians as leading to remilitarization of the Federation and perpetuating the partition of Bosnia and Herzegovina. However, at the time of the signature of the Agreement there were considerable discrepancies between the military capabilities of the Republika Srpska, which possessed armaments in excess of the agreed levels, and the military capabilities of the Federation of Bosnia and Herzegovina, which possessed armaments well below the agreed levels. In 1999, after the start of NATO's bombing of Yugoslavia, the government of the Federal Republic of Yugoslavia suspended the implementation of the Florence Agreement, but resumed it a few weeks after the withdrawal of its armed forces from Kosovo. A temporary suspension also took place in 2000, at the time of international ostracism of the Yugoslav President Milosevic's regime.

Annex 1-B of the Dayton Agreement contains the Agreement on Regional Stabilization, which in Article II provides for the following CSBMs in Bosnia and Herzegovina: restrictions on military deployments and exercises in certain geographical areas; restraints on the reintroduction of foreign forces; restrictions on locations of heavy weapons; withdrawal of forces and heavy weapons to cantonment/barracks areas or other designated locations; notification of disbandment of special operations and armed civilian groups; notification of certain planned military activities; identification and monitoring of weapons manufacturing capabilities; exchange of data on the holdings of the five weapon categories as defined in the CFE Treaty; and establishment of military liaison missions between the chiefs of the armed forces of the Federation of Bosnia and Herzegovina and the Republika Srpska.

The Dayton Agreement, tolerating the existence of three armies in Bosnia and Herzegovina, cannot guarantee lasting stability in the Balkans. The separatist movement of Croats of the Federation of Bosnia and Herzegovina, as well as the secessionist rebellions of the ethnic Albanians of Kosovo and of the Former Yugoslav Republic of Macedonia (FYROM), cast doubts on the future of the Dayton Agreement, especially its arms limitation clauses.

14.5 Arms Limitations in Latin America

Attempts to bring about limitations on armaments in Latin America have a long history.

The 1923 Convention Regarding Central America

The Convention for the Limitation of Armaments, adopted at the Conference on Central American Affairs, was signed in February 1923 and entered into force on 24 November 1924. Considering their population, area, extent of frontiers and other factors of military importance, the five Central American states agreed that, except in case of civil war or impending invasion by another state, their standing armies and national guards would not exceed certain, rather low, levels. These levels were as follows: 5,200 men for Guatemala, 4,200 for El Salvador, and 2,000–2,500 for the remaining three countries – Costa Rica, Honduras and Nicaragua. Each state was allowed to possess up to ten aircraft but was not allowed to acquire war vessels. The parties agreed not to export or permit the exportation of arms or munitions from one state to another. They also agreed to furnish one another with full reports on the measures adopted for the execution of the Convention. (In 1953 Honduras denounced the Convention, but it remains in force for the other parties.)

The 1974 Declaration of Ayacucho

In the 1974 Declaration of Ayacucho, the six members of the Andean Group, created in 1969 for the purpose of sub-regional economic integration (Bolivia, Chile, Colombia, Ecuador, Peru and Venezuela), plus two non-members (Argentina and

Panama), undertook to create conditions permitting an effective limitation of armaments and putting an end to their acquisition for offensive purposes. The stated aim
of these measures was to devote all possible resources to the economic and social
development of the countries of Latin America.

After the signing of the Declaration of Ayacucho, several consultative meetings of
the Andean countries took place with a view to translating its provisions into an
internationally binding instrument. In 1978 a conference was convened in Mexico
City, the first of this kind in the history of Latin America, to deal exclusively with
the problem of conventional arms control in the region. This conference was
attended by representatives of Argentina, Bolivia, Colombia, Costa Rica, Cuba, the
Dominican Republic, Ecuador, El Salvador, Guatemala, Haiti, Honduras, Jamaica,
Mexico, Nicaragua, Panama, Peru, Suriname, Trinidad and Tobago, Uruguay and
Venezuela. The participants recommended, *inter alia,* the initiation of studies and
talks concerning possible limitations on the transfer of certain types of conventional
armaments to Latin America and among the countries in the area, as well as limitations or prohibitions on conventional weapons considered to be excessively injurious or indiscriminate in their effects.

The 1985 Contadora Act

In 1985 a group of Latin American countries – Colombia, Mexico, Panama and
Venezuela – known as the Contadora Group (after the Panamanian island where it
first met in 1983) put forward proposals for stopping and reversing the militarization
of the Central American Isthmus. The proposed Contadora Act on Peace and Cooperation in Central America contained the following arms control stipulations.

The parties would be required to provide advance notification of national military
manoeuvres held in areas less than 30 kilometres from the territory of another state.
International manoeuvres – those involving armed forces of two or more countries
on the territory of one country or in an international area – would also require prior
notification but would eventually be prohibited. Observers would be invited to both
national and international manoeuvres.

A set of criteria would have to be observed in establishing limits on military
developments in Central America. These criteria, of potential application also in
other regions, included: internal and external security needs; area; population; distribution of economic resources; extent and characteristics of land and sea boundaries; military expenditure as a proportion of gross domestic product; military budget in relation to public expenditure; and level of military technology. The Contadora Act also contained bans on the introduction of new weapon systems that would
alter the quality and quantity of current inventories of *matériel,* as well as on the
introduction, possession or use of weapons which are excessively injurious or produce indiscriminate effects. The parties would undertake to stop all illegal flows of
arms – meaning the transfer by governments, individuals or regional or extra-
regional groups of weapons intended for irregular forces or armed bands seeking to
destabilize governments in the region. The Contadora Act could have provided an
equitable basis for a peaceful settlement of the Central American problems. It failed,
however, to obtain the approval of the main protagonists.

The 1999 Inter-American Convention

On 7 June 1999 the General Assembly of the Organization of American States (OAS) approved the text of the Inter-American Convention on Transparency in Conventional Weapons Acquisitions. The stated objective of this Convention is to contribute to regional openness and transparency in the acquisition of conventional weapons by exchanging information regarding such acquisitions for the purpose of promoting confidence among states in the Americas.

The list of weapons covered by the Inter-American Convention includes battle tanks, armoured combat vehicles, large-calibre artillery systems, combat aircraft, attack helicopters, warships, missiles and missile launchers. The term 'acquisition' is defined as obtaining conventional weapons through purchase, lease, procurement, donation, loan or any other method, whether from foreign sources or through national production.

The parties to the Convention committed themselves to report annually to the Depositary (the General Secretariat of the OAS) on their imports and exports of conventional weapons during the preceding calendar year. In addition, the parties must notify the Depositary of their acquisitions of conventional weapons, through imports or through national production, no later than 90 days after incorporation of the acquired weapons into the inventory of the armed forces. Any state that is not a member of the OAS may contribute to the objective of this Convention by providing information annually to the Depositary on its exports of conventional weapons to the parties.

The 2002 Lima Commitment

On 17 June 2002 the ministers for foreign affairs and defence of Bolivia, Colombia, Ecuador, Peru and Venezuela signed the Lima Commitment, establishing the Andean Charter for Peace and Security and for the Limitation and Control of External Defence Spending. The member-states of the Andean Community undertook, *inter alia*, to: work out a common policy of Andean security; establish a zone of peace in the area of the Andean Community; take necessary measures to combat terrorism; limit defence expenditures, prohibit or restrict the use of certain conventional weapons and promote transparency in armaments; work towards declaring a zone free of strategic missiles in Latin America; strengthen the bans on nuclear, chemical and biological weapons; destroy all stocks of anti-personnel mines; and eradicate the illicit traffic in firearms, munitions, explosives and related materials. The signers of the Lima Commitment appealed to other governments to adhere to the Andean Charter.

14.6 The 1997 Anti-Personnel Mines Convention

The first category of conventional weapons whose possession has been prohibited under a multilateral treaty is the anti-personnel mine (APM). The reason for the special attention devoted to APMs lies in the fact that they are particularly cruel and directly affect civilian populations.

APMs kill or inflict wounds that usually result in surgical amputation. Survivors require extended hospital stays and prolonged rehabilitation treatment. The amputees need prosthetic devices to lead a normal life, and some are so disfigured that they need psychological counselling to cope with the trauma. Landmines, which are designed to attack persons and vehicles, render whole regions unsuitable for human habitation, deny cropland to farmers and impede safe repatriation of refugees and displaced persons long after the cessation of hostilities.

Owing to their simple design, low production cost (in certain countries the cost is as low as US$3 per mine) and long shelf life, APMs are easily available, whereas the cost of mine-clearing operations is exorbitant (US$300–1,000 per mine) and mine clearers are often seriously injured or killed. APMs are used on a large scale by regular and irregular forces as offensive weapons. They may also have a defensive role when used to protect national borders or vital installations, or as an impediment to the deployment of enemy troops in locations advantageous to an attacker. However, their military utility is outweighed by the humanitarian impact on civilians.

Legal restrictions on the use of APMs have been in force since the entry into force of the 1981 Inhumane Weapons Convention but have proved insufficient (see Chapter 17.2). Under the pressure of non-governmental organizations, especially the International Committee of the Red Cross (ICRC) and the International Campaign to Ban Landmines (ICBL), subsequently awarded a Nobel Peace Prize, several governments promulgated moratoria or bans on exports of APMs. Some governments went even further by stopping their production of APMs and starting the destruction of their stocks. However, unilaterally contracted obligations have a limited value; they do not carry the force of international law and can be easily reversed. This was recognized in the 1996 UN General Assembly resolution which urged states to pursue an effective, legally binding international agreement to ban the use, stockpiling, production and transfer of anti-personnel landmines.

The realization that nothing short of outlawing APMs would suffice prompted the Canadian government to sponsor a conference to discuss the issue. This conference, held with the support of some 50 governments, adopted, in October 1996, the Ottawa Declaration committing the participants to carry out a plan of action to ensure that a treaty banning APMs be concluded at the earliest possible date. The Canadian Foreign Minister invited all governments to come to Ottawa to sign the treaty. This is how the 'Ottawa Process' for the abolition of APMs was set in motion. States which were opposed to a comprehensive ban on APMs or saw it only as a distant goal were opposed to the Process. They wanted to discuss the matter piecemeal (beginning with a prohibition on the transfer of APMs) at the Conference on Disarmament, where – as experience had shown – negotiations on any subject can go on for years, and where, because of the requirement of consensus, any participant can block progress. These attempts to derail the Ottawa Process failed.

The follow-up to the 1996 Ottawa conference took place in June 1997 in Brussels, where 97 governments signed the Brussels Declaration launching formal negotiations for a comprehensive APM ban. The negotiations were conducted at the Diplomatic Conference in Oslo and were successfully concluded in September 1997.

The Convention on the Prohibition of the Use, Stockpiling, Production and Transfer of Anti-Personnel Mines and on their Destruction (APM Convention) was opened for signature on 3 December 1997. It entered into force on 1 March 1999.

Main Obligations

The APM Convention prohibits the use of APMs as well as their development, production, acquisition by other means, stockpiling and retention or transfer (involving not only the physical movement of mines into or from national territory but also the transfer of title to and control over mines) to anyone, directly or indirectly. Assistance, encouragement or inducement to any prohibited activity is equally prohibited. The prohibitions are valid 'under any circumstances', which means that they are valid in time of peace and in time of armed conflict, whether international or non-international (including internal disturbances), both for offence and for defence. Thereby, APMs have been declared illegitimate weapons.

Each party is under the obligation to destroy or ensure the destruction of all stockpiled APMs which it possesses, or which are under its jurisdiction or control, not later than four years after entry into force of the Convention for that party. APMs in mined areas under the jurisdiction or control of the parties (that is, also in occupied territories) are to be destroyed within ten years, but in the meantime they must be perimeter-marked, monitored and protected by fencing or other means to ensure an effective exclusion of civilians. An extension of the deadline to complete the destruction of APMs in mined areas may be requested for a period of up to ten years. (Indeed, for certain poor mine-affected countries, such as Afghanistan, Angola, Cambodia or Mozambique, ten years may not suffice.) Each such request must contain a detailed explanation of the reasons for the proposed extension and specify its humanitarian, social, economic and environmental implications. The request may be granted at the Meeting of States Parties or at the Review Conference (see below) by a majority of parties present and voting. Further extension may be granted upon the submission of relevant additional information.

The APM Convention prohibitions cover mines designed to be exploded by the presence, proximity or contact of a person and that will incapacitate, injure or kill one or more persons. Mines designed to be detonated by the presence, proximity or contact of a vehicle as opposed to a person, that are equipped with anti-handling devices, are not considered anti-personnel mines as a result of being so equipped. 'Anti-handling device' is defined as a device intended to protect a mine and which is part of, linked to, attached to or placed under the mine and which activates when attempt is made to tamper with or otherwise intentionally disturb the mine. Some people interpret the above proviso as allowing the protection of anti-vehicle mines with devices producing the same effects as anti-personnel mines. The fact that the term 'vehicle' has not been defined could facilitate the circumvention of the ban. However, in the opinion of the ICRC, any mine – whatever its primary purpose – capable of being detonated by the innocent passage of a person over or near the mine or through inadvertent or accidental contact with the mine itself, is *eo ipso* an anti-personnel mine prohibited by the APM Convention.

Each party must take national implementation measures – legal, administrative and others – including the imposition of penal sanctions to prevent and suppress any prohibited activity undertaken by persons or on territory under its jurisdiction or control.

Exceptions

The APM Convention permits the retention or transfer of a 'number' of APMs for the development of and training in mine detection, mine clearance or mine destruction techniques. The number has not been specified. It is only stated that it shall not exceed the minimum number absolutely necessary for the purposes mentioned above. Transfer of APMs for the purpose of destruction is also permitted, again without restrictions on numbers. These are serious gaps. However, the obligation of the parties to file reports with the UN Secretary-General, the Depositary of the Convention, on mines retained or transferred may reduce the risks of abuse.

Cooperation and Assistance

The parties have undertaken to facilitate and have the right to participate in the exchange of equipment, material, and scientific and technological information concerning the implementation of the Convention. Those in a position to do so should provide assistance (through various international channels or on a bilateral basis) for the care and rehabilitation and social and economic reintegration of mine victims and for mine awareness programmes. Assistance for mine clearance and for the destruction of stockpiled APMs is also envisaged. Parties may request the United Nations, regional organizations, other parties or other competent intergovernmental or non-governmental bodies to assist them in the elaboration of national demining programmes.

Meetings of Parties

The nature of the banned weapon – its small size and the ease with which it can be manufactured – made it impossible for the drafters of the APM Convention to provide for a monitoring and inspection system as elaborate as the verification mechanism of certain other treaties. However, the parties are to meet regularly to consider matters regarding the application or implementation of the Convention. A review conference is to be convened five years after entry into force of the Convention (2004). Further such conferences may be held if so requested by one or more parties, provided that the interval between them is in no case less than five years.

An Amendment Conference must be convened if a majority of the parties agree to consider a proposal for an amendment. Any amendment can be adopted at such a conference by a majority of two-thirds of the parties present and voting. It will enter into force for all parties which have accepted it upon the deposit with the Depositary of instruments of acceptance by a majority of parties. Thereafter, it will enter into force for any remaining party on the date of deposit of its instrument of acceptance. The Convention can thus be adapted to new situations and, in particular, to new technologies.

States not parties to the Convention, as well as the United Nations, other intergovernmental organizations or institutions, the ICRC and other relevant non-governmental organizations may be invited to attend the Meetings of the States Parties, Special Meetings of the States Parties, Review Conferences and Amendment Conferences as observers.

Final Clauses

The APM Convention entered into force on the first day of the sixth month after the month in which the 40th instrument of ratification, acceptance, approval or accession had been deposited. This delay for the entry into force of the Convention is also applicable to states depositing the relevant instrument at a later date.

The articles of the Convention are not subject to reservations. The Convention is of unlimited duration, but each party has the right to withdraw from it. The notification of withdrawal, to be transmitted to other parties, to the UN Secretary-General and to the UN Security Council, must include an explanation of the reasons motivating such action, and the withdrawal itself may take effect six months after the receipt of the instrument of withdrawal by the UN Secretary-General. Such a relatively easy exit from the Convention would not be permitted if, on the expiry of the six-month period, the withdrawing state were engaged in an armed conflict; the withdrawal could not take effect before the end of the conflict. Otherwise, the treaty's protection would lapse precisely when it was most needed.

Assessment

The conclusion of the APM Convention confirmed the widely held belief that mere restrictions on the use of weapons – nearly always hedged with exceptions or reservations – are not good enough, and that they are useful only in so far as they constitute a prelude to an unconditional ban on use and lead to the abolition of the banned weapons. In this respect, the APM Convention followed the example of the 1972 Biological Weapons Convention (BWC) and the 1993 Chemical Weapons Convention (CWC). In several other respects, however, the APM Convention is unique.

Unlike the BWC or the CWC, the APM Convention does away with weapons that have been in widespread use, both in international and internal conflicts. It contains an obligation, unprecedented for an arms control treaty, to provide assistance for the care and rehabilitation of victims. Furthermore – again unlike the BWC or CWC – the APM Convention resulted from negotiations carried out in a record time of less than a year by a group of like-minded nations outside the Conference on Disarmament, and it entered into force in spite of the opposition of China, Russia and the United States.

Non-governmental organizations have played a key role in the generation of popular support for a ban on APMs. After entry into force of the APM Convention, the ICBL started publishing the *Landmine Monitor Report* to monitor the implementation of and compliance with the Convention, and to assess the efforts of the international community to resolve the landmine crisis. The Monitor complements the parties' reporting required by the Convention. A new type of multilateral diplomacy has emerged based on a partnership between state authorities and the civil society.

The APM Convention has shortcomings as well. As noted above, its articles dealing with definitions and exceptions may give rise to divergent interpretations affecting its implementation. There are no provisions regulating anti-vehicle mines, which are frequently used indiscriminately along roads and railways, killing and wounding civilians, impeding the delivery of humanitarian assistance and rendering it difficult to reconstruct war-devastated areas. Although a procedure is envisaged to clarify suspicions of breaches, no organization has been set up to oversee the operation of the Convention on a continuous basis; meetings of the parties and the assistance to be provided by the UN Secretary-General may not suffice to ensure compliance.

The above shortcomings can be removed or attenuated at the appropriate time through the amendment procedure envisaged in the Convention, or through agreed understandings worked out at review conferences. Nevertheless, to become fully effective, the Convention must be adhered to by all states, or at least by all the major producers, exporters and users of mines. Achieving the desirable degree of universality will probably take a long time, but the pressure of world and domestic public opinion on the hold-out governments may prove irresistible. For, from the humanitarian point of view, elimination of APMs is a necessity, whereas, from the military point of view, the value of APMs is considered by many as marginal. Even when used on a massive scale, APMs have usually little or no effect on the outcome of hostilities. For the purposes of defence there exist alternatives to APMs; protective fences in combination with electronic sensing devices for early warning are often mentioned as a possibility. Creating zones free of APMs in the mine-affected parts of the globe could facilitate observance of the global ban. The growing stigma on the use of APMs may discourage their deployment by non-parties to the Convention.

The most urgent task for the international community is to strengthen and accelerate the demining operations in the most heavily mined countries and to provide assistance to mine victims. Several governments have pledged to devote funds or other resources for these purposes. Switzerland established an international centre for humanitarian demining. The United Nations has assumed the role of coordinator of the relevant activities.

At the first Meeting of the States Parties, held in 1999, the participants started to put the Convention into practice. They agreed on a format for annual country reports and set out a work programme for meetings of standing committees of experts on: mine clearance; victim assistance and mine awareness; stockpile destruction; and the operation of the Convention. At the second and third Meetings of the States Parties, held in 2000 and 2001, respectively, the participants expressed concern about the continued use of APMs but welcomed the progress made in the implementation of the Convention. They noted that large areas of mined land had been cleared, that mine casualty rates had dropped in several of the world's most mine-affected states and that assistance to mine victims had improved.

15

Constraints on Conventional Arms and Technology Transfers

The attempts to control international transfers of arms prior to World War II failed (see Chapter 2). It was only during the 1960s, when the United Nations contemplated restrictive measures with regard to transfers of both weapons of mass destruction and conventional weapons, that these attempts resumed. Weapons of mass destruction subsequently became the subjects of bilateral and multilateral negotiations aimed at formally prohibiting their proliferation (see Chapters 6 and 8), whereas conventional weapons proved difficult to control. Many states, including the developing countries, continued unimpeded to trade in conventional arms of ever-increasing sophistication and destructiveness or spread them by other means.

The wars waged on the African and Asian continents in the 1970s and 1980s, especially the Iraqi invasion of Kuwait in 1990, demonstrated the nefarious consequences of excessive accumulation of armaments. Once again, the idea of controlling the flow of arms among nations through openness and transparency in deliveries became the subject of intensive debates, which are reviewed in this chapter. (Arms embargoes imposed by the United Nations or other international bodies are not discussed here.)

The international transfer of arms covers imports and exports – under terms of grant, credit, barter or cash – of military equipment, including weapons of war, parts thereof, ammunition, support equipment and other items designed for military use. 'Arms transfer' can cover dual-use equipment having both military and civilian application, when its primary mission is identified as military. Military transfer agreements may also include the building of defence production facilities and licensing fees paid as royalties for the production of military equipment.

15.1 The CAT Talks

In 1977–78 the United States and the Soviet Union – at that time the leading arms suppliers – were engaged in bilateral talks on conventional arms transfers (CAT). One incentive for initiating these talks may have been the fear that an unconstrained flow of arms could lead to an unwanted armed confrontation between the two superpowers.

Divergent Positions

The United States aimed at developing norms of restraint that would include such undertakings as: no-first-introduction of advanced weapon systems into a region; restrictions on co-production and retransfer; development of norms for recipient restraint; establishment of consultative mechanisms to enhance the exercise of restraint; integration of restraint efforts with diplomatic efforts to resolve regional disputes; and reduction of possibilities for the substitution of suppliers where others have exercised restraint. The discussion was to focus on Latin America and sub-Saharan Africa, the two regions where superpower competition was practically non-existent.

The Soviet Union stressed its determination to continue supporting liberation movements with armaments but suggested that limits should be introduced on arms sales to racist regimes, states recognized as aggressors, states conducting militaristic policies, states having unjustified territorial claims and states rejecting disarmament efforts.

Reasons for Failure

The CAT talks collapsed ostensibly because of regional approaches to arms transfer control and, especially, because of the Soviet focus on West Asia and East Asia – regions of particular interest to the United States. In fact, however, the very state of US–Soviet relations at that time made it unlikely that the two powers would restrain the use of such an important tool as arms transfers in their competition for influence in the rest of the world. Moreover, an arms transfer limitation regime would have required the cooperation of the major West European suppliers. This would have been difficult to achieve, because the European states exported a considerably larger proportion of their armaments production than did the superpowers and feared that multilateral export restrictions would threaten their defence industries.

15.2 Guidelines, Principles and Codes of Conduct

Only many years later, after the 1991 Gulf War had revived concerns over international arms transfers, were all the five great powers ready to discuss ways to control transfers of conventional arms.

The 1991 Five-Powers Communiqué

Meeting in Paris on 8–9 July 1991, the permanent members of the UN Security Council issued a communiqué acknowledging that the transfer of conventional weapons, when conducted in a responsible manner, should contribute to the ability of states to meet their legitimate defence, security and national sovereignty requirements, and that it should enable them to participate effectively in collective measures requested by the United Nations for the purpose of maintaining or restoring international peace and security. At the same time, the participants in the Paris meeting recognized that indiscriminate transfers of military weapons and technology carry the risk of regional instability. They said that they were conscious of their special responsibility for ensuring that such risk is avoided and of the special role they were called upon to play in promoting confidence and transparency in this field. They were also aware that close consultation with recipient countries was needed.

The five powers pledged to observe restraint in conventional arms transfers, They also expressed the intention to develop modalities of consultation and information exchanges concerning arms transfers to the region of the Middle East.

The 1991 Five-Powers Guidelines

At a follow-up meeting, held in London on 17–18 October 1991, the permanent members of the UN Security Council adopted the Guidelines for Conventional Arms Transfers aimed at creating a 'serious, responsible and prudent attitude of restraint'. They undertook to consider carefully whether the conventional arms transfers which they envisaged would promote the recipients' capabilities to meet the needs for legitimate self-defence; whether they would serve as an appropriate and proportionate response to the threats confronting the recipients; and whether they would enhance the capability of the recipients to participate in collective arrangements consistent with the UN Charter.

They agreed to avoid transfers likely to prolong or aggravate an existing armed conflict, increase tension or introduce destabilizing military capabilities in a region, contravene embargoes or other relevant internationally agreed restraints, be used for purposes other than the security needs of the recipient, such as the support or encouragement of international terrorism, or be used to interfere with the internal affairs of sovereign states or seriously undermine the recipient's economy.

The five powers expressed the hope that other arms-exporting countries would adopt similar guidelines. As a matter of priority, they committed themselves to inform each other about transfers to the Middle East of tanks, armoured combat vehicles, artillery, military aircraft and helicopters, naval vessels, and certain missile systems.

The interpretation of the Guidelines was left to each state concerned. There is no procedure by which one government could question a specific export deal concluded by another government.

OSCE Principles

In a declaration of 1 December 1993, the members of the Organization for Security and Co-operation in Europe (OSCE) reaffirmed the principles governing their conventional arms transfers, namely: the need to ensure that arms transferred are not used in violation of the UN Charter; adherence to the requirement of transparency and restraint in the transfer of conventional weapons and related technology; the belief that excessive and destabilizing arms build-ups pose a threat to national, regional and international peace and security; the need for effective national mechanisms to control the transfer of conventional arms and related technology; and the commitment to provide data and information, as required by the UN resolution that established the Register of Conventional Arms (see below).

To further the aim of a cooperative and common approach to security, the OSCE members undertook, in considering transfers of conventional arms and technology, to take into account, among others, the respect for human rights and fundamental freedoms in the recipient country; the record of compliance by the recipient country with international commitments; the nature and cost of the arms to be transferred in relation to the circumstances of the recipient country; whether the transfers would contribute to an appropriate and proportionate response by the recipient country to the security threats confronting it; the legitimate domestic security needs of the recipient country; and the requirement that the recipient country should be able to participate in international peacekeeping measures.

UN Guidelines

On 3 May 1996 the UN Disarmament Commission adopted the Guidelines for International Arms Transfers. They included, to a large extent, the principles adopted earlier by some non-UN bodies, but they also identified ways and means of dealing with the problems of transfers.

In particular, at the national level the UN Guidelines stipulate that states should ensure that they have an adequate system of laws and/or regulations and administrative procedures to exercise effective control over the export and import of arms, in order, among other goals, to prevent illicit arms trafficking. States should scrutinize their national arms control legislation and procedures and, where necessary, increase their effectiveness in preventing the illegal production, trade in and possession of arms on their territory. They should intensify their efforts to prevent corruption and bribery in connection with the transfer of arms and make efforts to identify, apprehend and bring to justice those involved in illicit arms trafficking. They should establish and maintain an effective system of export and import licences for international arms transfers with the requirement for full supporting documentation. They should seek to obtain an import certificate from the receiving state covering the exported items, whereas the receiving state should seek to ensure that a certified licence of the authorities in the supplying state covers the imported arms. They should provide for adequate numbers of customs officials adequately trained to enforce the regulations on the export and import of arms. They should define, in accordance with their national laws and regulations, which arms are permitted for civilian use and which may be used or possessed by the military and police forces. They should take into account and apply, as appropriate, the relevant recommendations of the International Criminal Police Organization (Interpol).

At the international level, all arms transfer agreements and arrangements should be designed so as to reduce the possibility of diversion of arms to unauthorized destinations and persons, the requirement of import licences or verifiable end-use/end-user certificates being especially important. States should share relevant customs information on trafficking in and detection of illicit arms and coordinate their intelligence efforts. They should strengthen international cooperation in the relevant field of criminal law and make efforts to develop and enhance the application of compatible standards in their legislative and administrative procedures for regulating the export and import of arms. They should report all relevant transactions in their annual reports to the UN Register of Conventional Arms (see below) and maintain strict regulations regarding the activities of private international arms dealers. Sanctions and arms embargoes imposed by the UN Security Council must be strictly complied with.

In its 1999 Guidelines on Conventional Arms Control/Limitation and Disarmament, the UN Disarmament Commission listed some practical disarmament measures to be taken in post-conflict situations. These included collection, control, disposal and destruction of arms, conversion of military facilities, demining, demobilization and integration of former combatants.

The EU Code of Conduct

In June 1991 and June 1992 the European Union (EU) adopted a set of common criteria for the export of conventional arms. These criteria included, among others: respect for human rights in the country of final destination; preservation of regional peace, security and stability; national security of the member-states and security of the territories whose external relations are the responsibility of a member-state, as well as the security of friendly and allied countries; behaviour of the buyer country with regard to terrorism and respect for international law; existence of a risk that the exported equipment will be diverted within the buyer country or re-exported under undesirable conditions; and compatibility of the arms exports with the technical and economic capacity of the recipient country. On the basis of these criteria the EU worked out a detailed Code of Conduct on Arms Exports and adopted it on 8 June 1998.

According to the operative provisions of the Code of Conduct, EU member-states should circulate through diplomatic channels details of licences refused together with an explanation. Before any member-state grants a licence which has been denied by another member-state or states for an essentially identical transaction within the preceding three years, it should first consult states which issued the denial. If, after consultations, the member-state nevertheless decides to grant a licence, it should notify the states that issued the denial and explain the reasons.

The decision to transfer or deny the transfer of any item of military equipment remains at the national discretion of each state. The denials and relevant consultations are to be kept confidential and not to be used for commercial advantage.

15.3 The Wassenaar Arrangement

In July 1996 representatives of 33 states met in Vienna to establish an Arrangement on Export Controls for Conventional Arms and Dual-Use Goods and Technologies with the declared purpose of contributing to regional and international security and stability. This so-called Wassenaar Arrangement (after the city in the Netherlands where the initial elements of the Arrangement were negotiated) began operating in September 1996 and replaced the Co-ordinating Committee for Multilateral Export Controls (COCOM). The latter, set up during the Cold War and disbanded in 1994, was a regime for preventing exports of high-technology weapons and dual-use material to the Soviet bloc and other communist countries. The former does not explicitly target any particular state or region. Nevertheless, the United States was reported to have secured a political commitment from the states that joined the Arrangement not to supply arms and related technologies to certain countries.

States participating in the Arrangement are expected to share intelligence on potential threats, noting in particular dubious arms acquisition trends. They are to exchange, on a voluntary basis, information that enhances transparency and assists in developing common understandings of the risks associated with the transfer of arms and of sensitive dual-use goods and technologies. On the basis of the information received they may assess the scope for coordinating national control policies to combat these risks. The decision to transfer or to deny the transfer of any item is the sole responsibility of the participating state. The Arrangement has two pillars: one dealing with arms and another dealing with dual-use goods.

Regarding the first pillar, the information to be exchanged should cover battle tanks, armoured combat vehicles, large-calibre artillery systems, military aircraft/unmanned aerial vehicles, military and attack helicopters, warships and missiles or missile systems (including remotely piloted vehicles with the characteristics of missiles, but excluding ground-to-air missiles), and specify the quantity and the name of the recipient state and – except in the category of missiles and missile launchers – the details of model and type.

Regarding the second pillar, information to be exchanged should cover the items included in the List of Dual-Use Goods and Technologies and in the Munitions List. The List of Dual-Use Goods and Technologies (tier 1) has two annexes listing sensitive items (tier 2) and a limited number of very sensitive items (sub-set tier 2). The lists are to be reviewed regularly to reflect technological developments and experience gained by the participating states, including that in the field of goods and technologies which are critical for indigenous military capabilities.

States participating in the Arrangement should promptly notify licences denied to non-participants with respect to items on the List of Dual-Use Goods and Technologies, when the reasons for denial are relevant to the purposes of the Arrangement, and exert extreme vigilance for items included in the sub-set of tier 2. They should meet periodically to take decisions regarding the Arrangement, including the review of the lists of controlled items. The reporting requirements for armoured combat vehicles, aircraft and helicopters were subsequently expanded and the List of Dual-Use Goods and Technologies was updated and amended. At the annual meeting held in December 2001, the participants in the Arrangement amended its founding document by adding a new section in which they declared that they would continue to prevent terrorist organizations from acquiring conventional arms and dual-use goods and technologies that could be used for military purposes.

The factors to be taken into consideration in deciding on the eligibility of a state for participation in the Arrangement should include whether the state is a producer/exporter of arms or relevant industrial equipment, whether it adheres to arms non-proliferation regimes and whether it applies export controls. Decisions on the admission of new members as well as on the development of further guidelines must be taken by consensus. A secretariat established in Vienna facilitates the work of the Arrangement.

The Wassenaar Arrangement – the first multilateral export control regime covering both armaments and sensitive dual-use goods and technology – does not provide for binding agreements. This is its major weakness. This renders it difficult to achieve uniformity (and thereby also solidarity) in arms export policies.

15.4 The UN Register of Conventional Arms

In the Final Document of its 1978 Special Session, the UN General Assembly urged the major arms supplier and recipient countries to consult with each other on the limitation of all types of international transfer of conventional arms. Thirteen years later, on 9 December 1991, the General Assembly decided to establish a 'universal and non-discriminatory Register of Conventional Arms'. This decision was incorporated in the resolution on transparency in armaments, adopted without a dissenting vote. It called upon member-states to exercise restraint in exports and imports of conventional arms, particularly in situations of tension or conflict, to ensure that they had in place an adequate body of laws and administrative procedures regarding the transfer of arms and to adopt measures for their enforcement.

Transparency in Armaments

In addition to data on transfers of conventional weapons, states were encouraged to provide available background information regarding their military holdings, procurement through national production and relevant policies. 'Background information' was understood to consist of White Papers, policy statements and the like, whereas the term 'available' indicated that no special reports needed to be prepared for submission to the Register. This provision was included as a partial concession to those who argued that a register recording only transfers would be discriminatory: over a period of time, it would give a fairly accurate notion about the arms inventories of the importing countries not possessing a significant indigenous arms-manufacturing industry, whereas the major arms-producing countries, not importing much, would have considerably less to report and could therefore keep the actual state of their arms inventories undisclosed. It was agreed that the operation of the Register would be reviewed with a view to its expansion.

The Guidelines and Recommendations 'for objective information on military matters', as adopted by the UN Disarmament Commission in May 1992, made it clear that the Register of Conventional Arms was to be only one element of the regime of transparency in armaments, the other being the UN Standardized System of Reporting on Military Expenditures. In the longer run, states were also expected to provide information on nuclear weapons and other weapons of mass destruction, as well as on transfers of high technology with military applications.

Structure of the Register

An annex to the UN General Assembly resolution referred to above listed seven categories of armaments for which the imports and exports were to be reported. After some adjustments, this list included: (a) battle tanks; (b) armoured combat vehicles; (c) large-calibre artillery systems; (d) combat aircraft; (e) attack helicopters; (f) warships: vessels or submarines equipped for military use with a standard displacement of 750 metric tonnes or above, and those with lesser displacement if equipped for launching missiles with a range of at least 25 kilometres or torpedoes with similar range; and (g) missiles and missile launchers: guided or unguided rockets, ballistic or cruise missiles capable of delivering a warhead or weapon of destruction within a range of at least 25 kilometres, and means designed or modified specifically for launching such missiles or rockets, if not covered by categories (a) through (f). (Remotely piloted vehicles with the characteristics for missiles as defined above are included, whereas ground-to-air missiles are not.)

The standardized forms for exports and imports, elaborated by a panel of experts, consist of two parts. One part contains columns indicating the final importer state or states, the number of items, the state of origin (if not the exporter) and the intermediate location (if any). The other part contains columns for remarks regarding the description of the reported items and for comments on the transfers. If no imports or exports of any of the relevant categories of armaments have taken place, a so-called nil report is to be filed.

The Register of Conventional Arms was established with effect from 1 January 1992. States are asked to provide data on an annual basis by 30 April each year in respect of imports into and exports from their territory in the previous calendar year. The Register is open for consultation by representatives of member-states. The UN Secretary-General was requested to provide annually to the UN General Assembly a consolidated report of the registered data together with an index of other 'inter-related' information.

Assessment

A register of international transfers of conventional arms can only marginally increase transparency in armaments. Indeed, the UN Register does not make it significantly easier for countries to assess the military potential of their adversaries. There is a wide range of weapons that need not be reported. Even the seven categories of the Register are not fully covered. Lack of qualitative information on the type and model of weapons limits the value of the Register even further. Weapon sub-systems and dual-use items, which form a significant component of the trade in arms, are left out. Information on armed forces personnel is not provided either. Nor does the Register facilitate the evaluation of the economic aspects of the transfers recorded, as the value of the transactions concluded and the mode of financing them are not to be indicated. Certain arms contracts require confidentiality and may never be revealed. And since the reported data relate to arms that have already been delivered (there is no requirement for advance notification), the Register cannot give an early warning of military build-ups, as was hoped by some.

It is widely recognized that secrecy about military matters breeds fears and distrust and must, therefore, be reduced. Hence the multiplication of guidelines and codes of conduct regarding arms transfers, as described above. The Register of Conventional Arms could, even in its present form, perform a confidence-building function, if it were made legally binding, and if all the requested data were provided within the stipulated time frame by all major arms suppliers and recipients and in sufficient detail to render the information meaningful. There is no indication that this will happen; the provision of data remains voluntary and is not subject to verification.

Even if the above requirements were met, the Register, as well as other similar arrangements – whether global or regional – would not, by themselves, necessarily lead to limitations of arms transfers. There are no objective criteria for determining which types or quantities of armaments are excessive. In certain situations the Register could even stimulate, instead of dissuading, the acquisition of arms by states eager to redress what they perceive as military imbalances.

Effective prohibitions on transfers of arms are conceivable only in conjunction with prohibitions or severe restrictions on their production. This proved to be the case with weapons of mass destruction. Transfer of conventional weapons cannot be prevented or even significantly limited as long as their manufacture remains unlimited. In other words, there is a need to control arms production as well.

15.5   The Problem of SALWs

Since the end of the Cold War ever more attention has been focused on small arms and light weapons (SALWs), the easy availability of which contributes to the outbreak, intensity and duration of armed conflicts. According to UN definitions (which are not yet generally accepted), small arms are those which can be carried by an individual for personal use and include revolvers and self-loading pistols, rifles and carbines, sub-machine guns, assault rifles, as well as light machine-guns. Light weapons are those which can be handled by two or more persons serving as a crew and include heavy machine-guns, hand-held and mounted grenade launchers, portable anti-aircraft guns, portable anti-tank guns and recoilless rifles, portable launchers of anti-tank missile and rocket systems, portable launchers of anti-aircraft missile systems, as well as mortars of calibres of less than 100mm. Ammunition and explosives, forming an integral part of SALWs, include cartridges (rounds) for small arms, shells and missiles for light weapons, mobile containers with missiles or shells for single-action anti-aircraft and anti-tank systems, anti-personnel and anti-tank hand grenades, as well as landmines.

SALWs are particularly suitable for use in intra-state conflicts (in which most victims are civilians) because of their low cost, simplicity, durability, portability and concealability. Initiatives to control them have been taken by groups of countries both globally and regionally. Measures to prevent their unregulated spread are under continuous discussion.

Global Initiatives regarding SALWs

UN Reports on Small Arms. In August 1997 the UN Secretary-General issued a report on small arms, prepared with the assistance of a Panel of Governmental Experts. The recommendations made by the Panel included the following.

The United Nations should support, with the assistance of the donor community, all appropriate post-conflict initiatives related to disarmament and demobilization, such as the disposal and destruction of weapons, including weapon turn-in programmes sponsored locally by governments and non-governmental organizations.

Guidelines to assist the negotiators of peace settlements in working out plans to disarm the combatants, particularly as concerns light weapons, small arms and munitions, should be developed and include plans for the collection of weapons and their disposal, preferably by destruction. Consideration should be given to the establishment of a disarmament component in peacekeeping operations undertaken by the United Nations.

States and regional organizations, where applicable, should strengthen international and regional cooperation among police, intelligence, customs and border control officials in combating the illicit circulation of and trafficking in SALWs and in suppressing criminal activities related to the use of these weapons.

The establishment of mechanisms and regional networks for information sharing for the above-mentioned purposes should be encouraged.

All weapons which are not required for the purpose of national defence and internal security should be collected and destroyed by states as expeditiously as possible.

States should determine, in their national laws and regulations, which arms are permitted for civilian possession and under which conditions they may be used.

States should ensure that they have in place adequate laws, regulations and administrative procedures to exercise effective control over legally possessed SALWs and over their transfer in order to prevent illicit trafficking.

States emerging from conflict should, as soon as practicable, impose or re-impose licensing requirements on all civilian possession of SALWs on their territory.

States should exercise restraint with respect to the transfer of the surplus SALWs, manufactured solely for the possession of and use by the military and police forces; they should also consider the possibility of destroying all such surplus weapons.

States should ensure the safeguarding of such weapons against loss through theft or corruption, in particular from storage facilities.

The United Nations should urge the relevant organizations, such as Interpol and the World Customs Organization, as well as all states and their relevant national agencies, to cooperate in the identification of the groups and individuals engaged in illicit trafficking activities.

The United Nations should encourage the adoption and implementation of regional or sub-regional moratoria on the transfer and manufacture of SALWs.

The United Nations should initiate studies on the feasibility of establishing a reliable system for marking all SALWs from the time of their manufacture, the feasibility of restricting the manufacture and trade of such weapons to the manufacturers and dealers authorized by states, and the feasibility of establishing a database of such manufacturers and dealers.

In August 1999 the UN Secretary-General issued yet another report on small arms, prepared by a Group of Governmental Experts. The Group noted that progress was being made at various levels in the implementation of the recommendations of the previous report.

The above reports led to the UN General Assembly resolution on the convening in 2001 of a Conference on the Illicit Trade in Small Arms and Light Weapons in All Its Aspects (see below). In the understanding of the governmental experts, the phrase 'in all its aspects' included aspects of legal transfers insofar as they were directly related to illicit trafficking in and manufacturing of the arms in question.

UN Report on Ammunition and Explosives. In June 1999, following a request by the UN General Assembly, the Secretary-General issued a report prepared by a Group of Experts on the problems of ammunition and explosives. Having concluded that ammunition and explosives are an inseparable part of the problem of excessive accumulation, transfer and misuse of SALWs, the Group recommended the following preventive measures: adoption by states of rules, regulations and procedures for central national collections of information on the production, stocks and transfers of ammunition and explosives; collection and analysis of such data centrally in each country in a single database and linkage of databases on a regional and international basis; nomination by states of a national point of contact for regional and international exchanges of information and cooperation on all aspects of the problem of ammunition and explosives; creation of regional registers covering ammunition and explosives; pursuit of efforts to expand the scope of the UN Register of Conventional Arms so as to cover SALWs as well as ammunition and explosives; regional and international harmonization of laws and regulations relevant to the control of ammunition and explosives; international standardization of the form and content of end-use/end-user certificates; encouragement of states to register all the participants in the ammunition and explosives supply chain, including producers, brokers and shippers, and to deal only with those approved on a national and international level; and encouragement of states to promote regular meetings among the security community and intelligence agencies for the exchange of information on the activities of illegal actors in order to improve law enforcement strategies under the aegis of the United Nations.

To assist in the process of identification and tracing of ammunition and explosives, the Group recommended: encouraging the adoption of a common minimum standard for the marking of ammunition and explosives; including in the marking of small arms ammunition at least the three following elements in a standardized format – the factory of production, the year of production and the batch/lot of production; investigating and using new technologies to improve the marking of ammunition and the tracing and detection of explosives and explosive components; and encouraging regular international meetings of ammunition experts for the exchange of technical information regarding all aspects related to ammunition and explosives.

Firearms Protocol. In April 1998, the UN Economic and Social Council (ECOSOC) adopted a resolution calling for a legally binding international instrument to combat the traffic in firearms. Following that resolution, the ECOSOC Commission on Crime Prevention and Criminal Justice developed the Protocol against the Illicit Manufacturing of and Trafficking in Firearms, Their Parts and Components and Ammunition, often referred to as the Firearms Protocol. This Protocol, opened for signature in 2001, supplements the UN Convention against Transnational Organized Crime, signed in 2000 (two other protocols to the Convention aim at stopping the smuggling of migrants and the trafficking in persons, particularly women and children). The Firearms Protocol was adopted by the UN General Assembly on 31 May 2001 and became the first legally binding global measure to regulate international transfers of small weapons.

According to the Protocol, 'firearm' means any portable barrelled weapon that expels, is designed to expel or may be readily converted to expel a shot, bullet or projectile by the action of an explosive, excluding antique firearms or their replicas. The parties are to combat the illicit trade in firearms by criminalizing trafficking-related activities; seizing and destroying confiscated weapons; record keeping; marking weapons at the point of manufacture and at import; harmonizing the export, import and transit licensing requirements; taking security measures to prevent the theft, loss or diversion of weapons; providing for an exchange of information, experience, training and technical assistance; and, possibly, providing for registration and licensing of arms brokers. The Protocol does not apply to government transactions.

The Firearms Protocol is to enter into force after 40 countries have deposited their instruments of ratification, acceptance, approval or accession but not before the Convention has entered into force.

Regional Initiatives regarding SALWs

The 1997 Inter-American Convention. On 13 November 1997 the General Assembly of the Organization of American States adopted and opened for signature the Inter-American Convention Against the Illicit Manufacturing of and Trafficking in Firearms, Ammunition, Explosives, and Other Related Materials. The stated purpose of the Convention is not only to prevent, combat and eradicate illicit activities but also to promote and facilitate cooperation and exchange of relevant information and experience among the parties. Parties that have not yet done so are to adopt the necessary legislative or other measures to establish as criminal offences under their domestic law the illicit manufacturing of and trafficking in the prohibited items. The criminal offences should include participation in, association with or conspiracy to commit, attempts to commit, and aiding, abetting, facilitating and counselling the commission of the said offences.

The parties should require, at the time of manufacture of firearms, appropriate markings of the name of the manufacturer, place of manufacture and serial number. They should also require markings on imported firearms permitting the identification of the importer's name and address, as well as markings on any firearms confiscated or forfeited (as illicitly manufactured or trafficked) that are retained for official use. An effective system of export, import and international transit licences or authorizations must be established for transfers of firearms, ammunition, explosives and other related materials. States should exchange among themselves relevant information and guarantee its confidentiality if requested to do so by the party providing the information.

The offences to which the Convention applies should be deemed to be included as extraditable offences in extradition treaties among the parties. States may make reservations to the Convention, provided that the reservations are not incompatible with its object and purposes and that they concern its specific provisions. The Convention is to remain in force indefinitely, but any party may denounce it.

ECOWAS Moratorium. In October 1998, at the initiative of Mali, the heads of state and government of the Economic Community of West African States (ECOWAS), meeting in Abuja, Nigeria, approved the Moratorium on the Importation, Exportation and Manufacture of Light Weapons in West Africa. This politically binding document, known as the ECOWAS Moratorium, took effect from 1 November 1998. It gave the region an opportunity to address the problems caused by small arms proliferation, which fuels violence and prevents social and economic progress. West Africa became the first region in the world to announce a halt to the procurement of light weapons.

Subsequently, a Code of Conduct for the implementation of the ECOWAS Moratorium was agreed upon in 1999. The parties committed themselves to adopt and harmonize the regulatory and administrative measures necessary to exercise control of cross-border transactions with regard to light weapons and their components, as well as the ammunition relating to them. National commissions, made up of representatives of the authorities and civil society, were to promote and ensure the effective implementation of the Moratorium at the national level. Member-states undertook, with the assistance of the Programme for Co-ordination and Assistance for Security and Development (PCASED), ECOWAS and other organizations, to carry out a systematic collection and destruction of all light weapons that are not required for the purpose of national security and must, therefore, be considered as surplus weapons. States having valid reasons may request an exemption from the Moratorium in order to meet their legitimate national security needs. Such a request should be forwarded to the ECOWAS Executive Secretariat, which would make its assessment and circulate it to member-states. Provided there are no objections, the Executive Secretariat should issue a certificate confirming member-states' assent. Should a member-state object, the request for exemption must be referred to the ECOWAS Mediation and Security Council. Participation in the Moratorium may be extended to other interested African states.

EU Joint Action. Cooperation on small arms issues among the European Union member-states began in 1997 with an agreement on the Programme for Preventing and Combating Illicit Trafficking in Conventional Arms, which emphasized the particular relevance of the initiatives dealing with small arms. This cooperation was further developed in December 1998, when the EU Council of Ministers adopted a document, called the Joint Action, on the European Union's contribution to combating the destabilizing accumulation and spread of small arms and light weapons. This agreement built mainly on the EU Code of Conduct on Arms Exports, adopted on 8 June 1998 (see above). The EU states agreed to develop a cooperative policy, concentrating, among other undertakings, on ending the destabilizing accumulation and spread of small arms and reducing the existing accumulations to the levels consistent with the countries' legitimate security needs.

OSCE Document. In November 2000 the OSCE Forum for Security Co-operation (FSC) adopted the Document on Small Arms and Light Weapons. For the purpose of the Document, SALWs were defined as man-portable weapons made or modified to military specifications for use as lethal instruments of war. The OSCE participating states undertook, in particular, to combat illicit trafficking through the adoption and implementation of national controls on small arms; to contribute to the reduction and prevention of the excessive and destabilizing accumulation and uncontrolled spread of small arms; to exercise due restraint to ensure that small arms are produced, transferred and held only in accordance with legitimate defence and security needs and in accordance with appropriate international and regional export criteria; to build confidence, security and transparency through appropriate measures regarding small arms; and to develop appropriate measures on small arms at the end of armed conflicts, including their collection, safe storage and destruction linked to the disarmament, demobilization and reintegration of combatants.

The OSCE states agreed to ensure effective national control over the manufacture of small arms by issuing, regularly reviewing and renewing licences and authorizations for manufacture. Licences and authorizations should be revoked if the conditions under which they had been granted were no longer met. Those engaged in illegal production must be prosecuted under appropriate penal codes. The agreed criteria governing exports of small arms and related technology are based on the OSCE principles governing all conventional arms transfers (see above). National systems to be established to regulate the activities of international brokers in small arms should include measures such as: requiring registration of brokers operating within the territory of a given country; requiring licensing or authorization of brokering; or requiring disclosure of import and export licences or authorizations or accompanying documents, and of the names and locations of brokers involved in the transaction.

Each OSCE participating state should ensure that it has an effective capability to enforce its international commitments on small arms through its national authorities and judicial system. An information exchange among the participating states about their small arms exports to, and imports from, the other participating states during the previous calendar year is to be conducted on an annual basis, beginning in 2002. Information on small arms that have been identified as surplus and/or seized and destroyed is to be shared as well. The FSC has an obligation to regularly review the implementation of the norms, principles and measures specified in the Document and to consider specific small arms issues raised by the participating states.

Traceability of SALWs

One of the main problems in dealing with the illicit traffic in SALWs is the difficulty, often impossibility, of identifying the sources and lines of supply of such weapons, particularly in regions of conflict or tension. Many arms producers imprint on weapons their serial numbers and the name of the manufacturer, but the practice is not uniform and the identifying marks often can be easily removed. Hence the need for a global mechanism which would enable cooperation among states in identifying points where arms (and possibly also munitions) were diverted to illicit trade, a need reflected in reports, codes of conduct and statements, both governmental and non-governmental, dealing with SALWs. According to a 2001 Franco-Swiss proposal, an international agreement should be concluded by which states would commit themselves to assisting each other in tracing the supplies of SALWs that contribute to illicit arms trafficking or to excessive and destabilizing accumulation and flows of these weapons. National committees would be established to ensure minimum agreed standards of marking and record keeping of SALWs, whereas international bodies would help develop interstate cooperation, including technical cooperation. Confidentiality in the treatment of information provided to assist in the tracing efforts would be recognized.

Since the weapons recovered from conflict areas are commonly altered to disguise their origins, it was proposed that each SALW should be uniquely marked, using a combination of techniques, at the point of manufacture so as to enable individual weapons to be traced. The markings would have to contain sufficient information to allow the national investigation authorities to determine, at a minimum, the country and year of manufacture, the manufacturer and the serial number. Agreed minimum standards for record keeping would also be needed. The time periods over which reliable records need to be maintained would have to be commensurate with the lifetimes of SALWs. (Many SALWs have a lifetime of more than 50 years.)

The UN Conference on SALWs

On 9–20 July 2001, the United Nations Conference on the Illicit Trade in Small Arms and Light Weapons in All Its Aspects was held in New York. The Conference adopted a Programme of Action containing measures to be undertaken at the national, regional and global levels.

The National Level. The Conference agreed, *inter alia*, to: put in place adequate laws, regulations and administrative procedures to exercise effective control over the production of SALWs within their areas of jurisdiction and over the export, import, transit or retransfer of such weapons, in order to prevent the illegal manufacture of and illicit trafficking in these weapons, or their diversion to unauthorized recipients; adopt and implement, in the states that have not yet done so, the necessary legislative or other measures to establish as criminal offences the illegal manufacture, possession, stockpiling and trade in SALWs; establish national coordination agencies responsible for policy guidance, research and monitoring of efforts to prevent, combat and eradicate the illicit trade in SALWs, including aspects of illicit manufacture, control, trafficking, circulation, brokering, as well as tracing, finance, collection and destruction of SALWs; identify groups and individuals engaged in the illegal manufacture, trade, stockpiling, transfer, possession, as well as financing the acquisition of illicit SALWs and take action under appropriate national law against such groups and individuals; ensure that licensed manufacturers apply an appropriate marking on each SALW to identify the country of manufacture and enable the national authorities to identify the manufacturer; ensure that comprehensive and accurate records are kept on the manufacture, holding and transfer of SALWs; ensure responsibility for SALWs held and issued by the state; make every effort, without prejudice to the right of states to re-export SALWs that they have previously imported, to notify the original exporting state in accordance with their bilateral agreements before the retransfer of those weapons; develop national legislation or administrative procedures regulating the activities of those engaged in SALW brokering, including the registration of brokers, licensing or authorization of brokering transactions as well as penalties for illicit brokering activities; take measures against any activity that violates a UN Security Council arms embargo; ensure that all confiscated, seized or collected SALWs are destroyed, unless another form of disposition or use has been officially authorized; ensure that the armed forces, police or any other body authorized to hold SALWs establish standards and procedures relating to the management and security of their stocks of these weapons; develop public awareness of the consequences of the illicit trade in SALWs in cooperation with the civil society; develop and implement, where possible, effective disarmament, demobilization and reintegration programmes, including the effective collection, control, storage and destruction of SALWs, particularly in post-conflict situations, unless another form of disposition or use has been duly authorized and the weapons have been marked and registered; and address the special needs of children affected by armed conflict, in particular reunification with their family, reintegration into civil society and appropriate rehabilitation.

The Regional Level. The Conference agreed, *inter alia,* to: encourage regional negotiations with the aim of concluding relevant legally binding instruments to prevent, combat and eradicate the illicit trade in SALWs and, where such instruments exist, ratify and fully implement them; encourage the strengthening and establishing, where appropriate, of moratoria or the taking of similar initiatives in affected regions or sub-regions on the transfer and manufacture of SALWs; establish, where appropriate, sub-regional or regional mechanisms, in particular trans-border customs cooperation and networks for information-sharing among law-enforcement, border and customs control agencies; encourage states to promote safe, effective stockpile management and security, in particular physical security, for SALWs; and encourage regions to develop, on a voluntary basis, measures to enhance transparency.

The Global Level. The Conference agreed, *inter alia,* to: request the UN Secretary-General, through the UN Department for Disarmament Affairs, to collate and circulate data and information provided by states on a voluntary basis on the implementation of the Programme of Action; encourage, particularly in post-conflict situations, the disarmament and demobilization of ex-combatants and their subsequent reintegration into civilian life, including the provision of support for the effective disposition of collected SALWs; encourage states and the World Customs Organization, as well as other relevant organizations, to enhance cooperation with the International Criminal Police Organization (Interpol) to identify those groups and individuals engaged in the illicit trade in SALWs; develop common understandings of the basic issues and the scope of the problems related to illicit brokering in SALWs; and encourage the relevant international and regional organizations and states to facilitate the cooperation of civil society.

The Conference recommended that the UN General Assembly convene a conference no later than 2006 to review progress made in the implementation of the Programme of Action; to convene a meeting of states on a biennial basis to consider the national, regional and global implementation of the Programme of Action; and to undertake a UN study to examine the feasibility of developing an international instrument enabling states to identify and trace illicit SALWs in a timely and reliable manner.

Assessment

The Programme of Action adopted by consensus at the UN Conference was a modest but significant step towards the development of international norms for restrictions on SALWs. It has provided general guidance for national governments, regional and international organizations as well as the civil society in combating the illicit trade in SALWs. However, the Programme is not legally binding. Moreover, owing to the opposition of the United States, the Conference could not reach consensus on a statement recognizing the need to establish and maintain controls over private ownership of SALWs and the need to prevent sales of such arms to non-state groups – as demanded by most states represented at the Conference. The President of the Conference expressed his disappointment that there was no agreement on these important issues.

As with other categories of conventional armaments, measures intended to limit the transfers of SALWs and to control arms trading and brokering activities, rather than restrict and regulate the manufacture and the very possession of SALWs, cannot be expected to be really effective. Transparency, marking, record keeping and monitoring the activities of brokers are, of course, desirable, but they cannot suffice to combat illicit trafficking, especially in the absence of a generally acceptable definition of SALWs. Since the terms 'excessive' and 'destabilizing', used with reference to the accumulation of SALWs, lack clarity, and since no definite limits have been set for permitted activities, licit transfers are not always distinguishable from illicit ones, nor are they in every case less reprehensible.

16

Confidence Building in Europe, Asia and the Americas

Confidence building promotes communication and better understanding among states, as argued in Chapter 1.3. By introducing restraints in the military field it renders the use of force for settling international disputes less likely. It thereby facilitates negotiations for disarmament.

16.1 Openness and Constraints in Military Activities in Europe

In 1954, when Western efforts to bring the FRG into NATO were nearing fruition, the Soviet Union proposed that a conference on security in Europe should be convened to deal with what it then considered to be a threat to its security. The proposal called for the withdrawal of all occupation forces from Germany and for a treaty on collective security in Europe. The West rejected this proposal, among several other reasons because it did not provide for full US participation in the projected conference.

Only in the early 1970s, after the political situation in Europe had improved significantly, did a conference on European security become a possibility. The improvement was brought about by the 1970 Soviet–West German treaty, in which the two countries committed themselves to regarding the borders of all states in Europe as inviolable; by the 1970 treaty between Poland and West Germany, in which the latter formally abandoned its claim to the territories east of the Oder–Neisse line; by the 1971 quadripartite (French–Soviet–UK–US) agreement on Berlin, which allowed unhindered movement of people and goods between the western sectors of Berlin and West Germany; by the 1972 treaty between the two German states, reaffirming the inviolability of the border between them; and by the Soviet agreement to participate in talks on troops reduction in Europe.

The Conference on Security and Co-operation in Europe (CSCE) opened on 3 July 1973. Nearly all the participants were organized in three major groupings: NATO, the Warsaw Treaty Organization and the neutral/non-aligned countries. Proposals for measures to be negotiated by the Conference fell into three areas called 'baskets': political, including military security; economic; and cultural–humanitarian.

The 1975 Helsinki CBM Document

The first phase of the CSCE concluded on 1 August 1975 at Helsinki with the adoption of the Final Act, which contained a Document on Confidence-Building Measures and Certain Aspects of Security and Disarmament. The rationale for adopting this so-called Helsinki CBM Document was formulated as follows: 'to contribute to reducing the dangers of armed conflict and of misunderstanding or miscalculation of military activities which could give rise to apprehension, particularly in a situation when the participating states lack clear and timely information about the nature of such activities'.

Main Provisions. Much of the Document simply confirmed the practice existing among nations that maintain normal relations. One provision, however, introduced a new undertaking: to notify major military manoeuvres in Europe at least 21 days in advance or, in the case of a manoeuvre arranged at shorter notice, at the earliest possible opportunity before its starting date. The term 'major' meant that at least 25,000 troops were to be involved. Manoeuvres with fewer troops could also be considered as major if they involved 'significant numbers' of either amphibious or airborne troops, or both. Manoeuvres of naval and air forces, whether conducted independently or jointly, were not covered by the notification requirement. The following information was to be provided for each major manoeuvre: designation (code-name), if any; general purpose; the states involved; the types and numerical strength of the forces engaged; and the area and estimated time frame of its conduct. States could give additional information, particularly relating to the components of the forces engaged and the period of troop involvement, and could invite observers to attend the manoeuvres.

The area of application of the CBMs did not cover the territories of the non-European states participating in the CSCE – the United States and Canada. The territories of the Soviet Union and Turkey were included only in part – up to 250 kilometres from the frontiers 'faced or shared' with other participating European states. Notification of major military manoeuvres was politically binding, whereas other undertakings, including the undertaking concerning observers, rested on a 'voluntary basis'.

Shortcomings. The concept of advance notification of military manoeuvres was introduced into the international debate in the early 1960s as part of a larger programme to reduce the risk of war by accident, miscalculation, failure of communications or surprise attack. It was then discussed along with the proposed establishment of observation posts, mobile observation teams and exchange of military missions, or in conjunction with a proposed prohibition on certain types of military exercise.

It is clear that notification alone can do no more than contribute to minimizing the danger that detection of significant military activities might give rise to a misunderstanding, provoke a rapid, possibly disproportionate military response and initiate unpremeditated hostilities. However, to properly fulfil even such a relatively modest role, notification must be mandatory, not optional, apply also to small-scale military manoeuvres and be given well in advance of the start of the manoeuvres. It should also cover military movements other than manoeuvres, since transfers of combat-ready units outside their permanent garrison or base areas, especially over long distances and close to the borders of other states, may cause even greater concern than manoeuvres. Furthermore, the role of invited foreign observers was not clearly defined; the Helsinki Document left it entirely to the host country to determine whether the attendance of observers at manoeuvres was to be a meaningful or simply a ceremonial act. Many of these requirements were met in a document adopted by the CSCE participants as a result of their conference held in Stockholm from 1984 to 1986.

The 1986 Stockholm CSBM Document

The Document of the Conference on Confidence- and Security-Building Measures and Disarmament in Europe, the so-called Stockholm CSBM Document, was adopted on 19 September 1986. It followed the guidelines worked out at the 1983 Madrid CSCE Follow-up Meeting, which required the adoption of measures that were militarily significant, politically binding, verifiable and covering the whole of Europe. The addition of 'security' to the title of CBMs, making them CSBMs, indicated these new objectives.

Besides reaffirming their determination to respect the principles embodied in the Helsinki Final Act, the parties to the Stockholm Document stated that they would refrain from the threat or use of force against *any* state, regardless of that state's political, social, economic or cultural system and 'irrespective of whether or not they maintain with that State relations of alliance'. The latter commitment refuted the so-called Brezhnev doctrine, promulgated in 1968 to justify the invasion of Czechoslovakia – a member of the Warsaw Treaty Organization – by the forces of that Organization. The Stockholm Document was a considerable improvement over the Helsinki Document also in other respects, as can be seen from the summary of its provisions.

Notification. The Stockholm Document required that notification be given 42 days in advance of the start of notifiable military activities, that is, those involving at any one time during the activity at least 13,000 troops, including support troops, or at least 300 battle tanks, if organized into a divisional structure or at least two brigades/regiments not necessarily subordinate to the same division. Both exercises and movements were covered by the notification requirement.

Air force participation was to be included in the notification only if it was foreseen that 200 or more sorties by aircraft, excluding helicopters, would be flown. Military activities consisting in amphibious landing or in parachute assault by airborne forces in the zone of application of CSBMs were to be notified whenever the amphibious landing or the parachute drop involved at least 3,000 troops. Notifiable military activities carried out without advance notice to the troops involved (alert activities) were exempted from the 42-day advance notification requirement; notification could then be given at the time the troops commenced the said activities. All notifications were to contain both general and detailed information, including the envisaged area of the activity and time frame. Each participating state was to exchange with others an annual calendar of its military activities subject to prior notification, as forecast for the subsequent calendar year. The zone of application of the CSBMs was to cover the whole of Europe, from the Atlantic Ocean to the Ural Mountains, as well as the adjoining sea and ocean areas and air space.

Observation. Notified military activities were to be subject to observation by other participating states whenever the number of troops engaged reached or exceeded 17,000 (that is, more than the number of troops subject to notification), except for an amphibious landing or a parachute assault by airborne forces, which were to be subject to observation whenever the number of troops engaged was 5,000 or more.

Observers were to be provided with a general observation programme and also given opportunities to visit some units and communicate with commanders and troops. The host country was not obligated to permit observation of restricted locations, installations or defence sites. For notifiable military activities carried out without advance notice to the troops involved (alert activities), international observation was not required unless these activities lasted more than 72 hours and met the agreed thresholds for observation.

Constraints. By 15 November each year, each party was to communicate information concerning military activities subject to prior notification involving more than 40,000 troops and planned for the second subsequent calendar year. No military activities involving more than 75,000 troops were to be carried out unless they had been thus communicated; no military activities involving more than 40,000 troops were to be carried out unless they were included in the annual calendar.

Verification. Within the zone of application of CSBMs, each party was entitled to conduct inspections on the territory of any other party. The state requesting inspection had to give the reasons for its request, but in practice this requirement was not observed.

No state was obliged to accept on its territory more than three inspections per calendar year, nor was it obliged to accept more than one inspection per calendar year by the same state. The representatives of the inspecting state, accompanied by the representatives of the receiving state, were to be permitted access, entry and unobstructed survey, except for areas or sensitive points to which access is normally denied or restricted, military and other defence installations, as well as naval vessels, military vehicles or aircraft.

Subsequent CSBM documents, worked out in Vienna in 1990, 1992, 1994 and 1999, extended the scope of the obligations assumed by the parties to the Stockholm Document.

The 1990 Vienna CSBM Document

Under the document adopted in Vienna on 17 November 1990 by the participants in the CSCE, the main modifications of the CSBMs then in force concerned the exchange of military information, reduction of risks, military contacts, observation of military activities, constraints, verification, communications and assessment of implementation.

Exchange of Information. The parties to the 1990 Vienna Document agreed to exchange information on their military forces concerning organization, manpower, and major weapon and equipment systems. This information was to be provided in an agreed format not later than 15 December of each year.

There was also to be an annual exchange of information on the parties' plans for the deployment of major weapon and equipment systems and on their military budgets for the forthcoming fiscal year. Budgetary information was to itemize defence expenditures according to the categories set out in the 1980 UN Instrument for Standardized International Reporting of Military Expenditures; any party was free to ask for clarification of such information.

Risk Reduction. For the first time, the parties undertook to consult each other about any unusual and unscheduled activities of their military forces outside their normal peacetime locations within the zone of application of CSBMs and about which a party expressed security concerns. The party having such concerns could request an explanation from the party where the activities in question were taking place. The reply was to be transmitted within 48 hours. The requesting state could also ask for a meeting with the responding state or a meeting of all parties.

Hazardous incidents of a military nature were to be reported and clarified in order to prevent misunderstandings and mitigate the effects. Each party was to designate a point of contact in case of such incidents.

Contacts. Each party possessing air combat units was to arrange visits for representatives of all other parties to one of its peacetime airbases on which such units were located; the visit was to last at least 24 hours. No party was obliged to arrange more than one such visit in any five-year period.

The parties undertook to promote and facilitate: exchanges and visits between senior military/defence representatives; contacts between relevant military institutions; participation by foreign military representatives in instruction courses; exchanges between military commanders and officers of commands down to brigade/regiment level; exchanges and contacts between academics and experts in military studies and related areas; and sporting and cultural events between members of their armed forces.

Observation. The conditions for observation activities were to improve considerably. The host state was to provide observers with appropriate observation equipment. Observers were also allowed to use their own binoculars, maps, photo and video cameras, dictaphones and hand-held passive night-vision devices, but this equipment was to be subject to examination and approval by the host state. An aerial survey of the area of the military activity observed was envisaged to help the observers gain a general impression of the scope and scale of the disposition of forces engaged in that activity. Helicopters and/or aircraft could be provided by the host state or by another state at the request of and in agreement with the host state.

At the close of each observation, the observers were to be given an opportunity to meet with host state officials to discuss the course of the observed activity. The parties were encouraged to permit media representatives to attend observed military activities in accordance with accreditation procedures set down by the host state.

Constraints. Military activities involving more than 40,000 troops (instead of 75,000 as stipulated in the 1986 Stockholm CSBM Document) were not to be carried out unless they had been the object of communication concerning military activities planned for the second subsequent calendar year.

Verification. The modalities of inspection were further refined, and information on military forces and on plans for the deployment of major weapon and equipment systems, as provided by the parties, was to be subject to evaluation. According to this novel procedure, each party was obliged to accept evaluation visits (up to 15 per calendar year) in accordance with a specified quota, but no party was obliged to accept more than one-fifth of its quota of visits from the same state. A party with a quota of fewer than five visits was not obliged to accept more than one visit from the same state during a calendar year.

No formation or unit could be visited more than twice during a calendar year or more than once by the same state. Each party was entitled not to accept more than one visit on its territory at any given time. The evaluation measure applied only to active units and formations. A party had the right not to accept a visit if the formation or unit was unavailable for evaluation, but it would have to state the reason. This provision could be invoked up to five times per year.

Communications. To complement the existing diplomatic channels, the parties undertook to establish a network of direct communications between their capitals for the transmission of messages relating to agreed measures. They were to designate points of contact capable of transmitting and receiving such messages 24 hours a day.

Implementation Assessment. Each year a meeting of the parties was to be held to discuss the implementation of the CSBMs. The discussion was to cover clarification of questions arising from implementation, operation of agreed measures and implications of all information originating from the implementation of these measures for the process of confidence and security building.

The 1992 Vienna CSBM Document

The CSBMs described above were developed in the 4 March 1992 Vienna CSBM Document, which integrated new measures with those contained in previous CSBM documents. The area of application was extended to cover the territories of the non-Russian former Soviet republics in Asia, which had been admitted to the CSCE in January 1992.

Exchange of Information. The parties decided to provide information on the total number of their units, as well as on planned personnel increases of over 1,500 troops for each active combat unit and of over 5,000 troops for each active formation, for more than 21 days, and to exchange information on temporary activation of non-active combat units and non-active formations with more than 2,000 troops, for more than 21 days. Military information to be exchanged annually was to include data relating to major weapon and equipment systems, furnished together with photographs.

Risk Reduction. The risk reduction clauses of the 1990 Vienna Document were supplemented with a provision for voluntary hosting of visits. The purpose of these visits was to dispel concerns about military activities.

Contacts. According to new stipulations, a state invited to visit an airbase may decide whether to send military officers and/or civilians, including personnel accredited to the host state. Schedules for such visits for the coming year or years might be discussed at annual implementation meetings. It was also agreed that the first state to deploy a new type of major weapon or equipment system must arrange a demonstration for all other parties at the earliest opportunity.

Notification and Observation. The parameters for prior notification of certain military activities were changed. Thus, activities involving at least 9,000 troops (including support troops), or 3,000 troops in an amphibious landing/parachute drop, or at least 250 battle tanks, if organized into a divisional structure or at least two brigades/regiments, were now subject to notification. Activities involving at least 13,000 troops or 3,500 in an amphibious landing/parachute assault by airborne forces were subject to observation. A new provision allowed observation of activities engaging at least 300 battle tanks.

Constraints. Constraints on military activities subject to notification were strengthened. Thus, no party was allowed to carry out within two calendar years more than one military activity involving over 40,000 troops or 900 battle tanks. Nor was it allowed to conduct within a calendar year more than six military activities, each involving over 13,000 troops or 300 battle tanks but not more than 40,000 troops or 900 battle tanks. Of these six military activities, no party was allowed to conduct within a calendar year more than three activities, each involving over 25,000 troops or 400 battle tanks. Each party undertook not to carry out simultaneously more than three military activities, each involving over 13,000 troops or 300 battle tanks. No military activity involving over 40,000 troops or 900 battle tanks was to be carried out unless it had been the object of a communication concerning activities planned for the second subsequent calendar year, and unless it had been included in the annual calendar no later than 15 November the preceding year.

Verification. The evaluation provisions were complemented with the following requirement: non-active formations and combat units temporarily activated were to be made available for evaluation during the period of temporary activation and in the area/location of activation. In such cases the provisions for the evaluation of active formations and units were to apply, *mutatis mutandis.* Evaluation visits conducted under this provision were to be counted against the established quota.

The 1994 Vienna CSBM Document

The Vienna Document, adopted on 28 November 1994, incorporated a programme of military contacts and cooperation. Other major changes and additions were as follows.

Notification and Observation. The list of parameters determining the notifiability of a military activity was supplemented with provisions on the involvement of at least 500 armoured combat vehicles or at least 250 self-propelled and towed artillery pieces, mortars and multiple rocket launchers (100-mm calibre and above). Engagement of military forces in a heliborne landing was added to the provision on amphibious landings and parachute assaults subject to notification.

Compliance and Verification. States were encouraged to take, bilaterally, multilaterally or in a regional context, additional measures to increase transparency, such as notification of military activities carried out below the thresholds or close to the borders between them, as well as observation of non-notifiable exercises.

Communication. Agreement was reached on language, standard operating procedures and other ways to ensure the efficient use of the communication network.

The 1999 Vienna CSBM Document

The Vienna Document adopted on 16 November 1999 introduced further changes. A particularly important addition was a chapter which envisaged complementing the OSCE-wide CSBMs with voluntary binding measures tailored to specific regional needs. The principles upon which such measures had to be based were set out. Other important additions as well as modifications were as follows.

Exchange of Information. The limits for planned and reported increases in personnel strength for more than 21 days were now lowered to 1,000 troops for each active combat unit and to 3,000 troops for each active formation. As regards defence planning, the information to be exchanged annually was to include the date on which the military budget for the forthcoming fiscal year had been approved by the competent national authorities and to specify the identity of these authorities. The information was to be provided not later than three months after the approval of the military budget. The inability to meet this deadline had to be notified and explained, and the date envisaged for the actual submission had to be provided. States with no armed forces were to provide 'nil reports' together with their annual military information. If necessary, discrepancies between expenditures and previously reported budgets were to be clarified, and information was to be provided on the relation of the military budget to the gross national product as a percentage. States were also encouraged to hold periodic high-level OSCE military doctrine seminars.

Risk Reduction. New provisions regarding unusual military activities entitled both the requesting and the responding states to ask other states concerned about such activities to participate in meetings to discuss the matter. The requesting or the responding state, or both, may ask for a meeting of all participating states.

Military Cooperation. In addition to the provisions regarding visits to airbases, each participating state must arrange for representatives of all other participating states to visit one of its military facilities or military formations, or to observe military activities below the specified thresholds. Efforts should be made to arrange for one such visit or observation in any five-year period.

Notification and Observation. Military activities, 'including those where forces of other participating states were participants', had to be notified. The parameters for notification and observation were retained.

Compliance and Verification. Requests for an inspection were to be submitted at least 36 hours, but no more than five days, before the estimated entry of the inspectors into the territory of the receiving state. The inspection team, consisting of up to four inspectors, could include nationals from up to three participating states. Aircraft had to be provided by the receiving country, unless otherwise agreed. Reasons for a failure to carry out or accept an inspection, because of *force majeure*, were to be explained in detail. Requests for evaluation visits must be submitted not earlier than seven (and not later than five) days before the estimated entry into the territory of the receiving state. The team should consist of no more than three persons, unless otherwise agreed. The report of the team must be communicated to all participating states within 14 days.

Assessment

The process of confidence building in Europe started at a time when East–West confrontation carried the risk of a massive surprise armed attack that could initiate another world war. CSBMs made a genuine contribution to the relaxation of tensions between the two military blocs. Moreover, the Vienna Documents encouraged the OSCE participating states to complement the OSCE-area CSBMs with measures adapted to sub-regional needs. A number of CSBM-related agreements were entered into by neighbouring European states (some of them between historical adversaries) to increase transparency and openness. The conclusion in 1999 of the Stability Pact for South Eastern Europe was yet another achievement.

The dissolution of the Soviet Union and the consequent disappearance of one of the main protagonists of the Cold War have, to a certain extent, reduced the importance of CSBMs in Europe. Improving notification procedures and exchanges of information, lowering the thresholds for notifiable military activities and elaborating ever more intrusive verification measures will hardly produce new significant effects. Efforts would perhaps be better spent on tightening the undertakings that have already been assumed. Indeed, there are doubts as to whether the Vienna Document CSBMs are at all relevant in crisis situations. The impunity with which the authorities in Yugoslavia disregarded their political commitments may have highlighted the advisability of transforming CSBMs into legally binding obligations. Provision for rapid collective action against violators of such obligations would be particularly advisable.

16.2   Other Security-Related CBMs in Europe

New CSCE Institutions

At the conclusion of the summit meeting held in Paris on 19–21 November 1990, the CSCE participants signed the Charter for a New Europe, which laid down guidelines for the future activities of the CSCE. The signatories recognized that the changing political environment in Europe opened new possibilities for common efforts in the field of military security. They expressed their determination to cooperate in defending democratic institutions against activities that violated the independence, sovereign equality or territorial integrity of the participating states.

Several decisions were taken regarding the institutions of the CSCE. In particular, it was decided that heads of state or government should meet every two years, as a rule, and that ministers of foreign affairs should meet for political consultations, as a Council, at least once a year. Additional meetings of the representatives of the participating states could be convened to discuss questions of urgent concern. A Supplementary Document adopted together with the Charter for a New Europe set out the procedural and organizational modalities relating to the provisions of the Charter.

The CSCE summit meeting held at Helsinki on 9–10 July 1992 adopted a document called 'The Challenges of Change'. According to this document, the CSCE Committee of Senior Officials (CSO) was mandated to play a central role in early warning, crisis management, peaceful settlement of disputes and peacekeeping operations. It was then also decided to open the main CSCE meetings to Japan, to establish working relations with other European and trans-Atlantic organizations, and to strengthen relations with non-governmental organizations. A Forum for Security Co-operation (FSC) was set up, marking the end of the CSCE practice of elaborating only CSBMs while reserving arms limitation for negotiation between the military blocs. In December 1992 the CSCE decided that the Council of Ministers should appoint a secretary general whose tasks were to include management of the CSCE operations, preparation of meetings, implementation of CSCE decisions and liaison with other international institutions.

New institutional arrangements were needed for the proper functioning of the CSCE. They have not, however, helped to settle ongoing armed conflicts, to prevent further conflicts from breaking out and to stop breaches of the international humanitarian law, including genocide. As distinct from its 'competitors' – NATO or the European Union – the OSCE (successor of the CSCE) lacks operational powers, as well as the resources necessary to make states observe the proclaimed principles and implement the adopted resolutions. In the field of arms control its role remains marginal. Moreover, because of its considerable expansion, the OSCE is losing the character of a regional organization of states having common security interests.

The 1992 Open Skies Treaty

In 1955 US President Eisenhower advanced the concept of open skies for aerial photography as a way to build confidence between the United States and the Soviet Union. This concept was first rejected by the Soviet Union as 'legalized espionage', but a few decades later it became incorporated in a multilateral accord negotiated by NATO and former members of the WTO, which called on the signatories to submit their territories to unarmed surveillance flights. The Open Skies Treaty was signed on 24 March 1992 and entered into force on 1 January 2002.

Area of Application. The Treaty covers an area stretching from Vancouver in Canada eastwards to Vladivostok. It is thus the first major CSBM to include not only European territory but also the United States and Canada as well as the Asian portions of Russia. No part of these territories may be held off-limits to observation overflights; only flight safety considerations may restrict their conduct.

Quotas. The annual number of overflights which each party is obligated to accept – the so-called passive quota – depends on the country's geographic size. The number of flights over the territories of other nations, allocated annually to each party, constitutes that party's active quota. Each state has the right to conduct the same number of overflights over a particular state as that state may conduct over it. However, a country may be overflown by a state that it does not intend to overfly. The passive quotas of certain states need not be filled if there is no interest in overflying them. A state may also transfer a part of or its entire active quota to other parties, with the consent of the state to be overflown. No state may use more than half its active quota in flying over a single other state; nor may any one state use up more than half of the passive quota of another state.

Two or more states may form a 'group of state parties', the members of which may redistribute among themselves their active quotas. At the time of signature, Russia and Belarus, the Benelux nations (Belgium, the Netherlands and Luxembourg) and the other members of the Western European Union formed such groups. The Benelux group decided to operate at an exceptionally high level of integration: to establish a single point of entry for overflights over all the territories of the group; to have a single combined active and passive quota; and to conduct flights jointly. A flight over a group of parties is to be counted against the passive quota of each observed state. The number of group members overflown will be counted against the observing party's active quota. Once the regime is fully operational, the United States will have to accept up to 42 overflights, as will the Russia–Belarus group. The latter's quota may rise if other former Soviet republics decide to join the group.

Sensors. The Open Skies Treaty specifies four categories of sensor with which observation aircraft can be equipped: optical panoramic and framing cameras; video cameras with real-time display; infrared, line-scanning devices; and synthetic aperture radars. The combination of these sensors permits the acquisition of imagery in all types of weather, with a resolution sufficient to enable recognition of all significant types of military equipment (tanks, artillery pieces and armoured personnel carriers).

Parties must have equal commercial access to sensor systems capable of producing the maximum allowable image resolution. However, to enable all parties to participate on the basis of equality from the start, only the more affordable and accessible systems are to be allowed in the course of the Treaty's initial three-year phase-in period. During that period, the use of infrared imaging devices is prohibited, unless agreed to by both the observing and observed states. With the consent of all parties, the capabilities of the sensors employed may be upgraded and new sensor categories may be added. The latter could include air-sampling devices useful in detecting chemical and nuclear activities or in environmental monitoring.

The Treaty provides for raw-data sharing among the parties on a shared-cost basis. This should allow smaller states to have access to the entire data pool, rather than being limited to data from their own low number of overflights. Through a system of mission reports all parties will know where flights were undertaken and images collected.

Flight Procedures. The observed country may decide whose observation aircraft is to be used for overflights of its territory. This stipulation – which allows the country under observation to supply its own aircraft – was introduced at the insistence of Russia. The option may allay host-nation fears about concealed sensors, but it may also burden it with additional costs.

The observing party must provide notification of its intent to overfly another country no less than 72 hours prior to the estimated time of arrival of its personnel at a designated location. The observed party is required to acknowledge receipt of the notification within 24 hours; if that party exercises its right to provide the observation aircraft, it must supply this information to the observing party in its acknowledgement. Once it has arrived, the observing party must submit to the observed party a mission plan specifying the intended route as well as the distances involved. The two parties must agree on the plan no later than eight hours after receipt of the original mission plan. Unless otherwise agreed, the observation flight is to begin no less than 24 hours after the mission plan has been submitted and be completed within 96 hours of the observing party's arrival. The reason for establishing such tight schedules is to emphasize the short-notice character of the overflights, minimizing at the same time the expenses and logistical problems involved in hosting an observation flight.

Institutional Arrangements. The Open Skies Consultative Commission is to promote the objectives and facilitate the implementation of the Treaty. Its tasks are: to consider questions relating to compliance; to seek to resolve ambiguities and differences of interpretation; to take decisions on applications for accession to the Treaty; and to agree on necessary technical and administrative measures following the states' accession to the Treaty.

The Consultative Commission may also agree to improve the viability and effectiveness of the Treaty. Improvements relating to modification of the annual distribution of active quotas, to updates and additions to the categories or capabilities of sensors, to revision of the share of costs, to arrangements for the sharing and availability of data, and to handling mission reports, as well as to minor matters of an administrative and technical nature, are not to be considered as amendments to the Treaty. The Commission's decisions are to be taken only by consensus.

Final Clauses. The Treaty is of unlimited duration. A state deciding to withdraw from the Treaty must provide notice of its decision at least six months in advance of the date of its intended withdrawal. The depositaries must convene a conference of the parties no less than 30 days and no more than 60 days after they have received the notice, in order to consider the effect of the withdrawal. Any party has the right to propose amendments to the Treaty. An amendment must be approved by all parties. Unless requested to do so earlier by at least three parties, the depositaries must convene a conference to review the implementation of the Treaty three years after its entry into force and at five-year intervals thereafter.

Assessment. Whereas the great powers possess a variety of means to inspect each other, as well as other nations, the remaining countries are often in the dark as regards the military activities of their actual or potential adversaries. In most cases they have to rely on the superpowers for intelligence information. The Open Skies Treaty may attenuate this discrepancy by initiating a new phase in the process of enhancing transparency among states. On the other hand, since in recent years the international situation in Europe has changed, much of the original confidence-building rationale for open skies may have disappeared. Nonetheless, by providing information unavailable by other means, the Open Skies Treaty mechanism can help to verify compliance with the 1990 Treaty on Conventional Armed Forces in Europe (CFE Treaty) and the 1999 Vienna CSBM Document and serve as a tool for monitoring intra-state conflicts in the area covered by OSCE activities. With the consent of the parties, the Treaty could also be used in the field of environmental protection or natural disasters.

The 1994 Code of Conduct

Another contribution to the body of CSBMs in Europe is the Code of Conduct on Politico-Military Aspects of Security, adopted in Budapest at the 5–6 December 1994 meeting of the OSCE heads of state or government. This Code reaffirmed the validity of the guiding principles and common values embodied in the 1975 Helsinki CBM Document and specified the rights and duties of the OSCE members concerning both interstate and intra-state relations. Its provisions are politically binding. If requested, a state must provide clarification regarding their implementation.

The norms belonging to the category of interstate relations include: the duty to consult promptly with a state seeking assistance in realizing its individual or collective self-defence and to consider jointly the nature of the threat and actions that may be required in defence of the common values; the duty to prevent and combat terrorism in all its forms and to fulfil the requirements of international agreements regarding the prosecution or extradition of terrorists; the duty not to provide assistance to or support states that are in violation of their obligation to refrain from the threat or use of force against the territorial integrity or political independence of any state; the right to freely choose the security arrangements in accordance with international law and the commitments to OSCE principles and objectives; the right to belong or not to belong to international organizations and to be or not to be a party to bilateral or multilateral treaties, including treaties of alliance, as well as the right to neutrality; the duty to maintain only such military capabilities as are commensurate with individual or collective legitimate security needs (taking into account the obligations under international law) and not to attempt to impose military domination over any other state; the right to station armed forces on the territory of another state in accordance with a freely negotiated agreement, as well as in accordance with international law; the duty to cooperate in countering tensions generated by violations of human rights as well as manifestations of aggressive nationalism, racism, chauvinism, xenophobia and anti-semitism; and the duty to seek, in the event of armed conflict, to facilitate the effective cessation of hostilities and create conditions favourable to the political solution of the conflict, as well as to cooperate in support of humanitarian assistance to alleviate suffering among the civilian population.

The norms belonging to the category of intra-state relations include: the duty to consider the democratic political control of military, paramilitary and internal security forces, as well as of intelligence services and the police, to be an indispensable element of stability and security; the duty to provide for the legislative approval of defence expenditures and for transparency and public access to information related to armed forces; the duty to ensure that the armed forces as such are politically neutral without, however, restricting the civil rights of individual service members; the duty to provide and maintain measures to guard against accidental or unauthorized use of military means; the duty not to support or tolerate forces that are not accountable to or controlled by the constitutionally established authorities; the duty to ensure that the paramilitary forces refrain from the acquisition of combat mission capabilities in excess of those for which they were established; the duty to ensure that the recruitment or call-up of personnel for service in military, paramilitary and security forces is consistent with the commitments in respect of human rights and fundamental freedoms; the duty to reflect in internal laws or other relevant documents the rights and duties of armed forces personnel, and to consider introducing exemptions from or alternatives to military service; the duty to instruct the armed forces personnel in international humanitarian law, rules and conventions governing armed conflict, and to ensure that such personnel are aware that they are individually accountable for their actions; the duty to ensure that armed forces personnel vested with command authority exercise it in accordance with relevant national as well as international law and are made aware that they can be held individually accountable under those laws for the unlawful exercise of such authority, and that orders contrary to national and international law must not be given (the responsibility of superiors does not exempt subordinates from any of their individual responsibilities); the duty to provide appropriate legal and administrative procedures to protect the rights of the forces personnel; the duty to ensure that any decision to assign armed forces to internal security missions is arrived at in conformity with constitutional procedures and, if recourse to force cannot be avoided, to ensure that its use is commensurate with the needs for enforcement; and the duty not to use armed forces to limit the peaceful and lawful exercise of human and civil rights by persons as individuals or as representatives of groups, nor to deprive them of their national, religious, cultural, linguistic or ethnic identity.

No measures to enforce compliance with the above norms are included in the Code. This omission weakens its significance.

The 1996 OSCE Lisbon Document – A Framework for Arms Control

The Declaration on a Common and Comprehensive Security Model for Europe for the Twenty-First Century, which was adopted in Lisbon on 2 December 1996 by the heads of state or government of the states participating in the OSCE, contained the Framework for Arms Control.

The purposes of the Framework are as follows: to contribute to the further development of the OSCE area as an indivisible common security space by, *inter alia*, stimulating the elaboration of further arms control measures; to provide a basis for strengthening security and stability through tangible steps aimed at enhancing the security partnership among OSCE states; to enable OSCE states to deal with specific security problems in appropriate ways, not in isolation but as part of an overall undertaking to which all are committed; to create a web of 'interlocking and mutually reinforcing' arms control obligations and commitments that will give expression to the principle that security is indivisible for all OSCE states; to provide structural coherence to the interrelationship between existing and future agreements; and to provide a basis for the establishment of a flexible agenda for future arms control in the OSCE.

Considering the existing and possible future challenges and risks in the field of military security, the following issues should be addressed: military imbalances that may contribute to instabilities; interstate tensions and conflicts, particularly in border areas; internal disputes with the potential to lead to military tensions or conflicts between states; enhancing transparency and predictability as regards the military intentions of states; helping to ensure democratic political control and guidance of military, paramilitary and security forces by constitutionally established authorities and the rule of law; ensuring that the evolution or establishment of multinational military and political organizations is fully compatible with the OSCE's concept of security and is also fully consistent with arms control goals and objectives; ensuring that no participating state, organization or grouping strengthens its security at the expense of the security of others or regards any part of the OSCE area as a particular sphere of influence; ensuring that the presence of foreign troops on the territory of a participating state is in conformity with international law, the freely expressed consent of the host state, or a relevant decision of the UN Security Council; ensuring full implementation of arms control agreements at all times, including times of crisis; ensuring, through a process of regular review undertaken in the spirit of cooperative security, that arms control agreements continue to respond to the security needs in the OSCE area; and ensuring full cooperation, including cooperation in the implementation of existing commitments, in combating terrorism in all its forms and practices.

The following principles were developed to guide future negotiations: arms control regimes should contain measures designed to ensure that each participating state will maintain only such military capabilities as are commensurate with legitimate individual or collective security needs and will not attempt to impose military domination over any other participating state; a key element of an effective arms control regime is provision for complete, accurate and timely exchange of relevant information, including the size, structure, location and military doctrine of military forces, as well as their activities; the measures adopted should be combined, as appropriate, with verification that is commensurate with their substance and significance, including verification sufficiently intrusive to permit an assessment of the information exchanged and of the implementation of measures subject to verification; and limitations and, where necessary, reductions are an important element in the continuing search for security and stability at lower levels of forces.

The 1997 NATO–Russia Founding Act

Whereas the Warsaw Treaty Organization had dismantled its organs and structures by 1 April 1991, NATO engaged in enlarging its membership. The countries aimed at were, in the first place, those of Central and Eastern Europe, which feared the reconstitution of the Soviet empire, with its traditional drive westwards, and looked for reliable assurances that they would be effectively defended if their independence were once again placed in jeopardy. For Russia, eastwards expansion of NATO signified bringing closer to the Russian borders the armed forces of an organization which the Russian leaders, as well as the Russian public at large, considered a hostile alliance in spite of the apparently changing functions of this alliance. NATO ignored the Russian objections and, on 16 December 1997, its members signed the protocols of accession to NATO by the Czech Republic, Hungary and Poland. To attenuate the negative impact of this event on the relations between Russia and the Western powers, an agreement called the Founding Act on Mutual Relations, Co-operation and Security between NATO and the Russian Federation was signed on 27 May 1997 (that is, prior to the formal enlargement of NATO) between the President of Russia and the heads of state of the NATO countries.

The Founding Act noted that NATO and Russia did not consider each other as adversaries and cited the political and economic transformations on both sides that made possible this relationship. A new forum, called the NATO–Russia Permanent Joint Council, was created 'in order to enhance each other's security and that of all nations in the Euro-Atlantic area'. Disagreements were to be settled on the basis of goodwill and mutual respect within the framework of political consultations. The issues that NATO and Russia decided to discuss with each other included conflict prevention, peacekeeping operations, exchange of information on defence policies, arms control questions, prevention of the proliferation of weapons of mass destruction, possible cooperation in theatre missile defence and conversion of defence industries.

In the key provisions of the Founding Act, those describing the military dimensions of the NATO–Russia relationship, NATO reiterated its December 1996 statement of 'no intention, no plan and no reason' to deploy nuclear weapons on the territories of new members. Nor would nuclear-weapon storage sites be established on these territories. In the current and foreseeable security environment, NATO would carry out its collective defence and other missions through interoperability, integration and capability for reinforcement rather than by additional permanent stationing of substantial combat forces.

With all the above-declared restraints, NATO has kept the door open for further enlargement, to include even the former Soviet republics.

The NATO–Russia Council

The NATO–Russia Permanent Joint Council held its last meeting (at the level of foreign ministers) on 14 May 2002. Fourteen days later, in conjunction with the NATO–Russia summit meeting in Rome, Italy, the NATO heads of state and government agreed with the Russian President to establish a new NATO–Russia Council (NRC).

The NRC is to focus on specific projects in which NATO and Russia share a common goal. The initial workplan included projects in the following areas: assessment of the terrorist threat; crisis management; non-proliferation; arms control and confidence-building measures; theatre missile defence; search and rescue at sea; military-to-military cooperation; defence reform; civil emergencies; and new threats and challenges, including scientific cooperation and airspace management. Other projects may be added as the NRC develops.

The NRC does not affect NATO's existing responsibilities as a political and military alliance based on collective defence. It does not provide Russia a veto over NATO decisions or action. The NATO allies retain the freedom to act, by consensus, on any issue at any time. They will decide among themselves the issues they will address in the NRC, as well as the extent to which they will take a common position on these issues. NATO has established an information office in Moscow, where NGOs, academic institutions and interested Russian citizens can obtain information about NATO.

The creation of the NRC was regarded by many as an event ending the era of the Cold War. At the same time a question arose about the future of NATO – the alliance founded over five decades ago to defend Western Europe against a potential Soviet aggression – once Russia takes a seat at the NATO table as a partner and obtains a voice on a range of issues concerning the alliance.

16.3   CBMs in Asia

During the decade of the 1990s measures similar to European CSBMs – although less extensive – were adopted in relations among several Asian states to diminish military tensions along the disputed borders.

The 1996 Shanghai Agreement

On 26 April 1996 Russia and three Central Asian republics bordering on China – Kazakhstan, Kyrgyzstan and Tajikistan – constituting the Joint Party, and China signed, in Shanghai, the Agreement on Confidence Building in the Military Field in the Border Area. The signatories committed themselves not to attack the other party or carry out any military activity threatening the other party and disturbing the tranquillity and stability in the border area between Russia, Kazakhstan, Kyrgyzstan and

Tajikistan, on the one hand, and China, on the other. They decided to exchange information on the agreed components of the armed forces and the border troops; not to conduct military exercises directed against the other party; to limit the scale, geographical scope and number of military exercises; to give notification of any large-scale military activity and troop movements resulting from emergency situations; to give notification of the temporary entry of troops and weapons into the 100-kilometre geographical zone on both sides of the border between the Joint Party territories and China; to invite observers to military exercises on a reciprocal basis; to give notification of the temporary entry of the parties' river-going combat vessels of navies or naval forces into the 100-kilometre geographical zone on both sides of the eastern part of the Russian–Chinese border; to take measures to prevent hazardous military activity; to make inquiries about unclear situations; to strengthen friendly contacts between military personnel of the armed forces and the border troops in the border area and carry out other confidence-building measures agreed upon by the parties.

Moreover, the border troops should not use inhuman or rough treatment in dealing with border violators. The use of weapons by the border personnel would be determined by the domestic legislation of the parties and the corresponding agreements of Russia, Kazakhstan, Kyrgyzstan and Tajikistan with China.

The Agreement should not affect the obligations previously assumed by the parties. It was concluded for an indefinite period of time, but each party has the right to terminate it by notifying the other party of its decision to do so not less than six months in advance. Each state of the Joint Party has the right to withdraw from the Agreement, but it will remain in force as long as at least one state of the Joint Party and China remain parties to it.

The 1996 Sino-Indian Agreement

On 26 November 1996 China and India signed the Agreement on Confidence-Building Measures in the Military Field along the Line of Actual Control in the India–China Border Areas. It was a follow-up to their Agreement on the Maintenance of Peace and Tranquility along the Line of Actual Control in the India–China Border Areas, signed on 7 September 1993.

The signatories to the 1996 Agreement agreed that neither side should use its military capability against the other side. No armed forces deployed by either side in the border areas along the line of actual control, as part of their respective military strength, may be used to attack the other side or engage in military activities that threaten the other side or undermine the peace, tranquillity and stability in the India–China border areas. The two sides reiterated their determination to seek a fair, reasonable and mutually acceptable settlement of the boundary question. Pending an ultimate solution to this question, they reaffirmed their commitment to strictly respect the line of actual control. They also reaffirmed that they would reduce or limit their military forces, within mutually agreed geographical zones along the line of actual control in the border areas, to minimum levels compatible with the friendly and good-neighbourly relations and consistent with the principle of mutual and equal security. Reductions or limitations concerned the number of field army, border defence forces, paramilitary forces and any other mutually agreed category of armed force deployed in mutually agreed geographical zones. The major categories of armament to be reduced or limited were: combat tanks, infantry combat vehicles, guns (including howitzers) with 75-mm or larger calibre, mortars with 120-mm or larger calibre, surface-to-surface missiles, surface-to-air missiles and any other weapon system mutually agreed upon. Data were to be exchanged on the military forces and armaments to be reduced or limited. The ceilings on military forces and armaments to be kept by each side within mutually agreed geographical zones were to be determined with due consideration given to parameters, such as the nature of terrain, road communication and other infrastructure, as well as the time needed to induct troops and armaments.

China and India undertook to avoid holding large-scale military exercises, those involving more than one division (approximately 15,000 troops), in close proximity to the line of actual control in the border areas. However, if such exercises had to be conducted, the strategic direction of the main force involved was not to be towards the other side. If either side conducted a major military exercise, involving more than one brigade group (approximately 5,000 troops), in close proximity to the line of actual control in border areas, it would have to give the other side prior notification with regard to the type, level, planned duration and area of exercise, as well as the number and type of units or formations participating in the exercise.

Both sides must take adequate measures to ensure that air intrusions across the line of actual control do not take place. Combat aircraft (to include fighter, bomber, reconnaissance, military trainer, armed helicopter and other armed aircraft) may not fly within 10 kilometres of the line of actual control. If either side is required to undertake flights of combat aircraft within 10 kilometres, it must give the relevant information to the other side through diplomatic channels. No military aircraft of either side may fly across the line of actual control, except by prior permission. Unarmed transport aircraft, survey aircraft and helicopters are to be permitted to fly up to the line of actual control. Neither side is allowed to open fire, cause biodegradation, use hazardous chemicals, conduct blast operations or hunt with guns or explosives within 2 kilometres of the line of actual control.

Detailed implementation measures are to be decided through mutual consultations in the India–China Joint Working Group on the Boundary Question. The India–China Diplomatic and Military Expert Group is to assist the Joint Working Group in devising implementation measures under the Agreement.

The 1997 Moscow Agreement

On 24 April 1997, as a follow-up to the 1996 Shanghai Agreement, Russia, Kazakhstan, Kyrgyzstan and Tajikistan, constituting the Joint Party, and China signed, in Moscow, the Agreement on the Mutual Reduction of Armed Forces in the Border Area. The signatories agreed on the following measures.

The parties' armed forces stationed in the border area should not be used to attack another party or to conduct any military activity that threatened the other party or disturbed the tranquillity and stability in the border area.

The parties should reduce and limit the number of personnel and the quantities of basic types of armament and military equipment of the ground forces, air forces and air defence aviation, deployed within the geographical zone of application (GZA) of the Agreement, that is, in the geographical area extending to a distance of 100 kilometres from either side of the border between Russia, Kazakhstan, Kyrgyzstan and Tajikistan, on the one side, and China, on the other side. Certain limited areas within the GZA of the Agreement are to be considered sensitive areas. In the Eastern Sector (the eastern part of the state border between Russia and China), on the Russian side, these are the Khabarovsk sensitive area and the Vladivostok sensitive area.

Upon expiration of the reduction period (see below) the maximum level of personnel of ground forces, air forces and air defence aviation remaining for each party in the GZA of the Agreement should not exceed 130,400 persons, including 115,400 in ground forces, 14,100 in air forces and 900 in air defence aviation. The maximum level of personnel for the Eastern Sector should not exceed 119,400 persons; for the Western Sector (the western part of the state border between Russia and China, as well as the state borders between Kazakhstan, Kyrgyzstan, Tajikistan and China), 11,000 persons. From the date of entry into force of the Agreement, the maximum number of personnel of the border forces for each party within the GZA of the Agreement may not exceed 55,000 persons, including 38,500 for the Eastern Sector and 16,500 for the Western Sector.

The maximum levels of armaments and military equipment remaining for each party within the GZA of the Agreement should include armaments and military equipment located in combat units as well as in storage. Upon expiration of the reduction period, the maximum levels remaining for each party within the GZA should not exceed: 3,900 battle tanks, 5,890 armoured combat vehicles, 4,540 artillery systems, 96 tactical rocket launchers, 290 combat aircraft and 434 combat helicopters.

The reductions provided for in the Agreement must be brought about within 24 months from the date of its entry into force. The reduction of military personnel should be carried out by disbanding entire military formations, by reducing the staff size of military formations or by removing military formations from the GZA of the Agreement. The reduction of armaments and military equipment should be carried out by destroying, dismantling, converting to civilian purposes, placing on permanent display, using as ground or aerial targets, reclassifying into training *matériel*, or partially removing from the GZA.

In order to reinforce mutual confidence and ensure control over the implementation of the Agreement, the parties should exchange information about the troop formations, the number of personnel in these formations, and the quantity of main types of armament and military equipment deployed within the GZA of the Agreement. The information exchanged must be treated as confidential. Each party has the right to conduct and the obligation to accept inspections within the GZA, with the exception of the specified sensitive areas. The inspecting party should bear the expenses related to the transportation of the inspectors to the established entry/exit points. The inspected party should bear the expenses related to the visit of the inspectors. A Joint Control Group supervises the implementation of the Agreement.

The Agreement does not affect the obligations previously undertaken by the parties in relation to other states and is not directed against third states or their interests. Each party is allowed to terminate the Agreement by notifying the other party of its intention to do so at least six months before the date of the Agreement's expiration, which was set for 31 December 2020. In the absence of such notification, the duration of the Agreement is to be automatically extended for successive five-year periods. Each state belonging to the Joint Party may withdraw from the Agreement by notifying the other party and the other states of the Joint Party of its decision. After such notification the parties should conduct negotiations on the maximum levels of armed forces and border forces in the border area.

The 1997 Sino-Russian Statement

As a result of the meeting held in Beijing on 10 November 1997, the Presidents of China and Russia issued a statement on the development of relations between the two countries. In particular, the heads of state stated that all points of contention regarding the demarcation of the eastern section of the Sino-Russian border had been resolved and that the demarcation of the western section would be completed within an agreed period of time. Hope was expressed that a fair demarcation of the border would enhance friendship and good-neighbourly relations between the two countries and contribute to regional stability.

Exchanges of visits by heads of state, regular meetings between prime ministers and consultations between foreign ministers were found conducive to improving mutual communication and understanding, as well as to expanding and deepening cooperation between the two nations in various fields. It was noted that cooperation in the field of military technology was an important component of Sino-Russian relations and that it was not directed against a third country.

The 1998 Almaty Joint Statement

In the joint statement issued on 3 July 1998 at Almaty as a result of the five-nation meeting of China, Kazakhstan, Kyrgyzstan, Russia and Tajikistan, the participants undertook to take all the necessary measures to ensure the implementation of the 1996 Shanghai Agreement and the 1997 Moscow Agreement. They valued the positive impact of these agreements on the security in their region and the world at large, appreciated the initiative of the Central Asian countries for the establishment of a Central Asian nuclear-weapon-free zone and reaffirmed the importance of holding regular consultations among themselves.

The parties expressed concern over the tensions in Afghanistan and noted that greater effort should be made to promote a peaceful settlement of the conflicts in that country under the auspices of the United Nations and with the participation of the states concerned. They also expressed concern over the growing tension in South Asia following the nuclear test explosions in that region and called for stopping the nuclear arms race there.

The 1999 Lahore Memorandum of Understanding

On 21 February 1999 the Foreign Secretaries of India and Pakistan signed a Memorandum of Understanding identifying measures aimed at promoting an environment of peace and security between the two countries. The parties undertook to engage in bilateral consultations on security concepts and nuclear doctrines with a view to developing measures for confidence building in the nuclear and conventional fields;

to take national measures to reduce the risks of accidental or unauthorized use of nuclear weapons under their control; to notify each other immediately in the event of an incident that could create the risk of a fallout with adverse consequences for both sides or an outbreak of a nuclear war between the two countries; to adopt measures aimed at diminishing the possibility of such incidents being misinterpreted by the other side, and to identify or establish appropriate communication mechanisms for this purpose; to abide by their moratoria on nuclear test explosions, unless either side decided that extraordinary events had jeopardized its supreme interests; to conclude an agreement on the prevention of incidents at sea; to periodically review the implementation of the CBMs and, where necessary, set up consultative mechanisms; and to review the existing communication links with a view to upgrading them.

The 2001 Sino-Russian Good-Neighbourliness Treaty

On 16 July 2001 Russia and China signed, in Moscow, a Treaty of Good-Neighbourliness, Friendship and Cooperation. The contracting Parties reaffirmed their commitments not to be the first to use nuclear weapons against each other and not to target strategic nuclear missiles on each other; pledged to expand and deepen confidence-building measures in the military field so as to consolidate the security of both countries and strengthen regional and international stability; promised not to be members of any alliance or bloc nor embark on any action which compromises the sovereignty, security or territorial integrity of the other party, nor allow its territory to be used by third countries to the detriment of the other party; and undertook to cooperate in combating terrorism, separatism and extremism and in fighting organized crime, illegal trafficking in drugs, psychotropic substances and weapons. Should a situation arise which, in the view of either party, might endanger or undermine the peace or affect its security interests, or should either party face the threat of aggression, the parties shall immediately contact and consult each other with a view to averting the danger.

The ASEAN Undertakings

In the 1990s the Regional Forum of the Association of South East Asian Nations (ASEAN) adopted a series of CBMs covering, *inter alia*, military and defence-related issues. In particular, the Forum's members have developed bilateral exchanges on security perceptions; expanded high-level defence contacts and military exchange/training; submitted annual defence policy statements; prepared defence White Papers; invited observers to and provided notification of select military exercises on a case-by-case basis; and exchanged views on defence conversion programmes. None of these measures is mandatory.

Assessment

The CBMs in Asia have not helped to resolve the most controversial issue in Indian–Pakistani relations, namely, the territorial dispute over Jammu and Kashmir. The enmity between the two countries, including its religious dimensions, has remained unchanged. It has even intensified in recent years owing to the nuclear arms competition; armed clashes have continued. In contrast, the CBMs undertaken by China, Russia and the Central Asian republics, as described above, helped to set aside the disputes over large sectors of their common borders and to put off the final delineation of these borders for some unspecified period of time. Unlike those in Europe, the CBMs in Asia do not cover all the militarily important countries of the continent. Nor have they been followed by substantial, verifiable cuts in the military potential of the participating states.

16.4 CBMs in the Americas

On 9 November 1995 the governments of the member-states of the Organization of American States (OAS), meeting in Santiago, Chile, agreed to recommend a series of confidence- and security-building measures.

The Santiago Declaration called for agreements on prior notification of military exercises; participation by all member-states in the UN Register of Conventional Arms and the system for the standardized reporting of military expenditure; exchange of information on defence policies and doctrines; consideration of a process of consultations with the aim of making progress in conventional arms limitation and control; agreements on the invitation of observers to military exercises, visits to military installations and the exchange of civilian and military personnel for training; activities to prevent incidents and enhance security in land, sea and air traffic; cooperation to deal with or prevent natural disasters; development and introduction of communications between the civilian or military authorities of neighbouring countries; seminars and studies on measures of mutual trust and confidence-building policies, involving civilian and military personnel; holding a high-level meeting on the special security concerns of small island states; and peace education programmes.

17

Restrictions on the Methods of Warfare

Efforts to reduce brutality in war, motivated by humanitarian and religious as well as practical considerations, have a long history. Over the centuries, a body of rules and principles guiding the behaviour of belligerent states has developed as customary law. 'Custom' means a widespread repetition, over a long period, of a specific type of conduct in the belief that such conduct is obligatory and should be respected by all. It has thus been generally recognized that weapons and war tactics must, in their application, be confined to military targets; that they must be proportional to their military objectives as well as reasonably necessary to the attainment of these objectives; and that they should not cause unnecessary suffering to the victims or harm human beings and property in neutral countries.

From the second half of the 19th century, customary law began to be codified and supplemented by conventional law in the form of multilateral treaties. Rules prohibiting or regulating the use of weapons or methods of warfare now form part of the international humanitarian law applicable in armed conflicts.

17.1 Pre-World War II Agreements

The 1868 Declaration of St Petersburg

The Declaration of St Petersburg, adopted in 1868, was of special significance. It proclaimed that the only legitimate objective that states should endeavour to accomplish during war is to weaken the military forces of the enemy and that the employment of arms which uselessly aggravate the suffering of disabled men or render their death inevitable would be contrary to the laws of humanity. From this principle, the Declaration went on to forbid the use of a specific type of weapon – a projectile of a weight below 400 grams which is explosive or charged with 'fulminating or inflammable substances'. Owing to the physiological effects of shock, loss of body fluid and infections, the severe wounds created by such a projectile would inevitably result in death.

The 1874 Brussels Declaration

In 1874, representatives of 15 European states met in Brussels to examine a draft, submitted by the Russian government, of an agreement concerning laws and customs of war. The participants adopted the so-called Brussels Declaration, prohibiting *inter alia* the employment of poison or poisoned weapons, arms calculated to cause unnecessary suffering and projectiles already prohibited by the Declaration of St Petersburg. The Brussels Declaration was never ratified, however, because not all governments were willing to accept it as a binding agreement.

Nevertheless, in the year in which the Declaration was adopted, the Institute of International Law appointed a committee to study its contents and to submit proposals. The work of the Institute led to the adoption at Oxford, in 1880, of the Manual of the Laws and Customs of War. The Brussels Declaration and the Oxford Manual formed the basis of the relevant documents signed at the 1899 and 1907 Hague Conferences.

The 1899 and 1907 Hague Declarations and Conventions

Following the spirit of the St Petersburg Declaration, Declaration IV, 3 of the Hague Peace Conference, held in 1899 (see Chapter 2), prohibited the use of so-called dumdum bullets, which expand, flatten easily in the human body and cause more serious wounds than other bullets.

The Second Hague Conference, held in 1907, adopted several conventions. Convention IV, on laws and customs of land warfare, confirmed the principles of the St Petersburg Declaration. It stated that the right of belligerents to adopt means of injuring the enemy is not unlimited, and it prohibited the employment of arms, projectiles or material calculated to cause unnecessary suffering or the destruction of the enemy's property, unless such destruction is 'imperatively demanded' by the necessities of war. In particular, Convention IV prohibited the use of poison or poisonous weapons, the treacherous killing or wounding of individuals belonging to a hostile nation or army, or the killing or wounding of an enemy who had either laid down arms or surrendered. The Conference also restricted and regulated, in Convention VIII, the use of submarine mines. This Convention forbids the laying of unanchored automatic contact mines, except when so constructed as to become harmless one hour at most after the person who laid them ceases to control them. Also forbidden is the use of anchored automatic contact mines which do not become harmless as soon as they have broken loose from their moorings as well as torpedoes which do not become harmless when they have missed their mark. Upon the termination of hostilities, the parties to a conflict in which mines are used are obliged to remove the mines they have laid, each removing its own mines. With regard to mines laid by one of the belligerents off the coast of the other, their position must be made known to the other party by the power that laid them. Convention IX prohibited the bombardment by naval forces of ports, cities, villages, habitations or buildings which are not defended.

Regulations on Submarines and Noxious Gases

In 1922 the victorious powers of World War I signed in Washington, DC, the Treaty related to the Use of Submarines and Noxious Gases in Warfare. The aim was to make more effective the rules for the protection of the lives of neutrals and non-combatants at sea as well as to reaffirm in a treaty the customary law prohibiting the use of gases in war. Strict restrictions were to be imposed on the employment of submarines in naval warfare, and their use as 'commerce destroyers' was to be entirely prohibited. However, France did not ratify the Treaty, which consequently never came into effect.

The problem of submarines was again raised at the 1930 Naval Conference in London (see Chapter 11.1). In Part IV, Article 22, of the 1930 Naval Treaty, signed at that Conference, the participants agreed that, in their action with regard to merchant ships, submarines must conform to the rules of international law applicable to surface vessels. When the Treaty expired in 1936, its Article 22 was recast as a separate protocol and signed by many nations.

The 1925 Geneva Protocol

The problem of gases was dealt with in the 1925 Geneva Protocol, which prohibited the use of asphyxiating, poisonous or other gases, and of all analogous liquids, materials or devices, as well as the use of bacteriological methods of warfare. The Protocol ratified the prohibition declared in the 1899 Hague Declaration IV, 2, under which the contracting powers had agreed to abstain from the use of projectiles for the diffusion of asphyxiating or deleterious gases, as well as the prohibition on the use of poison, contained in the 1907 Hague Convention IV referred to above. Of the direct antecedents of the Geneva Protocol, as far as the prohibition of chemical weapons is concerned, the 1919 Treaty of Versailles and other peace treaties of 1919–20 were applicable only to the vanquished countries (Germany, Austria, Bulgaria and Hungary), while the 1923 Convention for the Limitation of Armaments, adopted at the Conference on Central American Affairs, was binding only on the countries of this region (Costa Rica, El Salvador, Guatemala, Honduras and Nicaragua). (For the origin of the 1925 Geneva Protocol, see Chapter 2.3, and for an analysis of the Geneva Protocol, see Chapter 8.1.)

17.2 Post-World War II Agreements

When World War II broke out, the following agreements for the protection of war victims were in force: the Convention for the Amelioration of the Condition of the Wounded and Sick in Armies in the Field (which replaced the Red Cross Geneva Conventions of 22 August 1864 and 6 July 1906) and the Convention relative to the Treatment of Prisoners of War, both signed on 27 July 1929. Since the relevant international instruments failed to provide adequate humanitarian safeguards during the war, the international community considered it necessary to reinforce the existing rules, as well as to establish new rules once the war was over.

The 1948 Genocide Convention

The mass murder of millions of people during World War II led to the adoption, on 9 December 1948, of the UN Convention on the Prevention and Punishment of the Crime of Genocide, known as the Genocide Convention. Genocide, a term coined in 1943, was defined as the commission of acts intended to destroy, in whole or in part, a national, ethnic, racial or religious group, as such. It was declared to be a crime punishable under international law.

Persons charged with genocide are to be tried by a competent tribunal of the state on the territory of which the act was committed or by an international penal tribunal. The parties may call upon the competent organs of the United Nations to take such action under the UN Charter as they consider appropriate for the prevention and suppression of genocide. The Genocide Convention entered into force in 1951. Initially effective for a period of ten years, it was to remain in force for successive periods of five years for such contracting parties as have not denounced it at least six months before the expiration of the current period.

The Nuremberg and Tokyo International Tribunals

The International Military Tribunal, known as the Nuremberg trial, established in 1945 for the trial and punishment of the major World War II criminals, defined the crimes falling within its jurisdiction as follows:

(a) Crimes against peace: planning, preparation, initiation, or waging a war of aggression, or a war in violation of international treaties, agreements or assurances, or participation in a common plan or conspiracy for the accomplishment of any of the foregoing.

(b) War crimes: violations of the laws or customs of war; such violations include, but are not limited to, murder, ill-treatment or deportation to slave labour or for any other purpose of civilian population of or in occupied territory, murder or ill-treatment of prisoners of war or persons on the seas, killing of hostages, plunder of public or private property, wanton destruction of cities, towns or villages, or devastation not justified by military necessity.

(c) Crimes against humanity: murder, extermination, enslavement, deportation and other inhumane acts committed against any civilian population, or persecutions on political, racial or religious grounds.

The trial was held in Nuremberg. The verdict rendered in 1946 found all but three of the 22 defendants guilty, of whom 12 were sentenced to death.

The Japanese war criminals were tried by the International Military Tribunal for the Far East, established in 1946. The verdict rendered in 1948 found 25 defendants guilty, of whom seven were sentenced to death.

Other Ad Hoc International Tribunals

In 1992, when the breakup of Yugoslavia was followed by the atrocities of the 'ethnic cleansing' campaign, the Secretary-General of the United Nations, the depositary of the Genocide Convention, appointed a commission to examine and analyse the evidence of grave breaches of international humanitarian law committed on the territory of the former Yugoslavia after 1991. In May 1993 the UN Security Council considered the report of the Secretary-General and unanimously resolved that the International Criminal Tribunal for the Former Yugoslavia (ICTY) be established – as an independent, temporary organ, with its seat in The Hague – for the purpose of prosecuting persons responsible for these breaches, including genocide. The judges are elected by the UN General Assembly from a list submitted by the UN Security Council, whereas the prosecutor is appointed by the Security Council on nomination by the Secretary-General. The trial may be held only in the presence of the accused person.

In 1995, after hundreds of thousands of people had been killed in internal strife in Rwanda, the UN Security Council decided to establish the International Criminal Tribunal for Rwanda. The purpose of this tribunal, which has its seat in Arusha, Tanzania, is to prosecute persons responsible for genocide, crimes against humanity and other serious violations of international law in Rwanda in 1994, including violations of legal norms relating to the protection of victims of non-international armed conflicts.

Although their statutes differ, the two tribunals have some organizational and institutional links to ensure a uniform legal approach as well as the economy and efficiency of resources. They have issued a number of indictments and arrest warrants. By 1 March 2002, 26 persons had been convicted by the Tribunal for the Former Yugoslavia and eight by the Tribunal for Rwanda. The trial of Slobodan Milosevic, former President of Yugoslavia, by the ICTY began on 12 February 2002.

Further tribunals are envisaged to try those who committed atrocities in Cambodia (in the 1970s) as well as in Sierra Leone and East Timor (in the 1990s). However, as distinct from the tribunals for Yugoslavia and Rwanda, these are to be national courts with international participation. The degree of international participation is to be negotiated between the United Nations and the governments of the respective countries. In the case of Cambodia, the UN Secretary-General decided, in 2002, to interrupt negotiations that had been held for several years with the Cambodian authorities because the independence, impartiality and objectivity of the envisaged court could not be guaranteed.

It is remarkable that new frontiers are being crossed in international law. After World War II it was the victors who set the rules for punishing the vanquished, whereas now it is the international community, as a whole, that brings to justice perpetrators of war crimes and crimes against humanity.

The Standing International Tribunal

An international conference, convened by the UN General Assembly and held in Rome from 15 June to 17 July 1998, adopted by a large majority the Statute of a permanent International Criminal Court (ICC), often referred to as the Rome Statute. The seat of the ICC is in The Hague, but the Court may sit elsewhere whenever it considers this to be desirable.

The jurisdiction of the ICC covers crimes of genocide, crimes against humanity and war crimes, listed in Article 8 of the Statute, but only those committed after the entry into force of the Statute. (Unlike chemical weapons, biological weapons are not listed as weapons whose use constitutes a war crime.) The Statute also stipulates that the Court shall exercise jurisdiction over crimes of aggression once a provision is adopted defining such crimes. (The definition of aggression contained in 1974 UN General Assembly Resolution 3314 is regarded by many as out of date.) If a state becomes a party to the Statute after it has entered into force, jurisdiction for that state may be exercised only with respect to crimes committed after the Statute has entered into force for that state.

A state may refer to the ICC a situation in which crimes within the jurisdiction of the Court appear to have been committed and request the prosecutor to investigate the situation. The prosecutor may also initiate investigations, and the UN Security Council – acting under Chapter VII of the UN Charter – may refer situations to the prosecutor. An important principle to be applied by the ICC is that concerning the irrelevance of official capacity. This means that the capacity as head of state or government, member of government or parliament, or government official shall in no case exempt a person from criminal responsibility under the Statute, nor shall it constitute a ground for a reduced sentence.

It should be noted that the ICC is to be 'complementary to national criminal jurisdiction'. It must therefore declare a case inadmissible if it is being investigated or prosecuted by a state having jurisdiction, where such a state has already decided not to prosecute or where the accused person has already been tried. This provision is not to apply where the ICC determines that the state concerned is unwilling or unable to genuinely carry out the investigation or prosecution and, in particular, where such investigation, prosecution or trial has been carried out for the purpose of shielding the accused, has been unduly delayed or is not being conducted independently and impartially.

To resolve a major legal difficulty that arose during the Nuremberg war crimes trial, the Statute includes specific provisions regarding the responsibility of commanders and other superiors. The rule is that a military commander shall be criminally responsible for crimes committed by forces under his or her command and control as a result of a failure to exercise control properly over such forces, where the commander knew or should have known that the forces were committing such crimes and where the commander failed to take all reasonable and necessary measures to prevent the crimes or to submit the matter to the competent authorities for investigation and prosecution. On the other hand, a person is to be relieved of criminal responsibility if he or she was under a legal obligation to obey orders or did not know that the order was unlawful, or if the order was not manifestly unlawful. Orders to commit genocide or crimes against humanity are manifestly unlawful. The ICC may impose on a convicted person an imprisonment which should not exceed 30 years or a term of life imprisonment, but not a death sentence. In addition, the Court may order a fine and forfeiture of proceeds, property and assets derived from a crime.

No reservations are allowed by the Statute. However, any state that becomes a party to it may make a declaration that, for a period of seven years after entry into force of the Statute for that state, it does not accept the jurisdiction of the Court with respect to war crimes specified in the Statute when a crime is alleged to have been committed by its nationals or on its territory. Only after the expiry of seven years from the date of entry into force of the Statute may amendments be formally submitted. Moreover, except in the case of a Security Council referral, the state in which the crime took place, or whose nationals committed it, must be a party to the Statute in order for such an act to be prosecuted.

On 6 May 2002 the US State Department informed the depositary of the Rome Statute that the United States did not intend to become a party to the Statute and that, consequently, it had no legal obligations arising from its signature on 31 December 2001. The State Department requested that the US intention not to become a party be reflected in the depositary's status lists relating to the Statute.

In spite of the shortcomings of the Rome Statute, the most important of which are mentioned above, there is a widespread belief that the ICC will eventually put an end to impunity and further deter international crimes. The Statute entered into force on 1 July 2002.

The 1949 Geneva Conventions

Important rules of international humanitarian law were worked out at a conference held in Geneva in 1949. They were included in the following four conventions: Convention I for the Amelioration of the Condition of the Wounded and Sick in Armed Forces in the Field; Convention II for the Amelioration of the Condition of the Wounded, Sick and Shipwrecked Members of Armed Forces at Sea; Convention III relative to the Treatment of Prisoners of War; and Convention IV relative to the Protection of Civilian Persons in Time of War.

The Geneva Conventions of 12 August 1949 were conceived chiefly as a code of behaviour in wars of the traditional type – those conducted between states and between regular armed forces. However, most armed conflicts since World War II have been civil wars or have started as such. Guerrilla warfare – a frequent type of conflict – has complicated application of the principle that a distinction must be observed between civilians and the military. As a result, the protection of civilians has become considerably weakened. Furthermore, the laws of war which relate directly to the methods and means of warfare – as distinct from rules designed to accord protection to certain persons, places or objects in armed conflicts – had not developed since the 1907 Hague Conventions, with the sole exception of the 1925 Geneva Protocol banning the use of chemical and bacteriological means of warfare. In particular, air warfare had remained largely uncodified and new weapons of an especially cruel nature were not specifically prohibited. To deal with these matters, a Diplomatic Conference on the Reaffirmation and Development of International Law Applicable in Armed Conflicts was convened in Geneva in 1974. At the end of the fourth session of the Conference, two protocols were agreed; they were opened for signature on 12 December 1977.

Protocols Additional to the 1949 Geneva Conventions

1977 Protocol I. Protocol I, which relates to international armed conflicts, reaffirms the basic rules that the right of the parties to an armed conflict to choose methods or means of warfare is not unlimited and that it is prohibited to use weapons, projectiles, and material and methods of warfare of such a nature as to cause superfluous injury or unnecessary suffering. In addition, the parties are under an obligation to determine in their study, development, acquisition or adoption of a new weapon, means or method of warfare whether its employment would in some or all circumstances be prohibited by the Protocol or other rules of international law. In the understanding of the great powers, the rules of warfare established by Protocol I do not regulate or prohibit the use of nuclear weapons.

Protocol I also reiterates and expands the traditional rules regarding the protection of the civilian population. The prohibition against indiscriminate attacks now covers attacks by bombardment via any methods or means which treat as a single military objective a number of clearly separated and distinct military objectives located in a city, town, village or other area containing a similar concentration of civilians or civilian objects, as well as attacks which may be expected to cause incidental loss or injury to civilians, excessive in relation to the direct military advantage anticipated. Reprisals against the civilian population are forbidden.

Furthermore, it is prohibited to destroy or render useless foodstuffs, agricultural areas for the production of foodstuffs, crops, livestock, drinking water installations and supplies, and irrigation works, for the specific purpose of denying the civilian population those objects which are indispensable for its survival. However, in recognition of the vital requirements of any party to a conflict to defend its national territory against invasion, derogation from these prohibitions may be made by a party within such territory under its own control where required by 'imperative military necessity'. Dams, dykes and nuclear electricity-generating stations have been placed under special protection and are not to be attacked if such attack may cause the release of dangerous forces and consequent severe losses among civilians. However, this protection may cease if the installations in question are used in significant and direct support of military operations and if an attack on them is the only feasible way to terminate such support (see also Chapter 9.2). The Protocol urges the parties to conclude further agreements to provide additional protection for objects containing dangerous forces. (Proposals have been made to expand the list of protected installations by adding to it oil rigs and pipelines.)

Detailed precautionary measures are prescribed to spare the civilian population and civilian objects in the conduct of military operations. There is a prohibition on attacking, by any means, non-defended localities, declared as such by the appropriate authorities of a party to the conflict, or extending military operations to zones which the parties agreed to consider as having the status of demilitarized zones. Members of civil defence organizations and journalists engaged in dangerous professional missions in areas of armed conflicts must also be protected.

Several articles dealing with relief action in favour of the civilian population have strengthened the corresponding clauses of the 1949 Geneva Convention IV. The duties of the occupying power include providing, 'to the fullest extent of the means available', supplies essential to the survival of the civilian population of the occupied territory.

A special provision concerns the protection of the natural environment against 'widespread, long-term and severe damage'. It includes a prohibition on the use of methods and means of warfare that are intended or may be expected to cause such damage to the natural environment.

Protocol I, which brought about the convergence of the 'Hague rules' affecting the conduct of hostilities and the 'Geneva rules' for the protection of victims of war, is applicable not only to interstate armed conflicts but also to conflicts in which peoples – in the exercise of their right to self-determination – are fighting against colonial domination, alien occupation and racist regimes. In this way, guerrilla fighters are now covered by international protection. In particular, they have the right to prisoner-of-war status if they belong to organized units subject to an internal disciplinary system and under a command responsible to the party concerned. They also have to carry their arms openly during each military engagement and during such time as they are visible to the adversary before launching an attack. Any combatant, as defined in this Protocol, who 'falls into the power' of an adversary is entitled to prisoner-of-war status. On the other hand, mercenaries, as defined in the Protocol, have no right to combatant or prisoner-of-war status.

1977 Protocol II. This Protocol develops and supplements Article 3, which appears in all four Geneva Conventions of 1949 and deals with armed conflicts not of an international character. It prescribes humane treatment for all the persons involved in such conflicts, care for the wounded, sick and shipwrecked as well as protection of civilians against the dangers arising from military operations. It does not apply to internal disturbances, such as riots, sporadic acts of violence and similar acts.

The Protocols of 1977 constitute a step forward in the development of the humanitarian law of armed conflict. In particular, Protocol I broke new ground by making the protection of the natural environment, as such, a component of international law.

Their main shortcoming is that they have not restricted or forbidden the use of any specific weapon.

The 1981 'Inhumane Weapons' Convention

To fill the gaps left by the 1977 Protocols, a special UN conference was convened in 1979 to discuss the problem of so-called inhumane weapons. No weapon can be considered as 'humane', but there are substantial differences in the effects which different types of weapon produce on individual combatants or civilians – in particular as regards the magnitude and severity of the wounds and the duration of the injury caused as well as the extent of the area covered and the degree of control that can be exercised by a user.

In 1980, at the conclusion of its second session, the UN conference adopted the text of the Convention on Prohibitions or Restrictions on the Use of Certain Conventional Weapons which may be deemed to be Excessively Injurious or to have Indiscriminate Effects. Signed in 1981, this so-called CCW Convention, also referred to as the Inhumane Weapons Convention, applies to international conflicts in the same way as Protocol I to the 1949 Geneva Conventions. It has the format of an 'umbrella treaty', under which specific agreements are subsumed in the form of protocols. Three protocols were agreed in the first instance.

Protocol I. This Protocol prohibits the use of any weapon whose 'primary' effect is to injure by fragments which in the human body escape detection by X-rays. Prohibiting a weapon whose design makes medical treatment very difficult has a clear humanitarian appeal. In this particular case, however, it is of little consequence, as the weapon prohibited does not exist and there does not seem to be any serious military interest in developing it. The use of fragmentation weapons that do exist has not been banned.

Protocol II. This Protocol restricts the use of mines, booby traps and other devices. 'Mines' are defined as any munitions placed under, on or near the ground or other surface area and designed to be detonated or exploded by the presence, proximity or contact of a person or vehicle. 'Booby traps' are defined as any devices or materials which are designed, constructed or adapted to kill or injure, and which function unexpectedly when a person disturbs or approaches an apparently harmless object or performs an apparently safe act. 'Other devices' are defined as manually emplaced munitions and devices designed to kill, injure or damage, actuated by remote control or automatically after a lapse of time.

The use in international armed conflicts of mines, booby traps and other devices against the civilian population as such, or against individual civilians, is prohibited in all circumstances, whether in offence or defence, or by way of reprisal. All 'feasible' precautions, defined as those 'practicable or practically possible', must be taken to protect civilians from the effects of such weapons. Also prohibited is the indiscriminate use of the weapons in question against military objectives in conditions which may be expected to cause incidental loss of civilian life, injury to civilians or damage to civilian objects that is excessive in relation to the concrete and direct military advantage anticipated. Booby traps designed to cause superfluous injury or unnecessary suffering are prohibited in all circumstances. In addition, the Protocol bans the use of remotely delivered mines (both anti-personnel and anti-vehicle), those delivered by artillery, rocket, mortar or similar means, or dropped from an aircraft, unless such mines are used only within an area which is itself a military objective, or which contains military objectives, and unless the location of mines can be accurately recorded or a mechanism is used to render a mine harmless or cause it to destroy itself when it no longer serves the military purpose for which it was emplaced.

Guidelines for the recording of the location of minefields (areas in which mines have been emplaced), mines and booby traps are contained in an annex to Protocol II. International cooperation in the removal of these devices after the cessation of hostilities is envisaged in a separate article, but no specific obligation is imposed on the parties to remove or otherwise render these devices ineffective.

In practice, the restrictions described above proved patently inefficacious. All kinds of landmines continued to be used, uncharted minefields continued to exist in a number of countries, and in most cases no precautions were taken to safeguard against harm to non-combatants. The CCW Convention does not apply to the use of anti-ship mines at sea or in inland waterways; in this respect, the 1907 Hague Convention VIII adopted over 90 years ago is still valid (see above).

The CCW Convention provides that conferences may be held in order to review the operation of both the Convention and its annexed Protocols, as well as to consider additional protocols. At the request of the parties, such a review conference was convened in 1995, following two years of meetings of governmental experts. The main purpose of this conference was to reinforce the constraints under Protocol II regarding landmines. (Another purpose was to consider a ban on blinding laser weapons.) On 3 May 1996, the Conference adopted an amended text of Protocol II to the Convention.

Main Amendments to Protocol II. The Amended Protocol II prohibits the use of anti-personnel mines (APMs) which are not detectable, such as plastic mines. Those produced after 1 January 1997 must incorporate in their construction a material or device that enables the mine to be detected by commonly available technical mine-detection equipment and provides a response signal equivalent to a signal from 8 grams or more of iron in a single coherent mass. APMs produced before 1 January 1997 may have attached to them, prior to their emplacement and in a manner which makes them not easily removable, a material or device making them detectable, instead of having such material or device incorporated in their construction. At the time of its notification of consent to be bound by the Amended Protocol, each party is free to declare that it will defer compliance with the latter obligation for a period of up to nine years from the entry into force of the Protocol – a period presumably needed for the acquisition of the capability to introduce the necessary changes to mines. Moreover, it is prohibited to use *any* mines which employ a mechanism or device specifically designed to detonate the munition by the mere presence of a standard mine detector, or to use a self-deactivating mine equipped with an anti-handling device that is designed in such a manner that the anti-handling device is capable of functioning even after the mine has ceased to be capable of functioning. 'Self-deactivating' means automatically rendering a mine inoperable by means of the irreversible exhaustion of a component, such as a battery, that is essential to the operation of the mine. 'Anti-handling device' is defined as a device intended to protect a mine; it is part of, linked to, attached to or placed under the mine, and activates when an attempt is made to tamper with the mine. The above prohibitions are intended to facilitate the detection of both APMs and anti-vehicle mines and to reduce the risks to mine-clearance personnel.

The presence of metal in a mine does not guarantee that the mine will be detected and can be safely removed, even if its location has been properly recorded. This is especially true for mines laid in soil rich in iron or on former battlegrounds that contain large numbers of metal fragments, including spent cartridges. To deal with this and other uncertainties, the Amended Protocol requires that all remotely delivered anti-personnel mines be designed and constructed so that no more than 10% of activated mines will fail to self-destruct within 30 days of emplacement, and that each mine have a back-up self-deactivation feature designed and constructed so that, in combination with the self-destruction mechanism, no more than one in 1,000 activated mines will function as a mine 120 days after emplacement. All non-remotely delivered anti-personnel mines used outside marked areas shall also comply with the above requirements. However, each party may declare, at the time of its notification of consent to be bound by the Amended Protocol, that it will, with respect to mines produced prior to the entry into force of the Protocol, defer compliance with the requirements of self-destruction and self-deactivation for a period of up to nine years from the entry into force of the Protocol. During the period of deferral, the parties will have to minimize 'to the extent feasible' the use of anti-personnel mines not corresponding to the specified requirements. With respect to remotely delivered anti-personnel mines, the parties will then have to comply with either the requirement of self-destruction or the requirement of self-deactivation and, with respect to other anti-personnel mines, with at least the requirement of self-deactivation.

The Amended Protocol II applies not only to international armed conflicts between the parties to the Protocol, as does its original version. It also applies to armed conflicts which are not of an international character and to those which occur on the territory of one of the parties. It was thus recognized, at least with respect to APMs, that what is inhumane and therefore prohibited in wars between states must also be considered inhumane and prohibited in civil strife.

After the cessation of active hostilities, all mines (both anti-personnel and anti-vehicle mines) must be either cleared, removed and destroyed without delay by the parties to the conflict or maintained in controlled fields.

Transfer of non-detectable mines, the use of which is banned, is not allowed, even if compliance with the requirement of detectability has been deferred. As regards mines, the use of which is only restricted, the parties must 'exercise restraint' in their transfer to states. Transfer to recipients other than a state or an authorized state agency is prohibited.

The legislative measures to be taken by the parties to prevent and suppress violations must include penal sanctions. These should be applied against persons who, in relation to an armed conflict and contrary to the provisions of the Protocol, wilfully kill or cause serious injury to civilians. An annual conference is to be held by the parties to consult and cooperate with each other on issues related to the operation of the Protocol.

The Amended Protocol II is an improvement over its original version in that it aims at further reducing civilian casualties and the loss of land for civilian purposes. It also provides in greater detail for the protection from the effects of mines of UN forces or missions, missions of the International Committee of the Red Cross and other humanitarian missions, as well as missions of enquiry. However, in several important respects, the improvements are illusory.

'Anti-personnel mine' is defined in the Amended Protocol as a mine 'primarily' designed to incapacitate, injure or kill persons. This implies that a mine not primarily designed to incapacitate, injure or kill persons is not subject to agreed restrictions, even if another purpose is to do so. The long-lived so-called 'dumb' APMs may continue to be produced and used if they are placed in fenced, marked and guarded minefields, whereas self-destructing and self-deactivating so-called 'smart' APMs may be laid without any specific restriction on their placement. The employment of APMs that are technically more sophisticated than those currently in use has thus been legitimized, and the creation of a new lucrative branch of the armaments industry for manufacture of a new generation of mines has been encouraged, primarily in industrialized countries. Less developed countries may be unable to manufacture such mines or buy them from others.

Moreover, since no international verification mechanism to check compliance with the requirements of self-destruction and self-deactivation has been set up, it will be practically impossible to reliably ascertain whether a mine was designed according to the agreed specifications and whether it was produced before or after entry into force of the Amended Protocol. Nor is it clear how the envisaged failure rate of the self-destructing and self-deactivating mechanisms can be ensured. The provision for long transition periods for the implementation of the new obligations allows the parties to continue their present practices for many years. Hence, the need to prohibit the very possession of anti-personnel mines is widely recognized (see Chapter 14.6).

Protocol III. This Protocol refers to the use of incendiary weapons. They are defined as weapons or munitions primarily designed to set fire to objects or to cause burn injury to persons through the action of flame, heat or a combination thereof, produced by a chemical reaction of a substance delivered on the target – for example, flame-throwers, fougasses, shells, rockets, grenades, mines, bombs and other containers of incendiary substances. Munitions which may have only incidental incendiary effects, such as illuminants, tracers, smoke or signalling systems, are excluded from the scope of the Protocol. So are munitions designed to combine penetration, blast or fragmentation effects with an additional incendiary effect, such as armour-piercing projectiles, fragmentation shells, explosive bombs and similar combined-effects munitions in which the incendiary effect is not specifically designed to cause burn injury to persons but to be used against military objectives, such as armoured vehicles, aircraft and installations or facilities.

The prohibitions and restrictions introduced by Protocol III aim only at the protection of civilians. Thus, it is prohibited in all circumstances to make the civilian population as such – individual civilians or civilian objects – the object of attack by incendiary weapons. It is also prohibited to make a military objective situated within a concentration of civilians the object of attack by air-delivered incendiary weapons. However, even the protection of civilians is qualified: military objectives located within populated areas but separated from the concentration of civilians are excluded from the restriction in respect of ground-delivered incendiary weapons. The Protocol stipulates that all feasible precautions should be taken in order to limit the incendiary effects to the military objective and to avoid or minimize incidental loss of civilian life, injury to civilians and damage to civilian objects.

The Protocol also prohibits attacks by incendiary weapons on forests or other kinds of plant cover, except when these are used to cover, conceal or camouflage combatants or other military objectives, or are themselves military objectives. However, plant cover is most likely to be attacked precisely when it is being used as cover or camouflage. The attempt to reconcile the postulate of environmental protection with claims of military necessity resulted here in banning only some preventive or vindictive tactics. A general prohibition on the use of incendiary weapons in war, as advocated by a group of non-aligned and neutral nations, could not be agreed.

The most serious deficiency of Protocol III is that it does not protect combatants. Evidently, incendiary weapons are still considered by certain countries as too valuable from the military point of view to be outlawed. Thus the scope of the ban on the use of incendiary weapons is rather narrow.

Protocol IV. On 13 October 1995 the parties to the CCW Convention adopted a Protocol on Blinding Laser Weapons as Protocol IV of the Convention. The Protocol prohibits the use of laser weapons that are specifically designed to cause permanent blindness to unenhanced vision, that is, to the naked eye or to an eye with corrective eyesight devices, such as prescription glasses or contact lenses.

The term 'weapons specifically designed' means those weapons having as their sole combat function, or as one of their combat functions, to cause permanent blindness. 'Permanent blindness' is defined as irreversible loss of vision that cannot be corrected and which is seriously disabling with no prospect of recovery. 'Serious disability' is described as visual acuity of less than '20/200 Snellen' measured using both eyes. This means that a disabled person cannot see at 20 feet (approximately 6 metres) what a person with normal vision can see at 200 feet (approximately 60 metres). (This language was introduced at the insistence of the US Delegation.)

Protocol IV forbids the transfer of blinding laser weapons to any recipient. Blinding as an incidental or collateral effect of military employment of other laser systems, including lasers used against optical equipment, is exempt from the prohibition.

Although the military utility of blinding laser weapons is limited, the adoption of Protocol IV of the CCW Convention was an achievement. Blinding is a particularly abhorrent way of wounding the enemy and is more debilitating than most battlefield injuries because sight provides 80–90% of a person's sensory stimulation. Protection against the threat of blinding laser weapons is virtually impossible. Blinding cannot be considered a military necessity and therefore belongs to that category of generally condemned methods of warfare which cause superfluous injury or unnecessary suffering.

It is remarkable that the blinding weapon, a weapon developed and reportedly also tested, has been prohibited before ever being used on the battlefield. However, Protocol IV suffers from several weaknesses. The production of blinding laser weapons has not been outlawed, and the blinding of persons using optical devices has not been banned. Such devices, including binoculars, magnify the intensity of the laser beam and increase the potential for blindness. For example, a member of a tank crew looking through a periscope could be permanently blinded by an anti-*matériel* laser that has been designed mainly to destroy or damage optical devices.

The parties are required to take all feasible precautions – which include training of the military and other 'practical measures' – to avoid causing blindness with laser systems other than those specifically designed to inflict damage to vision. The term 'feasible' can lend itself to different interpretations, and the relevant paragraph of Protocol IV should have unequivocally established the rule that blinding as a method of warfare is prohibited. Such wording would have banned *all* practices that are intended, or which can be expected, to cause blindness. An exception for laser systems for targeting and range-finding purposes could be justified, since any blindness caused by them would be accidental rather than the result of intentional use. Laser systems aimed at destroying optical equipment should not be exempted. From the humanitarian point of view they can hardly be considered legitimate because they are expected to destroy human eyesight in most cases, unless and until effective means are universally used as standard equipment to prevent such injuries. Differentiation between intentional and accidental blinding may be difficult; it would be possible if a consistent pattern of violations were discerned.

Protocol IV applies only to international armed conflicts. Extension of its scope to cover non-international armed conflicts received general support but was not included in the final text.

Strengthening the Convention. The Review Conference of the parties to the CCW Convention, held on 11–21 December 2001, decided to amend Article 1 of the Convention. The main elements of the amended Article are as follows.

The Convention and its annexed Protocols apply in the situations referred to in Article 2 common to the Geneva Conventions of 12 August 1949 for the Protection of War Victims, including any situation described in paragraph 4 of Article 1 of Additional Protocol I to these Conventions. Article 2 of the 1949 Conventions stipulates that, in addition to the provisions which are to be implemented in peacetime, each Convention applies to all cases of declared war or of any other armed conflict which may arise between two or more of the parties, even if the state of war is not recognized by one of them. The Convention also applies to all cases of partial or total occupation of the territory of a party, even if the occupation meets with no armed resistance. While one of the powers in conflict may not be a party to the Convention, the powers that are parties to the Convention remain bound by it in their mutual relations. Moreover, they are bound by the Convention in relation to a non-party if the latter accepts and applies the provisions of the Convention. The situations described in paragraph 4 of Article 1 of Additional Protocol I to the 1949 Conventions include armed conflicts in which peoples are fighting against colonial domination and alien occupation and against racist regimes in the exercise of their right of self-determination. The CCW Convention and its annexed Protocols also apply to situations referred to in Article 3 common to the 1949 Geneva Conventions, namely, situations of armed conflict not of an international character occurring in the territory of one of the parties. The CCW Convention does not apply to situations of internal disturbances and tensions, such as riots, isolated and sporadic acts of violence, and other acts of a similar nature, as not being armed conflicts. Nothing in the CCW Convention or its annexed Protocols may be invoked as a justification for intervening, directly or indirectly, for any reason whatever, in the armed conflict or in the internal or external affairs of the party in the territory of which that conflict occurs.

The decision to extend the scope of application of the CCW so as to cover not only interstate but also intra-state conflicts was a major contribution to the development of international humanitarian law. The amendment is to enter into force six months after the deposit of the 20th instrument of ratification, acceptance, approval or accession.

The Review Conference also decided to establish an open-ended group of governmental experts in order to discuss ways and means of addressing the issue of explosive remnants of war and further explore the issue of mines other than anti-personnel mines. Moreover, it invited the interested parties to convene experts to consider issues related to small-calibre weapons and ammunition, such as military requirements, scientific and technical factors, medical factors, legal/treaty obligations/standards and financial implications.

17.3 The Legality of Nuclear Weapons

Attempts to establish a rule of law expressly banning the use of nuclear weapons have been made for several decades. In 1961, by a vote of 55 to 20, with 26 abstentions, the UN General Assembly adopted a resolution stating that the use of nuclear weapons was contrary to the 'spirit, letter and aims' of the United Nations and, as such, a direct violation of the UN Charter. The resolution went on to proclaim the use of nuclear weapons to be a 'crime against mankind and civilization'. The United States and other NATO countries opposed this resolution, contending that in the event of aggression the attacked nation must be free to take whatever action with any weapons not specifically banned by international law. In addition to the pronouncement of the illegality of nuclear weapons, the General Assembly asked the Secretary-General to ascertain the views of the governments of UN member-states on the possibility of convening a special conference for signing a convention on the prohibition of the use of these weapons. The Secretary-General's consultations proved inconclusive, and the requested conference was never convened. Resolutions advocating an unconditional ban on the use of nuclear weapons were also considered at subsequent sessions of the UN General Assembly. In particular, the Final Document of the Tenth Special Session of the UN General Assembly, held in 1978, recommended that efforts be made to bring about conditions in international relations that would preclude the use or threat of use of nuclear weapons. This and other similar recommendations have not been followed up.

Applicability of Existing Law to Nuclear Weapons

There exists a body of opinion that there is no need to create a legal norm to ban the use of nuclear weapons because such a ban is already covered by the humanitarian law of armed conflict. The arguments are as follows.

The use of nuclear weapons can be deliberately initiated either in a surprise pre-emptive attack aimed at disarming an adversary, who may or may not be nuclear-armed, or in the course of escalating hostilities started with non-nuclear means of warfare. The first situation, usually referred to as 'first strike', is covered by the fundamental rule of international law enshrined in the UN Charter, namely, that the threat or use of force against the territorial integrity or political independence of any state is prohibited unconditionally, irrespective of the type of weapon employed – nuclear or non-nuclear. The second situation, usually referred to as 'first use', involves the right of self-defence, which is also enshrined in the UN Charter: all states may defend themselves, individually or collectively, until the UN Security Council has taken the necessary measures to restore and maintain international peace and security. The Charter does not specify which weapons may or may not be used by states in such a situation, but the right of self-defence is not unlimited.

In discussing the limitations on the right of self-defence, one should start from the rule (embodied in the 1907 Hague Convention IV) which prohibits the employment of arms causing 'unnecessary' suffering or the destruction of the enemy's property, unless such destruction is 'imperatively demanded' by the necessities of war (see above). Since nuclear explosions could cause massive injury to people and massive damage to property, and since mass destruction can hardly be a necessity, it would be nearly impossible to observe the relevant rule in a nuclear war.

Modern weapons are capable of precise targeting. It is conceivable, therefore, that a low-yield tactical nuclear weapon might be used against an isolated military objective without causing indiscriminate harm to other objectives. However, once the nuclear threshold has been crossed, there can be no guarantee that a high-yield nuclear weapon will not be used. There will always be a risk of nuclear escalation on the part of the attacker as well as on the part of the attacked nation if the latter also possesses nuclear weapons. Thus, irrespective of motivation, a single use could provoke a nuclear war, which is impossible to contain in either space or time. Indeed, it is not the targeting that should be decisive in determining the legality of nuclear weapons, but rather the enormous destructive potential of these weapons and the uncontrollable effects of their use. Even the 1925 Geneva Protocol, which deals with less devastating weapons than nuclear weapons, does not differentiate between targets or between more or less severe effects caused by the use of the banned weapons.

Under customary international law, reiterated in the 1949 Geneva Conventions for the protection of war victims, the belligerents are under strict obligation to protect civilians not taking part in hostilities against the consequences of war. The indiscriminate nature of nuclear weapons renders this norm very difficult to comply with. Even if exclusively military targets were aimed at, civilian casualties could be an important by-product; in many cases they might outnumber the military casualties. Yet another iniquitous aspect of nuclear warfare is the inability of the belligerents to comply with the requirement to respect the inviolability of the territory of neutral states. It is impossible to confine the effects of nuclear explosions, particularly radioactive contamination, to the territories of states at war.

Although the primary effects of nuclear explosions are blast and heat, the nuclear radiation and radioactive fallout which they produce inflict damage on the biological tissue of humans, animals and plants. Nuclear weapons can, therefore, for the purpose of the international humanitarian law, be compared to poison, the use of which as a method of warfare is prohibited by the Hague Declarations and the Geneva Protocol, discussed above. In addition, since nuclear explosions may also be expected to cause widespread, long-term and severe damage to the natural environment, their use would contravene Protocol I Additional to the 1949 Geneva Conventions.

Finally, it should be noted that, in placing limitations on the conduct of hostilities, the 1907 Hague Convention IV included the so-called Martens Clause (named after Fedor Martens, professor of international law and member of the Russian Delegation to the Hague Peace Conference), which was subsequently reaffirmed in several treaties. This Clause makes usages established among civilized peoples, the laws of humanity and the dictates of the public conscience obligatory, even in the absence of a specific treaty prohibiting a particular type of weapon. It was this legal yardstick that the Nuremberg International Military Tribunal applied in concluding that the law of war is to be found not only in treaties but also in customs and practices of states and that, by its continual adaptation, this law follows the needs of a changing world. Thus, also weapons and tactics which may be resorted to in the exercise of legitimate self-defence must not be violative of the existing norms, whether or not these norms are spelled out in formal international agreements.

The cumulative effect of the generally accepted restraints on the use of all weapons is such that nuclear war can hardly be initiated in compliance with the rules of customary international law. It should be noted that, in its judgement of 1986 in the case concerning military and paramilitary activities in and against Nicaragua, the International Court of Justice (ICJ) confirmed that customary law has the same standing as treaty law. Nonetheless, in view of the special character of nuclear weapons, a ban on their use cannot simply be deduced from restrictions regarding other types of weapon. This reasoning must have guided those who in 1925 decided to sign the Geneva Protocol banning the use of chemical means of warfare, even though the use of these means had already been condemned by the 'general opinion of the civilized world', as stated in the Protocol. In other words, prohibitions concerning specific weapons ought to be incorporated in positive law, as they are in the case of chemical and biological weapons as well as in the case of anti-personnel mines.

In its advisory opinion of 8 July 1996, the ICJ, the judicial organ of the United Nations, declared its inability to rule that the use or threat of use of nuclear weapons is prohibited unconditionally. At the same time, the ICJ declared the existence of an international obligation to achieve nuclear disarmament 'in all its aspects'. However, nuclear disarmament is not achievable without a prior undertaking by states not to use nuclear weapons under any circumstance. Mere cuts in nuclear arsenals will not necessarily lead to their abolition.

Under UN Security Council Resolution 984 of 1995, the non-nuclear-weapon states parties to the NPT obtained from France, Russia, the United Kingdom and the United States so-called negative assurances that nuclear weapons would not be used against them. These assurances are conditional. They would cease to be valid in the case of an invasion or any other attack on these powers, their territories, their armed forces or other troops, their allies, or on a state towards which they have a security commitment 'carried out or sustained by a non-nuclear-weapon state in association or alliance with a nuclear-weapon state', whereas China's assurances not to use nuclear weapons against non-nuclear-weapon states are unconditional (see Chapter 6). In 1998, after its nuclear test explosions, India stated that it would not be the first to use nuclear weapons and expressed its readiness to enter an international agreement banning such use. Assurances of non-use of nuclear weapons are also contained in protocols to the nuclear-weapon-free-zone treaties. They have been given to parties to these treaties but are understood by the nuclear-weapon powers (again with the exception of China) to be subject to the same or similar conditions as the assurances given to parties to the NPT (see Chapter 13). The use of nuclear weapons against non-parties to the above-mentioned treaties, or between nuclear-weapon powers, is not formally prohibited.

Consequences of No-Use Commitments

Only a formal unconditional undertaking not to use nuclear weapons against any country, whatever its status – nuclear or non-nuclear, aligned or non-aligned, party or not party to the NPT or a nuclear-weapon-free-zone treaty – appears to have real significance. As a corollary to such an undertaking, tactical nuclear weapons would have to be totally eliminated because of their first-strike characteristics: once deployed close to the front lines – as they must be to have military value – they are likely to be employed very early in armed conflict to avoid capture or destruction by the adversary's conventional forces.

To become even more credible, the non-use commitments would have to be backed up by taking nuclear strategic forces off alert. Continuously monitored de-alerting could reduce the risk of a surprise attack and of an unauthorized or accidental launch of nuclear weapons. Once the use of nuclear weapons is prohibited, the very threat of such use will become unlawful.

According to the doctrine of belligerent reprisals, a retaliatory (second) use of nuclear weapons to make a violator of the ban on use desist from further illegitimate actions would not be considered a breach of the ban, if it were proportionate to the violation committed and to the injury suffered. Thus, countries possessing nuclear weapons would be committed only to no first use, it being understood that attacks on the civilian population and objects protected by international law could not be tolerated under any circumstance. Some people argue that the doctrine of belligerent reprisals is inapplicable in the context of nuclear warfare because the particularly inhumane nature of nuclear weapons makes their second use as illegal as their first use. Several countries adopted such an attitude with regard to biological and/or chemical weapons, even before the possession of these weapons had been banned; they have thereby recognized that the prohibition on use is absolute, not subject to exceptions. It is doubtful, however, whether those possessing nuclear weapons would be willing, in case of a nuclear aggression, to give up the right to respond in kind.

Responses to CBW Attacks

Once the right of legitimate self-defence, individual or collective, is restricted to the use of non-nuclear means of warfare, a nuclear response to an aggression committed with chemical or biological weapons will be prohibited as well. Breakouts from the 1925 Geneva Protocol and the 1993 Chemical Weapons Convention, which ban the employment of chemical and biological weapons (CBW), could be countered with modern conventional weapons, if the situation could not be remedied with non-military means. Moreover, parties may withdraw from arms control treaties when some extraordinary events have jeopardized their interests. A proven violation would justify withdrawal, even without the required several months' notice. The withdrawing treaty-abiding country would not have chemical or biological weapons readily available for immediate response to a violation of the relevant bans, but reconstitution of CBW stocks would not be a very complicated or overly lengthy process. Besides, the use of these weapons is unlikely to be decisive for the outcome of an armed conflict.

Although widely considered as weapons of mass destruction along with nuclear weapons, biological and chemical weapons have several important distinctive features. Under certain exceptional circumstances, the use of biological weapons may produce fatalities comparable to those caused by nuclear weapons. However, since it would be difficult to recognize each unusual outbreak of a disease as an aggression, and since there would be no 'signature' of the user, BW attacks could hardly be deterred by a threat of nuclear retaliation. Attacks with chemical weapons, even on a large scale, could not reach the level of destructiveness caused by a nuclear attack. Moreover, there exist means of defence against biological and chemical weapons (vaccination, antidotes, masks, protective clothing and decontaminants) as well as warning systems with highly sensitive sensors capable of detecting the agents in question. There are no such means or systems against nuclear weapons. By excessively magnifying the dangers posed by biological and chemical weapons, the opponents of nuclear disarmament encourage nuclear-weapon proliferation.

Assessment

A global ban on the use of nuclear weapons, preferably included in a multilateral treaty rather than in easily reversible declarations, would reinforce the firebreak separating conventional and nuclear warfare. It would thereby diminish the risk of nuclear war and weaken the political force of explicit or implicit threats to initiate such a war. Indeed, the doctrine of nuclear deterrence, in so far as it consists in threatening a nuclear attack in response to a non-nuclear attack, would have to be declared invalid. Furthermore, in discarding the war-fighting functions of nuclear weapons, the non-use posture would minimize the importance of nuclear superiority, whether quantitative or qualitative. It would, therefore, clear the way towards the abolition of tactical nuclear weapons and towards new substantial reductions of strategic nuclear forces.

Given the attitudes of the majority of the de jure or de facto nuclear-weapon powers, the prospect of reaching a no-use treaty soon is not bright. However, only when such a treaty is signed will the pledges made by these powers to eventually bring about complete nuclear disarmament become credible.

17.4 Laws of Armed Conflict and Disarmament

All laws of armed conflict suffer from one common weakness: rules of conduct for belligerents, set in time of peace, may not resist the pressure of military expedience generated in the course of hostilities. War itself is caused by the breakdown of certain legal constraints, and attempts to 'humanize' war may prove futile. Indeed, there have been enough violations of international humanitarian law to justify widespread pessimism regarding its usefulness. Moreover, existing norms overemphasize military necessity and are subject to different interpretations and reservations.

Nevertheless, there is evidence that some basic rules of conduct in war are by and large observed, especially those which concern the treatment of prisoners of war, respect for the neutral status of non-belligerent states and the protection of certain objects. A measure of constraint has also been exercised in the use of weapons. However, the danger that, under certain circumstances, prohibited weapons may be resorted to, as has already occurred, will not disappear as long as these weapons remain in the arsenals of states. Hence the intrinsic link between the development of international humanitarian law and progress in the field of disarmament.

18

Prevention of Accidental War

In 1958 an exchange of letters between Soviet Premier Khrushchev and US President Eisenhower led to an agreement to convene a conference of experts for the study of measures that might be helpful in preventing a surprise attack. The conference opened on 10 November. Participants were, on the Western side, experts from Canada, France, Italy, the United Kingdom and the United States and, on the Eastern side, experts from Albania, Czechoslovakia, Romania, Poland and the Soviet Union.

The group of experts from the five Western countries viewed their task to be that of preparing a technical, military analysis of the problem of surprise attack and of evaluating the effects of various systems of inspection and observations. The five eastern experts, on the other hand, submitted proposals for a system of inspection and disarmament in Europe as a means of preventing surprise attack. The two groups were thus operating under different terms of reference. The conference was suspended in December 1958 and never reconvened.

In 1962 came the US–Soviet confrontation provoked by the deployment of Soviet nuclear missiles in Cuba. This confrontation clearly demonstrated the need for quick and reliable communications between heads of government to reduce the danger of war breaking out because of technical failure, misunderstanding or miscalculation. An agreement to set up such communications was concluded between the United States and the Soviet Union less than one year after the Cuban Missile Crisis.

18.1 The Hotline Agreements

So-called hotlines are intended for use in time of emergency, when normal consultative procedures appear insufficient or impossible.

The First Agreement

Signed in June 1963 and put into effect two months later, the US–Soviet Memorandum of Understanding Regarding the Establishment of a Direct Communications Link – the first hotline agreement – permitted rapid exchanges of printed messages between the parties. Each party was responsible for arrangements for the communications link on its own territory, including continuous functioning of the link and prompt delivery of communications to its head of government. According to the annex attached to the Memorandum, the link was to comprise: (a) two terminal points with teletype equipment; (b) a full-time duplex wire telegraph circuit routed Washington–London–Copenhagen–Stockholm–Helsinki–Moscow; and (c) a full-time duplex radiotelegraph circuit routed Washington–Tangier–Moscow. Should the wire circuit be interrupted, messages would have to be transmitted via the radio circuit.

To assure reliability, circuits were to be tested hourly, using various non-political texts; and to ensure privacy of communications, all messages, including test messages, were to be encoded for transmission and decoded upon receipt. Each side was to maintain at its terminal 24-hour capability to transmit messages and translate those received in the sender's language.

Modernizations

Advances in satellite communications technology offered a possibility to increase the reliability and survivability of the hotline. The US–Soviet Agreement on Measures to Improve the Direct Communications Link, signed in 1971, supplemented and modified the 1963 Memorandum of Understanding by providing for the establishment of two satellite communications circuits for transmission of printed messages. A system of multiple terminals in each country was to allow the leaders of both countries to have several points through which they could receive and send messages. Under this so-called Hot Line Modernization Agreement, the United States was to provide a circuit via the Intelsat system, and the Soviet Union a circuit via its Molnya system. The original circuits were to be maintained.

Further improvements of the hotline were made through understandings reached between US and Soviet technical experts in 1972, 1973 and 1976, as well as through the 1975 exchange of notes between the two governments amending the 1971 Hot Line Modernization Agreement. It took more than six years after the signature of this Agreement to work out all the necessary technical and procedural arrangements, including the construction of satellite earth stations. When the two satellite circuits became operational in 1978, the radio circuit provided for in the 1963 Memorandum was terminated, but the wire telegraph circuit was retained as a backup.

In July 1984 the governments of the United States and the Soviet Union agreed, in an exchange of notes, to upgrade their hotline by adding facsimile transmission facilities. This upgrade became operational in 1986. The two sides could then exchange all kinds of graphic material, such as maps, charts or drawings that could be essential in resolving international crises. The agreement provided for reviews to improve even further the Direct Communications Link.

Assessment

The establishment of the US–Soviet hotline proved a useful and timely measure. The line was used many times in communications between Moscow and Washington during military crises, such as the 1967 and 1973 Arab–Israeli wars, the 1971 war between India and Pakistan, the 1974 Turkish invasion of Cyprus, the 1979 Soviet intervention in Afghanistan and, most probably, on other occasions as well. International agreements establishing direct communications links with the Kremlin were also signed by France, the United Kingdom and the Federal Republic of Germany, whereas the United States established such links with Belarus, Kazakhstan and Ukraine.

18.2 The 1971 Nuclear Accidents Agreement

Whatever the precautions to ensure the safety of weapons, and whatever the sophis-tication of command-and-control procedures, accidental or unauthorized use of nuclear weapons cannot be ruled out. There have been many false alarms of possible missile attack, caused mainly by misleading or ambiguous information from sensors aboard satellites or from early-warning radars but also by computer malfunctions or failures in communications equipment. In several cases, intercontinental bombers and missiles were ordered to a higher state of alert for long enough to use up much of the time allotted for taking a final decision. Thousands of lesser alarms have been caused by atmospheric disruptions. In addition, dozens of accidents have occurred directly involving nuclear weapons. All this creates the risk of an unintended nuclear war breaking out, especially during deep international crisis or conventional war, when command centres may be threatened or destroyed. An attempt to deal with this dilemma was made in September 1971, when the United States and the Soviet Union signed the Agreement on Measures to Reduce the Risk of Outbreak of Nuclear War.

Main Provisions

This agreement – called the US–Soviet Nuclear Accidents Agreement – was negoti-ated in conjunction with the Strategic Arms Limitation Talks (SALT). The parties undertook to maintain and improve their organizational and technical arrangements to guard against an accidental or unauthorized use of nuclear weapons under their control. This introductory provision reflected a recognition of the need for the safe handling of nuclear weapons. Each party was to notify the other immediately in the event of an accidental, unauthorized or any other unexplained incident involving the possible detonation of a nuclear weapon that could create a risk of the outbreak of nuclear war. In the event of such an incident, the party whose nuclear weapon was involved was to take measures to render harmless or destroy the weapon without causing damage.

The parties were to notify each other immediately if their missile warning systems detected unidentified objects, or in the event of signs of interference with these sys-tems or with related communications facilities, if such occurrences could create a risk of outbreak of nuclear war between the two countries. Furthermore, advance notification was to be given of any planned missile launches, if such launches extended beyond the national territory of one party in the direction of the other.

In other situations involving unexplained nuclear incidents, each party was to act in such a manner as to reduce the possibility of its actions being misinterpreted by the other party. The hotline was to be used to transmit urgent information.

Implementation

In 1976, at the initiative of the United States, a confidential protocol to the Nuclear Accidents Agreement was signed in the Standing Consultative Commission, a body set up to promote the objectives of the SALT I agreements. This protocol contained specific guidelines for implementing the provision on notification of missile launches and established a coding system designed to speed up transmission of information necessary in a crisis situation.

In 1985 the parties agreed that, in case of a nuclear incident instigated by a third party or an unauthorized group of individuals who had obtained a nuclear weapon, they would use appropriate prepositioned hotline messages.

Assessment

It is difficult to judge whether the Nuclear Accidents Agreement actually helped to avert the outbreak of nuclear war by accident or miscalculation. One can argue that the United States and the Soviet Union would – in their common interest – take all possible measures to avoid an unintended nuclear war, irrespective of any treaty obligations. Nevertheless, since the parties formally committed themselves to meet together a challenge to the security of both of them, the Agreement rendered it easier for them to take the necessary action. The need for further such agreements among nuclear-weapon-states was recognized when the French–Soviet and British–Soviet Nuclear Accidents Agreements were concluded in July 1976 and October 1977, respectively. They were patterned after the US–Soviet Agreement.

18.3 Agreements on the Prevention of Nuclear War

The 1973 US–Soviet Agreement

In June 1973, as a follow-up to the 1971 Nuclear Accidents Agreement, the United States and the Soviet Union signed the Agreement on the Prevention of Nuclear War.

The parties to the 1973 Agreement pledged themselves to act in such a manner as to prevent the development of situations capable of causing a dangerous exacerbation of their relations, to avoid military confrontations, and to exclude the outbreak of nuclear war. Since the two powers obviously shunned a nuclear confrontation, and since the Agreement also aimed at preventing a nuclear war 'between either of the parties and other countries', it was noteworthy even more for its multilateral implications than for its bilateral implications. The parties agreed to proceed from the premise that each would refrain from the threat or use of force 'against the allies of the other Party and against other countries'.

The Agreement provided for action to be taken if the risk of nuclear conflict appeared. The two parties would have to enter immediately into urgent consultations and make every effort to avert the risk. Although the UN Security Council is charged with the main responsibility for international peace and security, it would not be involved in handling such situations. Each party was 'free' (not obliged) to inform the Security Council, as well as the UN Secretary-General and the governments of allied or other countries, of the progress and outcome of the said consultations. This two-power exclusive consultation procedure was to be set in motion even when relations between countries not parties to the Agreement appeared to involve the risk of nuclear war either between the United States and the Soviet Union or between either of them and other countries. These stipulations were most probably motivated by a desire to prevent a local conflict from turning into a major great-power confrontation.

In signing the 1973 Agreement, the United States and the Soviet Union formally expressed their intention to minimize the probability of a nuclear war started by design – not only by accident. However, the new code of nuclear behaviour applied only obliquely to relations between either of the two superpowers and other nuclear-weapon powers. The very tenor of the Agreement gave rise to suspicions that the two powers accorded absolute priority to their bilateral relations over their multi-lateral alliance commitments and that they arrogated to themselves the role of refer-ees in matters relating to the security of others. In fact, neither the United States nor the Soviet Union had formally consulted its allies before signing the Agreement. This caused considerable dissatisfaction, especially among NATO member-states. Moreover, in the context of the main provisions of the Agreement, the clause that reiterated the right of self-defence implied that the parties continued to consider themselves free to employ nuclear weapons against an adversary that had used only conventional weapons, that is, against a non-nuclear-weapon state as well.

Other Agreements

In July 1976 France and the Soviet Union reached an agreement containing the fol-lowing undertakings.

Each party was to maintain and possibly improve the existing organizational and technical arrangements to prevent an accidental or unauthorized use of nuclear weapons under its control. The two parties would notify each other immediately of any accidental occurrence or any other unexplained incident that could lead to an explosion of their nuclear weapons and could be construed as likely to have harmful effects on the other party. In the event of an unexplained nuclear incident, each party would act in such a manner as to avoid, as far as possible, the possibility of its actions being misinterpreted by the other party.

A similar agreement was concluded in October 1977 between the Soviet Union and the United Kingdom.

18.4   The Nuclear Risk Reduction Centers

In the pursuit of their common objective of avoiding an accidental nuclear war, the United States and the Soviet Union decided to set up an institution specifically dedi-cated to this purpose. Thus, in September 1987, the two powers signed the Agree-ment on the Establishment of Nuclear Risk Reduction Centers. According to this Agreement – which is of unlimited duration – each party was to establish in its capi-tal a Nuclear Risk Reduction Center (NRRC) to operate under the control of its gov-ernment.

The NRRCs were to be used for the transmission of notifications of ballistic mis-sile launches under the 1971 Nuclear Accidents Agreement (see above), notifica-tions of ballistic missile launches under the 1972 Incidents at Sea Agreement (see Chapter 11.4) as well as other notifications.

The parties agreed to establish, via satellite circuits, a special facsimile communi-cations link between their national NRRCs. Each party was responsible for the pur-chase, installation, operation and maintenance of its own terminals. Direct facsimile messages from the Russian NRRC to the US NRRC were to be transmitted and received in Russian, and those from the US NRRC to the Russian NRRC in English. Transmission and operating procedures were to be in conformity with procedures

employed in the Direct Communications Link (see above). The NRRCs opened in
Washington and Moscow in March 1988. Regular meetings, at least once a year, are
to be held to consider matters related to their functioning.

18.5 Notifications of Missile Launches and Strategic Exercises

The regime of notifications of US and Soviet/Russian ballistic missile launches, as
initiated by the 1971 Nuclear Accidents Agreement and the 1972 Incidents at Sea
Agreement, was strengthened by the US–Soviet Agreement on Notifications of
Launches of Intercontinental Ballistic Missiles and Submarine-launched Ballistic
Missiles, signed in May 1988. Under this Agreement each party assumed the obligation to notify the other party, no less than 24 hours in advance, of the planned date,
launch area and area of impact for any launch of an ICBM or SLBM. For launches
from land, the area from which the launch was planned to take place had to be indicated. For launches from submarines, notification was to specify the general area
from which the missile would be launched. For all launches, notification must contain the geographic coordinates of the planned impact area or areas of the re-entry
vehicles. The notification was to be provided through the NRRCs and remain valid
for four days counting from the indicated launch date.

In September 1989 the Agreement on notification of US and Soviet ballistic
missile launches was complemented by the US–Soviet Agreement on Reciprocal
Advance Notification of Major Strategic Exercises. Each party undertook to notify
the other party, no less than 14 days in advance, about the beginning of a major
strategic forces exercise which included the participation of heavy bomber aircraft.
The United Kingdom told the Russian government that it would give five days'
notice of any British submarine-launched ballistic missile test firing.

18.6 The 1989 Dangerous Military Activities Agreement

To ensure the safety of the personnel and equipment of their armed forces operating
in proximity to one another in peacetime, the Chairman of the US Joint Chiefs of
Staff and the Chief of the General Staff of the Soviet Armed Forces signed, in June
1989, the Agreement on the Prevention of Dangerous Military Activities – the DMA
Agreement, which entered into force in 1990.

Definitions

The following activities were to be considered as 'dangerous military activities':
(a) entry by personnel and equipment of the armed forces of one party into the
national territory of the other party owing to circumstances brought about by *force
majeure*, or as a result of unintentional actions by such personnel; (b) using a laser
in such a manner that its radiation could cause harm to personnel or damage to
equipment of the armed forces of the other party; (c) hampering the activities of the
personnel and equipment of the armed forces of the other party in a Special Caution
Area; and (d) interfering with command and control networks in a manner that could
cause harm to personnel or damage to equipment of the armed forces of the other
party.

For the purposes of the Agreement, 'personnel' was defined as any individual, military or civilian, serving in or employed by the armed forces of the parties; 'equipment' meant any ship (warship or auxiliary ship), military aircraft (excluding spacecraft) or ground hardware (designed for use on land) of the armed forces.

'Laser' was defined as any source of intense, coherent, highly directional electromagnetic radiation in the visible, infrared or ultraviolet regions that is based on the stimulated radiation of electrons, atoms or molecules. 'Special Caution Area' was described as a region, designated mutually by the parties, in which the personnel and equipment of their armed forces are present and in which – owing to circumstances in the region – special measures must be taken. 'Interfering with command and control networks' related to actions that hamper, interrupt or limit the operation of the signals and information transmission means and systems providing for the control of personnel and equipment.

Main Obligations

Under the DMA Agreement, any incident arising from dangerous military activities is to be terminated and resolved by peaceful means, that is, without resort to the threat or use of force. Each party must exercise caution and prudence while operating near the national territory of the other party. If, however, the personnel and equipment of one party enter into the national territory of the other party, such personnel are to follow the requirements spelled out in Annexes 1 and 2 to the Agreement. Annex 1 sets out procedures for establishing and maintaining communications between the parties' armed forces, including specified radio frequencies, visual signals and English phrases for use in particular contingencies. Annex 2 sets out procedures for the resolution of incidents. If the personnel of one party intend to use lasers (such as range finders) that could cause harm to the personnel or damage to the equipment of the other party, the side intending such use must notify the other side. In any case, appropriate safety measures are to be observed. Personnel of the armed forces of the parties present in a designated Special Caution Area must establish and maintain communications and undertake such other measures as might be agreed to prevent dangerous military activities. Cases of interference with the command and control networks may be notified to the other party.

Since, in practice, it might be unclear whether an entering ship or aircraft is acting intentionally or as result of an error or *force majeure*, the parties made an agreed clarifying statement: the procedures set forth in the Annexes to the DMA Agreement would apply regardless of whether a party had been made aware of the circumstances of the entry into its national territory by the personnel and equipment of the other party. Information on instances of dangerous military activities is to be conveyed by the Chairman of the US Joint Chiefs of Staff through the Russian defence attaché in Washington, DC, and by the Chief of the General Staff of the Soviet Armed Forces through the US defence attaché in Moscow.

The rights of individual or collective self-defence are to remain unaffected, as are the rights of overflight and navigation under international law, including the right of warships to exercise innocent passage. A Joint Military Commission (JMC), meeting at least once a year, is to consider, among other matters, compliance with the obligations assumed, as well as ways to improve the effectiveness of the DMA Agreement.

Assessment

The DMA Agreement, aimed at preventing and resolving incidents that might occur between the superpowers' armed forces, grew out of actual experiences – such as US and Soviet military aircraft inadvertently crossing each other's national borders, aircraft crews being temporarily blinded by lasers of the other side, or military communications being interfered with or jammed. However, several provisions of the Agreement were vague, and certain crucial terms – such as 'unintentional actions' – were not defined. Moreover, intentional entries into the territory of one of the parties, such as the U-2 flights over the Soviet Union conducted in the 1950s, were not covered. Similarly, the conditions under which warships of one party may engage in innocent passage of the territorial sea of the other party remained unresolved. (For divergent opinions on this subject, see Chapter 11.5.) Proper functioning of the DMA Agreement depends, therefore, on the parties' willingness to tolerate such ambiguities and on the ability of the JMC to find practical solutions to possible disputes. The US–Soviet DMA Agreement served as a model for similar agreements concluded in the 1990s between Russia and a few other countries, including China.

18.7 OSCE Preventive Measures

The OSCE has developed mechanisms and procedures that, in cases requiring rapid action, facilitate prompt and direct contact between concerned parties (see Chapter 16). An exchange of information is envisaged regarding unusual and unscheduled activities of military forces of states outside their normal peacetime locations. The participating states are obliged to respond to requests for an explanation from other participating states and – if necessary – hold a meeting to discuss the matter.

Another mechanism requires states, in the event of hazardous incidents of a military nature, to cooperate by reporting and clarifying the incidents in order to prevent possible misunderstandings and to mitigate the effects on other states.

18.8 De-Targeting and Information Sharing

With the end of the Cold War, the threat of a deliberate full-scale exchange of nuclear missiles between Russia and the United States receded. However, the likelihood of a nuclear war between these two powers started by accident or misunderstanding became even greater than before. This was due, in the first place, to the degradation of the Russian early-warning system, which might lead to the acceptance by Russia of incomplete information as evidence of an incoming attack. (In 1995 a radar-detected scientific rocket, launched by Norway, was mistaken for a US Trident nuclear missile and triggered a nuclear alert in Russia.) Control over Russian nuclear weapons and materials was also considerably weakened. The emergence of two new nuclear-weapon powers – India and Pakistan – involved in territorial, political and religious conflicts with each other, heightened the risk of a nuclear catastrophe.

The awareness of this precarious situation prompted the two nuclear superpowers to adopt a 'de-targeting' posture. Consequently, on 14 January 1994 they agreed that, by 30 May of that year, neither country would be targeted by the strategic forces of the other country. Their missiles were to contain no targeting information or be targeted at broad high-sea areas.

On 15 February 1994 Russia and the United Kingdom declared that they would ensure the adoption of all necessary measures so that strategic nuclear missiles under their respective command would be de-targeted, also by 30 May 1994. A similar agreement was subsequently reached between Russia and China, and another between the United States and China.

De-targeting of missiles is not verifiable. Even if it were, it could be easily and very quickly reversed. Russia and the United States remain ready to launch massive nuclear strikes within minutes. Nevertheless, the agreed measure may contribute to confidence building – at least in a symbolic way – because the countries in question are now committed not to operate nuclear forces on a day-to-day basis in a manner that presumes that they are enemies.

On 2 September 1998 Russia and the United States agreed to share information on the launches of strategic and theatre ballistic missiles and space launch vehicles detected by their respective early-warning systems. The agreement reaffirmed their pledge to reduce the danger of nuclear war, including the danger that ballistic missiles could be launched inadvertently on the basis of false warning of attack. It was expected to strengthen strategic and regional stability and to help develop common responses to the threat posed by the proliferation of ballistic missiles.

On 4 June 2000, as a follow-up to the undertakings specified above, Russia and the United States signed a Memorandum of Agreement, which established a Joint Data Exchange Center (JDEC) for the exchange of information derived from each side's warning systems (space-based satellites, infrared systems and radars) on the launches of ballistic missiles of the parties and of third states, as well as of space launch vehicles. For ballistic missiles, the information to be provided includes the geographic area from which a launch has occurred, the time of launch, the class of missile, the launch azimuth, the geographic area of payload impact and the estimated time of payload impact. For space launches, the information should include the time of launch, the geographic area of the launch, the class of missile and the launch azimuth. The US–Russian Memorandum of Understanding, signed on 16 December 2000, established a pre- and post-missile launch notification system (PLNS). The JDEC is to be staffed with Russian and US personnel.

India and Pakistan also decided, in their Memorandum of Understanding, signed at Lahore on 21 February 1999, to provide each other with advance notification of ballistic missile flight-tests. Both countries further undertook to notify each other immediately in the event of any accidental, unauthorized or unexplained incident that could create the risk of a fallout with adverse consequences for both sides or an outbreak of a nuclear war between the two countries.

18.9 Assessment

All the agreements described above – if implemented fully and on a global scale – could reduce the dangers of a surprise nuclear attack and of an accidental, inadvertent nuclear war, but they could not remove these dangers altogether. This is why certain non-governmental organizations have suggested that all nuclear forces should be taken off alert and not be poised for immediate launch. Nuclear warheads would then have to be separated from their delivery vehicles – preferably under international observation – and safely stored at a significant distance from the launch site. This so-called missile de-mating would render the use of warheads physically impossible without a substantial delay, facilitating the detection of clandestine moves.

The main obstacle to carrying the de-alerting measures into effect is the belief in both the US and Russian military establishments that nuclear deterrence continues to be central to the US–Russian relationship and that deterrence must be based on the capacity for a quick retaliatory nuclear strike against the forces of the adversary.

19

Verification and Compliance

Generally, states are assumed to enter international treaties in good faith, intending to abide by their obligations. However, when such vital matters as national security are involved, special assurances are needed that the parties will not engage in violating or circumventing their contracted commitments. Hence the need for verification of compliance with arms control agreements.

19.1 Role and Functions of Verification

Insistence on verification measures which are obviously unacceptable to another party, or refusal to accept verification measures which are obviously indispensable, has often been used as a convenient excuse for blocking the conclusion of arms control agreements. Verification is only one among several criteria to be taken into consideration in deciding whether to sign or accede to an arms control treaty. It is easier, however, to argue about verification than about the political and military motives which, as a rule, are the real impediments to agreement.

Deterrence

For certain politicians, verification has been chiefly a means to open up closed societies and clear away suspicions of aggressive intent, irrespective of arms control obligations. For others, verification is conceivable exclusively in the context of specific measures because – they argue – the form and modalities of verification depend upon the nature, scope and military significance of the agreed constraints. There is consensus, however, that verification is necessary to deter cheating, for a government contemplating a violation may refrain for fear that detection would bring about an unwelcome response from the cheated state or states and perhaps even provoke an untoward reaction in its own country. On the other hand, deterrence of violations presupposes the ability to detect them. Timely detection is vital to enable the injured party to take corrective steps and redress the situation, especially in cases constituting an immediate military threat.

Confidence Building

Verification also has an important confidence-building function. By providing evidence that the parties are fulfilling their obligations and by confirming that the prohibited activities are not taking place, verification helps to generate an international belief in the viability of agreed arms control measures and to instil trust in the participating states that their national interests are protected. In addition, the existence of a verification mechanism makes it easier for a party unjustly accused of a breach to demonstrate its innocence. Charges that have not been disproved and misunderstandings that have not been clarified may negatively affect the international climate by weakening confidence in treaties and casting a shadow on arms control endeavours.

Conflicting Approaches

It is usually postulated that verification must be 'adequate', 'appropriate' or 'effective'. The meanings attached to these terms differ. Most people take the view that there will always be a limit to detecting violations, but that the threshold should be low enough to make the significance of undetected breaches negligible. The reasoning behind this pragmatic approach is that what matters most is not the fact of non-compliance, but the effect of non-compliance, and that, to significantly alter the military balance between states, cheating would have to be practised on such a scale as to render detection inescapable. Others, however, consider any deviation from the contracted obligations to be an offence that cannot be tolerated, regardless of its military significance, and insist on total verifiability. The reasoning behind this legalistic approach is that the principle *pacta sunt servanda* (contracts should be adhered to) must be observed unconditionally, under the threat of abrogation, even at the risk that disputes over trivial matters might undermine the treaty. Since fool-proof verification is not achievable, and since complete absence of violation can never be proved, only the first of the two approaches makes it possible to reach an arms control agreement. The parties must be prepared to take risks and judge whether the threat posed by undetected violations is greater than the threat posed by unconstrained military activity. In other words, each party must decide for itself how much cheating it could put up with – the degree of tolerable uncertainty being a judgement made by the state authorities, based on the impact of a possible violation on national security.

The Verification Process

The process of verification starts with collecting information regarding the application of the parties' undertakings. The correctness of this information is usually checked, and the performance of states is monitored by off-site and on-site inspection as well as observation of forces, weapons or activities covered by the agreement. Inspections can be conducted either in a systematic manner – continuously or periodically – or ad hoc, as decided by the verifying body, or upon challenge, as a result of a specific demand.

To assess the parties' performance, the monitoring stage of the verification process is followed by the interpretation of the data collected, processed and analysed, and the filtering out of possible false alarms. Verification procedures may include enquiries in cases requiring clarification.

Verification versus Intelligence

Arms control verification and military-related intelligence have much in common. What distinguish the two processes are the purpose and the degree of required accuracy. The mission of intelligence gathering is to determine, in the greatest possible detail, the numbers, characteristics and deployment of the opponent's forces and weapons, with a view to evaluating his overall military capability and intentions. Verification of arms control is directed only at certain forces, weapons or military activities, with a view to assessing whether their numbers and characteristics exceed the conditions imposed by the treaty. In making certain information easier to obtain, verification can constitute a complement to intelligence gathering, which is conducted irrespective of arms control. The exact dividing line is difficult to draw because it is frequently through military intelligence gathering that evidence of arms control violations is collected, whereas monitoring compliance with arms control obligations may provide military information that is unrelated to arms control.

19.2 Acquisition of Information

Basic information about the forces, weapons or military activities to be constrained, prohibited or eliminated by an arms control agreement is often provided by the parties themselves. For many treaties the provision of such information is an indispensable starting point for the implementation of the obligations. In addition to formal exchanges among the parties, pertinent information may also be acquired by national means.

International Data Exchanges to Enable Verification

Data exchanges may take place before and/or after the conclusion of the agreement – either directly among the parties or through an international body.

Bilateral. The 1974 Threshold Test Ban Treaty (TTBT) and the 1976 Peaceful Nuclear Explosions Treaty (PNET) require an exchange of detailed technical information regarding the sites of nuclear test explosions, in particular the geology of the testing areas. Such data are necessary to check whether the yield of any given explosion is in excess of the agreed threshold.

In the course of the Strategic Arms Limitation Talks, which resulted in the 1979 SALT II Treaty, the United States and the Soviet Union drew up a Memorandum establishing a database on the numbers of their strategic offensive arms, with the understanding that the data would be updated. Indeed, in 1979, at the time of signing the Treaty, the parties made two separate statements declaring the inventories they at that time possessed in all the ten categories of arms subject to limitations. They also undertook to notify each other of possible future changes in the database. This was the first time that an arms control agreement provided for such an exchange, but the Treaty never entered into force.

The Memorandum of Understanding attached to the 1987 Intermediate-Range Nuclear Forces Treaty, which established a database, listed all missiles and launchers – both deployed and non-deployed – as well as associated support structures and equipment (including their locations) that the United States and the Soviet Union undertook to eliminate. It also specified the technical characteristics of the items covered by the Treaty. Updating of the data took place when the Treaty entered into force.

A Memorandum of Understanding on the establishment of a database is also a part of the 1991 START I Treaty. It specifies for each party the numbers of warheads, of deployed intercontinental ballistic missiles (ICBMs) and submarine-launched ballistic missiles (SLBMs) and their associated launchers, of non-deployed ICBMs and SLBMs, of non-deployed launchers of ICBMs and SLBMs, of heavy bombers, and of ICBMs and SLBMs at space launch facilities, as well as technical data on heavy bombers and long-range nuclear air-launched cruise missiles.

The Memorandum of Understanding relating to the 1993 START II Treaty includes only those data used for the purposes of implementing this Treaty that differ from the data in the Memorandum of Understanding relating to the START I Treaty. The START I Treaty verification regime applies to the START II Treaty.

The conclusion of the 1990 US–Soviet Chemical Weapons Agreement was preceded by an exchange of data on the chemical weapon capabilities of the two sides as well as visits to the relevant facilities on their territories. This was done in accordance with the Memorandum of Understanding signed in 1989.

Multilateral. A well-developed system for multilateral exchange of information can be found in the 1959 Antarctic Treaty. Each party is obliged to inform the other parties of all expeditions to and within Antarctica, of all stations occupied there by its nationals, and of any military personnel or equipment intended to be introduced into Antarctica for scientific research or other peaceful purposes.

Parties to the 1979 Moon Agreement must provide information on the time, purposes, locations, orbital parameters and duration of each mission to the moon. In addition, the UN Secretary-General as well as the public and the international scientific community must be informed by the parties of any phenomena discovered in outer space, including the moon, which could endanger human life or health, as well as of any indication of organic life.

Accountancy is an important element of nuclear safeguards. The safeguards are administered by the International Atomic Energy Agency (IAEA), which is charged with monitoring compliance with the 1968 Non-Proliferation Treaty (NPT), the 1967 Treaty of Tlatelolco, the 1985 Treaty of Rarotonga, the 1995 Treaty of Bangkok and the 1996 Treaty of Pelindaba. States supply the IAEA with detailed information concerning nuclear material subject to controls and the features of facilities relevant to safeguarding such material. Inventory changes must be reported promptly.

Parties to the 1990 Treaty on Conventional Armed Forces in Europe (CFE Treaty) agreed to exchange information on the structure of their land forces and air and air defence aviation forces within the area of the Treaty's application; on the location, numbers and types of conventional armaments and equipment in service with their conventional forces; on the location and numbers of battle tanks, armoured combat vehicles, artillery, combat aircraft and attack helicopters within the area of application but not in service with conventional forces; on the location of sites from which conventional armaments and equipment have been withdrawn; on the changes in organizational structures or force levels; on the entry into and removal from service with the conventional armed forces of conventional armaments and equipment limited by the Treaty; on the entry into and exit from the area of application of conventional armaments and equipment limited by the Treaty; and on the conventional armaments and equipment in transit through the area of application. All information is to be provided in accordance with agreed procedures. The format for the exchange is specified.

Under the 1993 Chemical Weapons Convention (CW Convention), each party is obliged to declare whether it possesses any chemical weapons or whether there are chemical weapons located in any place under its jurisdiction or control and, if so, specify the precise location, aggregate quantity and detailed inventory of these weapons; report any chemical weapons on its territory that are possessed by another state and located in any place under the jurisdiction or control of another state; declare whether it has transferred or received, directly or indirectly, any chemical weapons since 1 January 1946 and specify the transfer or receipt of such weapons; provide a general plan for destruction of chemical weapons that it possesses or that are located under its jurisdiction or control; declare whether there are old or abandoned chemical weapons on its territory and provide all available information; declare whether it has, or has had at any time since 1 January 1946, a chemical weapons production facility in its possession or located in a place under its jurisdiction or control and, if so, specify; report any chemical weapons production facility in its territory that belongs to another state and that is located under the jurisdiction or control of another state; declare whether it has transferred or received, directly or indirectly, any equipment for the production of chemical weapons, also since 1 January 1946, and specify; provide a general plan for destruction of chemical weapons production facilities, specify actions to be taken for their closure and provide a general plan for their temporary conversion; and indicate any facility that has been designed, constructed or used since 1 January 1946 primarily for the development of chemical weapons. With respect to riot control agents, the parties must declare the chemical name, structural formula and Chemical Abstracts Service registry number, if assigned, of each chemical they hold for riot control purposes; this declaration is subject to updates. Initial as well as annual declarations are required for all industrial facilities producing more than certain amounts of chemicals.

Signatories to the 1996 Comprehensive Nuclear Test-Ban Treaty (CTBT) have undertaken to participate in the fullest possible exchange relating to technologies used in the verification of the Treaty. The aim of this exchange is to strengthen national verification measures and to benefit from their application for peaceful purposes.

The 1997 Convention Prohibiting Anti-Personnel Mines (APM Convention) requires that the parties report to the UN Secretary-General on their national implementation measures; on the total number and types of their stockpiled APMs; on the location of areas containing or suspected to contain APMs; on the number and types of APMs retained or transferred for allowed purposes; on the status of programmes for the conversion or decommissioning of APM production facilities; on the status of programmes for the destruction of APMs; on the number and types of APMs destroyed after entry into force of the Convention; on the technical characteristics of each type of APM produced; and on the measures taken to provide an effective warning to the population. The information is to be updated annually. The Treaty-required reporting by the parties is complemented by the annual reports of the unofficial publication *Landmine Monitor*, compiled by the International Campaign to Ban Landmines (see Chapter 14). These civil society-based reports cover all countries and provide information on the use, development, production, transfer and stockpiling of landmines, on stockpile destruction, on whether a state party has passed the national implementing legislation required by the Convention, and on the humanitarian mine action and assistance needs.

International Data Exchanges to Build Confidence

Exchanges of information, carried out as confidence-building measures not affecting the strength of the parties' armed forces or armaments, form the essence of several treaties analysed in the preceding chapters. These are, in particular, the 1999 Vienna CSBM Document (which has incorporated the 1975 Helsinki CBM Document, the 1986 Stockholm CSBM Document, as well as the 1990, 1992 and 1994 Vienna CSBM Documents), promoting openness and constraints in military activities in Europe (see Chapter 16); the 1975 Registration Convention, aimed at dispelling misunderstandings that may be caused by objects launched into outer space (see Chapter 10); as well as several agreements providing for notification of missile launches (see Chapter 18).

According to the provisions of the 1996 CTBT, each party is expected to contribute to the resolution of compliance concerns arising from possible misinterpretation of the verification data relating to chemical explosions. To this end, it may provide, on a voluntary basis, notification of any chemical explosion using 300 tonnes or greater of TNT-equivalent blasting material detonated as a single explosion anywhere on its territory, or at any place under its jurisdiction or control. Such notification should include details on the location, time, quantity and type of explosive used, as well as on the configuration and intended purpose of the blast. Moreover, each party may provide – also on a voluntary basis – as soon as possible after entry into force of the Treaty and at annual intervals thereafter, detailed information related to its national use of all other chemical explosions greater than 300 tonnes TNT-equivalent. These confidence-building measures are meant to assist in clarifying the origins of the detected explosions.

National Means

Information acquired through formal international exchanges cannot be considered fully reliable: it may be incomplete or misleading. However, its correctness can be checked by the parties using their own means.

National technical means (NTM) of verification are referred to in certain arms control treaties, as distinct from negotiated verification measures. They comprise nationally owned photographic, electronic and radar surveillance systems, seismic instrumentation and other specialized techniques.

Satellites. Photoreconnaissance with high-resolution cameras mounted on low-flying satellites can identify even small objects on the ground. Infrared sensors can track activities taking place under cover of night and detect camouflaged facilities and weapons. Satellites placed in geosynchronous orbit can ensure continuous surveillance of certain specific sites. Photographic satellites carrying scanners, which take pictures with light-sensitive lenses and infrared sensors, can provide wide-area surveillance. High-orbit satellites can be moved to lower orbits for close-look missions.

Aircraft. Aircraft can also conduct reconnaissance. Flying along borders and employing 'slant photography', they can photograph all types of military installation on the territory of the monitored country without its consent.

Electronic Eavesdropping. Radio receivers on land, in space and at sea can intercept telemetric data relayed from missiles to ground monitors during flight-tests, providing information about the size of the missile booster as well as the weight and range of missiles. Electronic intelligence satellite systems can also intercept telemetry from missiles during tests. Certain aircraft can carry out electronic reconnaissance as well.

Radar. Radar is particularly useful for monitoring areas often obscured by clouds. Certain types of ground-based radar can even monitor 'over the horizon' activities. Phased-array radar – a sophisticated technology, in which many small antennas can be individually adjusted electronically, allowing the beam to be switched rapidly from one direction to another – is used to track the initial and the final phases of ballistic missile flights. Air surveillance and control can be provided by airborne early warning vehicles.

Seismometers. Seismometers are employed to detect underground nuclear explosions. Often deployed in arrays of a defined configuration, they can – in combination with some other techniques – distinguish earthquakes from man-made blasts and help to determine the location as well as the magnitude of seismic events.

Non-Interference. Without the technological advances that have considerably facilitated verification of arms control agreements with states' own means, it would have been difficult to achieve certain important treaties, especially nuclear arms limitation and reduction treaties. Having accepted reconnaissance by satellites or aircraft as a legal activity contributing to the maintenance of international security, the superpowers have undertaken to refrain from interference with NTM. This implies, in particular, a ban on the use of anti-satellite systems against satellites serving verification purposes, as well as on the use of concealment measures impeding verification by NTM.

To enhance the effectiveness of the NTM, the 1987 INF Treaty included the following unprecedented provision: at the request of one of the parties, the other party was under the obligation, on six hours' notice, to open the roofs of those fixed structures for missile launchers not subject to on-site inspection and display them in the open for 12 hours to show that no missiles subject to elimination had been concealed. Following this example, each party to the START I Treaty is entitled to request the other party to display openly both mobile ICBMs and aircraft to make possible their observation with the help of satellites. Even certain testing activities must not be kept secret. For example, under the START I Treaty, it is prohibited to conceal the association between ICBMs or SLBMs and their launchers during testing. This means that access to telemetric information broadcast during flight-tests must not be denied by the use of encryption or jamming. The parties have even agreed to exchange tapes containing telemetric data after each test, along with the information necessary for interpreting the data.

None of the multilateral agreements in force stipulates that information acquired by NTM must be shared with parties not possessing them. Scanning the political, economic, technical and scientific literature may serve verification needs as NTM. Although it is unlikely that information indicating a violation would appear in open publications, thorough examination of certain specialized journals and national budgetary data may reveal clues regarding non-compliance.

Limitations. The available NTM are sophisticated enough to monitor with a high degree of reliability a large range of weapons and military activities covered by arms control agreements. There are, however, natural limitations to the accuracy of such monitoring, especially if the other party resorts to evasion or deception techniques. Moreover, qualitative weapon developments may outpace the technological capabilities for monitoring. As weapons become smaller, more mobile, more widely deployed and multi-purpose, the significance of restraints that can be verified by NTM alone will diminish. Even today, whole classes of weapon or military activity lack the distinctive features necessary for distant identification. Monitoring by NTM must therefore be complemented with on-site inspection, which requires cooperative efforts on the part of the participating states.

19.3 Systematic Inspection and Observation

Elaborate procedures for systematic, routine inspection and observation of relevant objects and activities have been included in several treaties controlling both weapons of mass destruction and conventional weapons.

Checking Limitations and Prohibition of Nuclear-Weapon Tests

To complement seismic methods used in verifying the yields of nuclear explosions, the 1974 TTBT and the 1976 PNET provide for on-site use of the so-called CORRTEX (continuous reflectometry for radius versus time experiments) technique. This yield-measuring technique is based on the following principle: since an underground nuclear explosion produces a shock wave that propagates radially outwards, a cable placed in the emplacement hole containing the nuclear device, or in one or more separately drilled 'satellite' holes adjacent to the emplacement hole, is crushed by over-pressure and shortened as the shock wave expands. A measuring instrument connected to the cable registers the rate at which the cable is short-circuited during the milliseconds it takes for the wave to reach the surface. The larger the explosion, the faster the wave will travel. Analysis of the wave expansion allows an estimation of the yield. The measurement operation must take place in the presence and under the surveillance of the personnel appointed by the verifying party, the so-called 'designated personnel'. The CORRTEX technique is known to be most efficient when applied to large explosions; low-yield explosions may provide too small a crush zone for accurate measurement. Parties to the 1996 CTBT, which prohibits all nuclear explosions, will – upon entry into force of the Treaty – make use of the International Monitoring System (IMS) comprising facilities listed in an annex to the Treaty. Data from the global seismic network, a key component of the IMS, make it possible to detect and locate seismic events and to distinguish between underground nuclear explosions and earthquakes. Hydroacoustic stations detect acoustic waves produced by natural and man-made phenomena in the oceans or on small islands. Infrasound stations detect very-low-frequency sound waves in the atmosphere, generated by natural and man-made events. Radionuclide stations use air samples to detect radioactive particles released from atmospheric explosions, as well as those vented from underground or underwater explosions. To permit the integration of all the contributing stations into the IMS, the host countries are required to sign an agreement with the Comprehensive Nuclear Test-Ban Organization (CTBTO). Data from all IMS facilities, including the results of analyses conducted at certified laboratories, are to be collected, processed, analysed and archived by the International Data Centre. Each party will have the right to participate in the international exchange of data.

Checking Reductions and Elimination of Nuclear and Chemical Weapons

In conformity with the stipulations of the 1987 INF Treaty, shortly after the Treaty entered into force the parties inspected each other's missile operating bases and missile support facilities to verify the number of missiles, launchers, support structures and support equipment covered by the Treaty, as well as other data declared by the parties. Inspections were also conducted to verify the elimination of the bases and facilities in question. The intermediate-range and shorter-range missiles (stripped of their nuclear warheads and guidance elements) as well as launchers of these missiles were removed from the deployment areas and taken to the elimination facilities specified in the Memorandum of Understanding or, in certain special cases, eliminated *in situ*. Inspectors observed the cutting, burning, crushing, flattening or destroying by explosion of the relevant items, in accordance with the Protocol on Elimination, as well as the launching of a restricted number of missiles for the purpose of their elimination. Some of these elimination operations were conducted in the presence of invited journalists and diplomats. Inspection of missile operating bases and missile support facilities was carried out – before and after their elimination – for 13 years after entry into force of the INF Treaty.

The 1991 START I Treaty provides for the following types of systematic on-site inspection: baseline data inspections at facilities to confirm the accuracy of the information specified in the initial exchange of data; data update inspections; new facility inspections; re-entry vehicle inspections to confirm that the deployed ballistic missiles contain no more re-entry vehicles than the number attributed to them; post-exercise dispersal inspections of deployed mobile launchers of ICBMs and their associated missiles to confirm that the number of such launchers and missiles located at the inspected ICBM base and those that have returned to it after dispersal does not exceed the number specified for that ICBM base; conversion or elimination inspections; close-out inspections to confirm that the elimination of facilities has been completed; formerly declared facility inspections to confirm that facilities, notification of the elimination of which has been provided, are not being used for purposes inconsistent with the Treaty; inspections during technical characteristics exhibitions of missiles and launchers; inspections during distinguishability exhibitions for heavy bombers, former heavy bombers and long-range nuclear ALCMs; and inspections during baseline exhibitions of heavy bombers equipped for non-nuclear armaments, training heavy bombers, and former heavy bombers.

Under the 1993 START II Treaty, the parties will have the right to conduct inspections in connection with the elimination of heavy ICBMs and their launch canisters, as well as in connection with the conversion of silo launchers of heavy ICBMs. As regards conversion, each party will be entitled to observe the entire process of pouring concrete into each silo launcher of heavy ICBMs, silo training launcher for heavy ICBMs, and silo test launcher for heavy ICBMs that is to be converted, and to measure the diameter of the restrictive ring to be installed in the upper portion of the silo launcher. Alternatively, inspection might take place only before the commencement and after the completion of the process of pouring concrete.

According to the 1993 CW Convention, inspection of a chemical weapons storage site may be arranged at 48 hours' notice. Inspectors have the right of 'unimpeded access' to all parts of the facility, including stored munitions and containers, as well as to any building or location there. Similar access must be accorded to inspectors at chemical weapons destruction sites. To monitor destruction, inspectors have the right to use continuous on-site monitoring devices, to carry out sample analysis during destruction, and to obtain samples from any container at the site. Conversion and destruction of chemical weapons production facilities are also subject to systematic on-site inspections and monitoring with on-site instruments.

Checking Conventional Arms Reductions

Each party to the 1990 CFE Treaty has the right to conduct and the obligation to accept inspections within the area of the Treaty's application. The inspections are intended to verify, on the basis of information previously supplied, compliance with the numerical limitations set forth in the Treaty; monitor the process of reduction of battle tanks, armoured combat vehicles, artillery, combat aircraft and attack helicopters, carried out at reduction sites in accordance with the Protocol on Reduction; and monitor the certification of recategorized multi-purpose attack helicopters and reclassified combat-capable trainer aircraft, carried out in accordance with the relevant Protocols. The CFE Treaty does not provide for permanent inspections or fixed remote sensors.

Checking Non-Production of Prohibited Items

Nuclear Arms Control. Each party to the 1987 INF Treaty was granted the right to inspect, by means of continuous monitoring, the portals of the facility of the other party, at which the final assembly was accomplished for ground-launched ballistic missiles having the first stage outwardly similar to the first stage of the missiles banned by the Treaty. The facility designated for inspection according to this criterion was the Votkinsk Machine Building Plant in the Soviet Union. To ensure that production of intermediate-range missiles has ceased, US inspectors were permitted to measure and weigh the canisters leaving the facility, as well as to open them randomly several times a year to check their contents. The United States had no comparable facility. Nevertheless, for the sake of symmetry, it allowed continuous monitoring at the portals of a plant at Magna, in the state of Utah, where rocket motors for the missiles covered by the Treaty were formerly produced. As stipulated by the Treaty, the US and Soviet/Russian factories were monitored only until 2001, although the Treaty is of unlimited duration.

As under the 1987 INF Treaty, parties to the 1991 START I Treaty may conduct continuous monitoring activities at certain production facilities. The Treaty specifies the number of containers or vehicles exceeding certain size criteria that may be inspected there each year. Two such facilities are located in Russia and one in the United States.

According to the terms of the 1968 NPT, the 1967 Treaty of Tlatelolco, the 1985 Treaty of Rarotonga, the 1995 Treaty of Bangkok and the 1996 Treaty of Pelindaba, non-nuclear-weapon states must accept IAEA safeguards to demonstrate the fulfilment of their obligation not to manufacture nuclear weapons. In conformity with the 1971 NPT Model Safeguards Agreement, the IAEA conducts ad hoc inspections to verify the information contained in the initial report of each party to the IAEA, as well as routine inspections to check that subsequent reports are consistent with the plant's operating records. Inspectors check the location, identity, quantity and composition of the nuclear material at the plant. Tamper-resistant photographic and television equipment is used to survey movements of nuclear material in nuclear plants in periods between inspections. (For the role of nuclear safeguards, see also Chapter 6.)

For years, routine inspections conducted by the IAEA had been formally confined to a limited number of so-called 'strategic points' of the nuclear fuel cycle, that is, those parts of the nuclear plant that are considered essential for safeguarding the flow of nuclear material. Moreover, the efficiency of the safeguarding operations was negatively affected by the requirement to obtain the consent of the inspected party to the designation of inspectors. Taking advantage of this requirement, many governments applied restrictions on nationality, linguistic qualifications or numbers of inspectors. This practice caused a waste of manpower, unnecessary expense and an unbalanced distribution of inspectors' nationalities. (In addition, under the START I Treaty the parties have the right to refuse an individual included in the proposed list of inspectors, but in a bilateral agreement such a clause does not impair the efficiency of inspections to the same extent as in a multilateral agreement.)

Since the adoption in 1997 of the Model Additional Safeguards Protocol, IAEA inspectors have the right to obtain from the parties to the Protocol more information than was previously required about, and ask for wider, more intrusive physical access to, all aspects of states' nuclear fuel cycles, from uranium mines to nuclear waste. They also have stronger authority to collect environmental samples for laboratory analysis from the air, water, vegetation, soil or building's surfaces for the purpose of assisting the IAEA in drawing conclusions about the presence or absence of undeclared nuclear material or nuclear activities at a specific location. (The leakage of fissile isotopes into the environment cannot be completely prevented in a nuclear-weapon programme.) New administrative arrangements will improve the procedures for the designation of inspectors and expand the capability of inspectors to communicate with their headquarters using modern means of communication. If there is a conflict between the original comprehensive safeguards agreement and the Additional Protocol, the Protocol overrules the safeguards agreement. However, by 2002 few countries had adhered to the Protocol.

Chemical Arms Control. Monitoring the non-production of chemical weapons is more complicated than monitoring the non-production of missiles or nuclear-weapon-usable material. This is so because of the vastness of the chemical industry in many countries, because of the relative ease with which chemical warfare agents can be fitted into existing delivery vehicles, and because of the generally accepted postulate that technical and commercial secrets of the chemical industry should not be revealed through inspection.

Under the 1990 US–Soviet Chemical Weapons Agreement, each party was to provide access to each of its chemical weapons production facilities for systematic on-site inspection to confirm that production of chemical weapons was not occurring at those facilities. This Agreement was superseded by the multilateral CW Convention of 1993.

Under the CW Convention, the stringency of inspections to check the absence of production of the prohibited weapons varies according to the degree of toxicity of the substances subject to verification. Chemicals in question are listed in three schedules, as described in Chapter 8. Production and possession of schedule 1 chemicals, which raise the greatest concern, is severely restricted, both quantitatively and qualitatively; systematic on-site inspections of facilities using or producing them are therefore envisaged to verify compliance. There are no limitations on production or possession of schedule 2 chemicals; however, to prevent their diversion for prohibited ends, the facilities producing, processing or consuming them may be subject to on-site inspection if they produce or use certain specified quantities. Schedule 3 chemical facilities are subject to on-site inspection only if they produce the relevant chemicals in excess of relatively high thresholds; those to be inspected are to be selected from among the declared facilities. So-called 'other facilities' – those that do not produce, process or consume the restricted chemicals, but have technical features permitting the production of chemical weapons – may also be inspected if they meet certain technical criteria; the selection of such facilities for inspection is to be done at random, as in the case of schedule 3 facilities. The main purpose of inspection of schedule 3 facilities and 'other facilities' is to ascertain the absence of schedule 1 chemicals.

Observation

In the context of verification, the rights of observers are more limited than the rights of inspectors. For example, the confidence- and security-building measures in Europe allow observation only of certain notified activities. The host country may even exclude from observation certain areas or installations. The purpose of observation, as provided for in the 1999 Vienna CSBM Document, is to confirm that the notified activity is not threatening in character, but the functions of observers are not clearly defined (see Chapter 16).

The original parties to the 1959 Antarctic Treaty, as well as those acceding states that conduct substantial scientific research in Antarctica, enjoy complete freedom of observation. All Antarctic areas, including stations, installations and equipment within these areas, and all ships and aircraft at points of discharging or embarking cargoes or personnel, are to be open at all times to observers; aerial observation is also permitted. In addition, scientific personnel may be exchanged between expeditions and stations, facilitating 'informal' observation.

Similar, although less sweeping, provisions are included in the 1967 Outer Space Treaty and in the 1979 Moon Agreement. All stations, installations, equipment and space vehicles on the moon and other celestial bodies are to be open to parties upon advance notice of a visit. However, objects in earth orbit are not subject to the regime of openness.

Observation is also allowed under the 1971 Seabed Treaty. It serves to verify the activities of states on the seabed and in the subsoil thereof beyond a 12-mile coastal zone. The possibility of 'appropriate' inspection of objects, installations or other facilities, reasonably expected to be of the kind prohibited by the Treaty, is envisaged as well.

19.4 Challenge Inspection

NPT Nuclear Safeguards

According to the 1971 NPT Model Safeguards Agreement, the IAEA may conduct challenge inspections, called special inspections (as distinct from systematic or routine inspections), if questions arise about the commitment of a party to the non-proliferation objectives of the NPT or about the safeguards coverage of nuclear fissionable material. An inspection is considered 'special' when it is either additional to the routine inspection or involves access to information or locations in addition to the access specified for routine inspections in the Agreement. For several decades the IAEA did not conduct inspections at locations or facilities other than those declared by the state concerned as containing safeguarded materials or equipment. Iraq – a party to the NPT – was therefore able to embark upon a substantial nuclear-weapon programme in undeclared facilities (often located in the vicinity of the routinely inspected objects) and thereby thwart the purpose of the Treaty without being hampered by IAEA controls.

According to the Model NPT Safeguards Agreement, only the IAEA – not the parties – may formally initiate visits to undeclared sites for storage of nuclear materials and installations in countries suspected of non-compliance with the non-proliferation commitments. To initiate such special inspections, the IAEA must first detect suspicious activities. However, it may not be able to do this, because indications of transgression collected in the course of routine inspections may be scant; information available through press and other media is often unreliable; and depositions of political defectors are not always credible. Since sending out people to blindly search the territories of NPT parties for evidence of secret nuclear activities is out of the question, the Agency can act responsibly only if it obtains information from the means of verification possessed by individual countries as to which sites and installations in countries under suspicion should be inspected. This is why, in 1991, the IAEA Director General proposed to set up, within his Secretariat and under his supervision, a special unit to receive and assess information from all sources, including intelligence sources, which could lead to the discovery of clandestine activities.

This proposal was not approved by the IAEA Board of Governors. Its opponents argued that information provided by the governments of certain countries with a view to triggering a special inspection could be discriminatory or deliberately misleading and that the IAEA Secretariat is not in a position to evaluate its trustworthiness. They also warned that, if the information upon which the Agency's request for special inspection was predicated came only or mainly from one source, the IAEA would risk becoming, or being seen as, an organization serving the interests of just one state or group of states, and that this would undermine its integrity. For these reasons, in 1992 the IAEA Board merely reaffirmed the Agency's right to undertake special inspections 'when necessary and appropriate', as described in the agreements providing for the application of safeguards to all nuclear materials in all peaceful nuclear activities within a state.

The IAEA Secretariat must be in a position to identify all instances in which available information about nuclear activities of a state appears inconsistent with that state's declaration and keep the IAEA Board adequately informed. However, formally requesting a special inspection – whatever the justification – constitutes a challenge of a political nature. This would better be made by governments of states parties to the NPT rather than an international technical organization. Such procedure would, of course, carry a risk of abuse, but this risk could be significantly reduced if the IAEA Board were entitled to rule that a requested inspection may not take place. A provision to this effect figures in several arms control agreements (see below). Whatever the arrangements, the IAEA cannot be given an absolute right to inspect at any time any place on the territory of the non-nuclear-weapon states parties. Such inspections would be unacceptable to many. The case of Iraq was exceptional, as the 1991 UN Security Council ceasefire resolution gave the Agency the authority to conduct an unrestricted countrywide search of illicit activities, an authority which under normal circumstances is not applicable to a sovereign state.

Denuclearized Zones

Special inspections to guard against breaches of obligations not to acquire nuclear weapons were also envisaged in the 1967 Treaty of Tlatelolco. Such inspections could be carried out either by the IAEA, in accordance with the nuclear safeguards agreements in force, or by the Agency for the Prohibition of Nuclear Weapons in Latin America and the Caribbean (OPANAL). In 1992, the parties decided to amend the Treaty so as to give the IAEA the exclusive prerogative of conducting special inspections.

Under the 1985 Treaty of Rarotonga, the Consultative Committee of parties may direct that a special inspection be made by a team of three qualified inspectors. The latter would be appointed by the Committee in consultation with the party complained of as well as the complaining party, provided that no national of either party shall serve on the team. If so requested by the party complained of, the special inspection team may be accompanied by representatives of that party. Special inspections must be given full and free access to all relevant information and places.

The African Commission on Nuclear Energy is entitled by the 1996 Treaty of Pelindaba to consider whether there is sufficient substance in the complaint of a party to warrant an inspection in the territory of another party. If there is, the Commission may request the IAEA to conduct such inspection as soon as possible. It may also designate its representatives to accompany the Agency's inspection team. If the party complained of so requests, the team may be accompanied by representatives of that party. Each party must give the team full and free access to all relevant information and places. The Commission may also establish its own inspection mechanisms.

The Comprehensive Nuclear Test Ban

To clarify whether a nuclear-weapon test explosion or any other nuclear explosion has been carried out in violation of the CTBT, and to gather information that might assist in identifying the violator, each party to the Treaty has the right to request an on-site inspection in the territory or in any other place under the jurisdiction or control of another party, or in an area beyond the jurisdiction or control of any state. The request must be based on information collected by the IMS, on any technical information obtained by NTM in a manner consistent with generally recognized principles of international law (which, in the understanding of some signatories, excludes information gathered through espionage), or on a combination thereof. It should be presented to the Executive Council of the CTBTO. If the request meets the requirements specified in the Protocol to the Treaty, the Technical Secretariat of the Organization begins preparations for the inspection, but the decision whether or not to actually carry out the inspection is to be taken by a majority of all members of the Executive Council. (A separate decision of the Council will be needed if, in the course of inspection, the inspection team finds it necessary to drill in order to obtain radioactive samples.) If the Council does not approve the inspection, the preparations must stop and no further action may be taken.

The Director-General of the CTBTO is empowered to determine the size of the inspection team and select its members. The inspected party is obliged to provide access within the inspection area, but it has the right to take measures to protect its national security interests and to prevent disclosure of confidential information not related to the purpose of the inspection. No national of the requesting state or the inspected state may be a member of the inspection team. The inspectors, on their part, must seek to minimize interference with the normal operations of the inspected state. Subject to the agreement of the inspected state, the requesting party may send a representative, who shall be a national of the requesting party or of a third party, to observe the conduct of the inspection.

If the Executive Council does not approve the inspection because it finds the request frivolous or abusive, or if for the same reasons the ongoing inspection is to be terminated, the following measures may be taken: requiring the requesting party to pay for the cost of any preparations made by the Technical Secretariat; suspending the right of the requesting party to request an on-site inspection for a period of time to be determined by the Executive Council; and suspending the right of the requesting party to serve on the Council for a period of time.

Bilateral Nuclear Arms Control

Short-notice inspections provided for in the 1987 INF Treaty and the 1991 START I Treaty possess certain characteristics which make them akin to challenge inspections. However, under the INF Treaty, no challenge inspection could be conducted at sites other than those specified in the Memorandum of Understanding. Under the START I Treaty, suspect-site inspections to confirm that covert assembly of missiles is not occurring may apply only to three agreed-upon sites; if suspect activity were to be detected at other facilities, the parties could raise the issue in the Joint Compliance and Inspection Commission (JCIC), which may then authorize special visits to undeclared sites, but only with the mutual consent of the parties.

Chemical Arms Control

The 1993 Chemical Weapons Convention stipulates that any party may request that an on-site challenge inspection be carried out of a facility or location in the territory or any other place under the jurisdiction or control of another party and that the requested party is obliged to accept such inspection. However, the Executive Council of the Organisation for the Prohibition of Chemical Weapons (OPCW) may decide by a three-quarters majority not to carry out an inspection if it considers the request to be frivolous or abusive. (Neither the requesting nor the inspected party may participate in such a decision.) Moreover, if the concern is about an undeclared facility, the requested party has the right to propose alternatives to full access to the suspected site; the perimeter within which the inspection is to take place may be negotiated, but it must include the whole of the requested perimeter and, as a rule, bear a close relationship to the latter. If no agreement is reached, the perimeter proposed by the inspected party is to be designated as final. In other words, the requested state may keep inspectors out of certain parts of the site to be inspected or delay access to them, but it cannot prevent a challenge inspection altogether.

In conducting the perimeter activities, the inspectors are entitled to use monitoring instruments and take air, soil and effluent samples. The requesting party may, subject to the agreement of the inspected party, send a representative to observe the conduct of the challenge inspection. Actual inspection may be restricted by so-called 'managed access' procedures, under which the challenged country is allowed to remove sensitive papers; shroud certain sensitive equipment; turn off computers and data-recording devices; allow inspectors into only a given percentage of the buildings or rooms, chosen at random; or, in 'exceptional cases', allow only one inspector from the team to view a particular area. Nevertheless, concealment of militarily significant amounts of the prohibited substances would be difficult: modern chemical detection equipment is able to detect traces of chemical agents even after extensive cleaning has taken place.

After entry into force of the Convention, the Technical Secretariat of the OPCW communicated to all parties a list of names, nationalities and ranks of the proposed inspectors and inspection assistants. Any person included in the list is to be regarded as designated unless a party, not later than 30 days after acknowledging the receipt of the list, has declared its non-acceptance in writing. The parties must provide each inspector and inspection assistant with multiple entry/exit and/or transit visas as well as other documents – valid for at least two years – necessary to carry out their duties.

Investigations of the alleged use of chemical weapons, as well as the use of riot control agents as a method of warfare, may also require challenge inspection. The inspection team, to be dispatched by the Director-General of the OPCW 'at the earliest opportunity', will have the right of access to any area which could be affected by the use of the prohibited weapons, and – subject to consultation with the inspected party – to hospitals, refugee camps and other locations deemed relevant to effective investigation. The team must be allowed to collect samples of chemicals and munitions, environmental samples, and biomedical samples from human and animal sources. It must also be free to interview and examine persons who may have been affected by the use of chemical weapons, eyewitnesses of such use or medical personnel. An inspection can be relatively easily arranged if the alleged use has taken place on the territory of the victim state that is interested in a speedy investigation, as in the case of Iran in the 1980s. In the case of alleged use of chemical weapons involving a state not party to the Convention, or in a territory not controlled by a state party, the OPCW shall closely cooperate with the UN Secretary-General and, if so requested, put its resources at his disposal. UN General Assembly resolutions adopted in the 1980s empowered the Secretary-General to carry out investigations, including on-site collection of evidence, in response to all reports that may be brought to his attention concerning the possible use of chemical and biological or toxin weapons.

European CSBMs

In the 1980s, when it was agreed that the CSBMs then under consideration in Europe must be verifiable, challenge inspection became an essential element of the negotiated measures. The breakthrough occurred when the 1986 Stockholm CSBM Document introduced an unconditional obligation of the parties to let foreign inspectors enter their territory upon simple request and on short notice.

The concept of non-refusable on-site inspections was further developed in subsequent documents of the Conference on Security and Co-operation in Europe and the Organization for Security and Co-operation in Europe, as described in Chapter 16. Such inspections may be set in motion whenever compliance with the agreed measures is in doubt. It is unclear, however, how doubts which may concern a variety of parameters of notifiable military activities could be dispelled within the relatively short period allotted to inspection, especially since 'sensitive' areas, places and equipment are to remain off-limits to inspectors. All that such inspections could do is to certify, just as in the case of observations, the 'benign' nature of the inspected activities.

The 1990 CFE Treaty envisages challenge inspections within specified areas, but the parties have the right to refuse them.

Assessment

As a means to deter violations, challenge inspections are more effective and less costly than routine inspections. However, their value must not be overestimated: they cannot replace the use of NTM of verification, nor can they provide proof that a breach has not been committed. A violator would scarcely permit inspection of areas or facilities in which he had conducted clandestine activities, whatever the consequences of his refusal. Rejection of inspection in response to a challenge to investigate a suspect event or activities may not necessarily amount to an admission of guilt, but it could result in a strong assumption by the accusing party that a violation has taken place.

19.5 Verification Institutions

In order to clarify problems which may arise among parties, most arms control agreements provide for consultation and complaint procedures, which form an integral part of the verification regime. In several cases, special verification bodies have been set up to oversee implementation of the parties' obligations and to resolve possible disputes. Under certain agreements disputes may be referred to the International Court of Justice.

Consultation and Complaints

Direct interstate consultation to resolve uncertainties regarding compliance is particularly important, and also relatively uncomplicated, in arms control agreements among technologically developed states. In those multilateral agreements where the verification tools are only in the hands of a few nations, consultation is of lesser use to countries that are technologically underdeveloped, do not possess the resources needed to ascertain the behaviour of others, and may be reluctant to seek foreign assistance in the conduct of verification.

International indirect consultation is envisaged in a few multilateral agreements. The procedure resorted to must be 'appropriate' and placed within the framework of the United Nations in accordance with its Charter. Given the vagueness of this language, the participants in the Biological Weapons Convention (BW Convention) Review Conferences agreed that the said procedure should include the right of any party to request that a 'consultative meeting', open to all parties, be convened promptly at the expert level. Such a meeting was held in August 1997 to discuss the suspicious appearance of an insect pest in Cuba, but it was unable to come to a formal conclusion.

Complaints of breaches of multilateral arms control agreements may be lodged with the UN Security Council, but they must contain evidence confirming their validity. A state not possessing such evidence for lack of reliable information may find its request for consideration rejected by the Council, especially if not all Council members are party to the treaty in question. Even if the Council agrees to discuss a charge not meeting the above requirement, there is always a danger that the case will not be given proper examination: great-power veto has often been used to block not only substantive decisions but also proposals for investigation or observation when the interests of the permanent members of the Security Council or their allies were involved. Certain countries may hesitate to embark on a procedure that extends the inequality of states under the UN Charter to relations under arms control agreements. It is not even clear to what extent parties to arms control agreements are obliged to cooperate if an investigation is initiated by the Security Council. In view of these uncertainties, it is essential that special bodies be set up to engage in verification operations and to deal effectively with suspicions regarding compliance.

US–Soviet/Russian Verification Bodies

The 1972 ABM Treaty, the 1972 SALT Interim Agreement and the 1979 SALT Treaty established a Standing Consultative Commission (SCC) to consider questions concerning compliance, to receive information considered necessary to assure confidence in the fulfilment of the obligations assumed by the parties, and to discuss questions involving interference with NTM or impeding verification by NTM. Most other responsibilities of the SCC went beyond the handling of compliance issues. The parties agreed to consider possible changes in the strategic situation that might have a bearing on the provisions of the agreements; to set procedures and dates for replacement, conversion and dismantling or destruction of arms, as provided for in the agreements, and to notify each other periodically of actions completed and those in process; to examine proposals for further increasing the viability of the agreements, including proposals for amendments; to maintain the agreed database of each side's arms subject to limitations; and to consider proposals for further measures aimed at limiting strategic arms.

Over the years, the SCC has considered a number of compliance-related issues raised by both sides. It has also been used to discuss possible amendments to the ABM Treaty. Several times during the Cold War, the parties disregarded the principle of confidentiality governing SCC proceedings, preferring to make public their suspicions rather than to seek clarification through bilateral consultation.

The 1974 TTBT set up the Bilateral Consultative Commission (BCC) to discuss questions related to the implementation of, or compliance with, the Treaty and its Protocol, as well as possible amendments. The BCC also has the authority to consider the implications of new verification technologies for the TTBT, and to seek agreement on verification costs and payment procedures. The 1976 PNET established the Joint Consultative Commission (JCC) with similar tasks.

In concluding the 1987 INF Treaty, the United States and the Soviet Union decided to bypass the existing institutions promoting the objectives and implementation of nuclear arms control agreements. A new body – the Special Verification Commission (SVC) – was set up to resolve questions relating to compliance and to agree on necessary measures for improving the viability and effectiveness of the Treaty. The SVC does not need to hold regular meetings; it can meet at any time at the request of either party.

Parties to the 1991 START I Treaty meet within the framework of the Joint Compliance and Inspection Commission (JCIC) at the request of either of them. The task of the JCIC is to resolve questions relating to compliance and to agree on such measures as may be necessary to improve the viability and effectiveness of the Treaty – both tasks being identical to those performed by the SVC – as well as to resolve questions related to the application of the relevant provisions of the Treaty to a new kind of strategic offensive arm. The parties to the 1993 START II Treaty would have had to establish the Bilateral Implementation Commission (BIC) with analogous duties if the Treaty had entered into force. The BIC is also envisaged in the 2002 Treaty on Strategic Offensive Reductions.

Regional Verification Bodies

The first attempts at creating a regional body with terms of reference similar to those of the SCC were made in the course of the Mutual and Balanced Force Reduction (MBFR) talks in 1982, when NATO participants submitted a draft treaty for the reduction of armed forces in Central Europe. The duties of the proposed international commission would have been: to ascertain compliance with the treaty; to clarify ambiguous situations through consultations; to exchange and collect information as provided for in the treaty; to arrange for observation of notified activities; to arrange for ground and aerial inspections; and to resolve problems referred to by the parties. The draft treaty put forward by the members of the Warsaw Treaty Organization made provision for a commission which would carry out identical tasks, but on an ad hoc basis rather than on a permanent basis. Owing to the differences concerning the substance of the negotiated treaty, proposals regarding an all-European verification institution did not at that time receive proper consideration.

ACA. For many years, the Agency for the Control of Armaments (ACA) of the Western European Union (WEU) was engaged in verifying compliance by the WEU member-states with arms limitations to which they had agreed under the 1954 Paris Agreements (see Chapter 2).

The ACA verification scheme was based on a detailed questionnaire, which had to be answered by states by a specific date each year. The states were obligated to give a full account of the armaments subject to control, stating their quantities and locations. A certain percentage of the weapons in question were regularly checked by international inspectors, and the units or depots to be inspected were selected by the

ACA on the basis of replies to the questionnaire. Data on production levels were cross-checked with defence budget figures and with information appearing in open sources. Intrusive inspection visits to relevant plants were also made. The system did not provide for challenge inspections, and only declared locations could be inspected; a carefully planned breach could therefore remain undetected. Control of conventional weapons was subsequently brought to an end, whereas weapons of mass destruction were eventually banned by other international instruments. The experience gathered by the ACA demonstrated that on-site inspections by an international control authority could be carried out without negatively affecting industrial processes or legitimate commercial interests. It should, however, be noted that ACA verification was conducted among allied states.

Euratom. The European Atomic Energy Community (Euratom) was set up in 1957 by the Treaty of Rome to promote the development of nuclear energy in Western Europe. One of its main tasks is to ensure that all member-states receive a regular and equitable supply of nuclear fuels. Euratom checks that nuclear materials are being used in conformity with the purpose stated by the user or with agreed supply conditions (military uses of nuclear energy are not prohibited). The safeguards which it applies are the responsibility of the Directorate General for Energy of the European Commission. Euratom has an agreement with the IAEA for joint application of safeguards to verify that there is no diversion of nuclear materials to nuclear weapons or other nuclear explosives in any of the European Union's non-nuclear-weapon states.

OPANAL. This Agency, set up under the 1967 Treaty of Tlatelolco, is responsible for holding periodic or extraordinary consultations among member-states on matters relating to the purposes, measures and procedures set forth in the Treaty and to the supervision of compliance with the obligations arising therefrom. The General Conference, composed of all parties to the Treaty of Tlatelolco, is the supreme organ of OPANAL. It holds regular sessions, as well as special sessions if circumstances so require. The Council – which can function continuously – is composed of five members elected by the General Conference. The General Secretary is the chief administrative officer of the Agency.

ABACC. The Argentine–Brazilian Agency for Accounting and Control of Nuclear Materials (ABACC) was established by the Bilateral Agreement for the Exclusively Peaceful Use of Nuclear Energy, signed by Argentina and Brazil and in force since 1991. It administers the Common System of Accounting and Control of Nuclear Materials (SCCC) to verify that nuclear materials in the nuclear activities of the parties are not diverted to purposes prohibited by the Agreement. The ABACC is made up of a Commission of four members (two for each party), which monitors the functioning of the SCCC, and a Secretariat. The Secretary is appointed by the Commission and is a national of one of the two states, alternating every year. Inspectors of Argentina control Brazilian facilities and vice versa. They are responsible only to the Secretariat.

JCG. Under the 1990 Treaty on Conventional Armed Forces in Europe (CFE Treaty), a Joint Consultative Group (JCG), composed of representatives designated by each party, was established. Within the framework of the JCG, the parties may: address questions relating to compliance with or possible circumvention of the provisions of the Treaty; seek to resolve ambiguities and differences of interpretation; consider and, if possible, agree on measures to enhance the viability and effectiveness of the Treaty; update the lists contained in the Protocol on Existing Types of Conventional Armaments and Equipment; resolve technical questions; work out the distribution of costs of inspections; work out measures to ensure that information obtained through exchanges of information among the parties or as a result of inspections is used solely for the purposes of the Treaty; as well as consider matters of dispute arising out of the implementation of the Treaty. The JCG may propose amendments to the Treaty for consideration by the parties. It must meet for regular sessions two times per year, and its decisions are to be taken by consensus.

JNCC. To implement the 1992 Joint Declaration on the Denuclearization of the Korean Peninsula, a South–North Joint Nuclear Control Commission (JNCC) was established. It is composed of seven members, including a chairman and a vice-chairman. The Commission's tasks include: exchanging information necessary to verify the denuclearization; deciding on the composition and operation of inspection teams; selecting objects of inspections; defining the equipment that may be used for inspections; discussing remedial measures based on the results of inspections; and settling disputes. The Commission is to meet, in principle, once every two months.

Commission for the Southeast Asia Nuclear Weapon-Free Zone. Established by the 1995 Treaty of Bangkok, and composed of all states parties, the Commission oversees the implementation of the Treaty with a view to ensuring compliance with its provisions. It meets as and when necessary, as far as possible in conjunction with the Ministerial Meeting of the Association of South-East Asian Nations. The Executive Committee, a subsidiary organ of the Commission – also composed of all parties – is responsible for ensuring the proper operation of the verification measures; considering and deciding on requests for clarification; setting up a fact-finding mission; deciding on the findings of such a mission; concluding agreements with the IAEA or other international organizations; and carrying out such other tasks as may be assigned to it.

African Commission on Nuclear Energy. Established by the 1996 Treaty of Pelindaba, the Commission is responsible for collating reports of the parties and exchanging information among them; arranging consultations and convening conferences of parties on any matter arising from the implementation of the Treaty; reviewing the application of IAEA safeguards; bringing into effect the complaints procedure; and promoting cooperation in the peaceful uses of nuclear science and technology. The Commission is to meet in ordinary session once a year, or in extraordinary session, as may be required.

Global Verification Bodies

IAEA. This Agency was the first specialized international organization to be directly involved in checking observance of multilateral arms control obligations. Its main task is to apply safeguards in non-nuclear-weapon countries to verify that they are abiding by the provisions of the 1968 NPT, the 1967 Treaty of Tlatelolco, the 1985 Treaty of Rarotonga, the 1995 Treaty of Bangkok, and the 1996 Treaty of Pelindaba. In fact, the IAEA was established many years prior to these treaties to ensure that international assistance in peaceful applications of nuclear energy was not used to further military purposes. Once nuclear-weapon proliferation had become the subject of international prohibition, the IAEA was given additional supervisory tasks. The General Conference of the IAEA, consisting of representatives of all member-states, meets in regular annual sessions. Its Board of Governors, consisting of a restricted number of member-states, meets at such times as it itself determines. The staff of the Agency is headed by a Director General.

OPCW. Established by the 1993 CW Convention, the OPCW ensures the implementation of the Convention and provides a forum for consultation and cooperation among the parties.

The Conference, composed of all parties, is the principal organ of the OPCW. It must meet annually in regular sessions, but special sessions may also be convened. Its main function is to oversee the operation of the Convention and review compliance with it. Other functions include the supervision of the activities of the Executive Council and of the Technical Secretariat, the appointment of the Director-General of the Technical Secretariat, and the approval of the budget of the Organization. Conference decisions on questions of procedure may be taken by a simple majority of the members present and voting. Decisions on matters of substance are to be taken as far as possible by consensus; if consensus is not attainable, the Conference may, after a period of deferment, take a decision by a two-thirds majority of the members present and voting. The issue of whether the question is one of substance or not is to be treated as a matter of substance (not of procedure), unless otherwise decided by the Conference by the majority required for decisions on matters of substance.

The Executive Council of the OPCW has a rotating membership of 41 states, based on equitable geographical distribution, with a certain number of seats in each region being designated for countries with the largest chemical industries. Members are elected by the Conference for a term of two years. The voting procedures of the Council are similar to those of the Conference, but the simple and two-thirds majorities apply to all members of the Council, not just to those present and voting. As an executive organ of the OPCW, the Council supervises the activities of the Technical Secretariat and performs other operational and administrative functions, such as preparing the budget of the OPCW and concluding agreements concerning the provision of assistance to parties that may be attacked or threatened with attack with chemical weapons. The Council must consider any issue within its competence affecting the Convention and its implementation, including concerns regarding compliance, as well as cases of non-compliance, including abuse of the rights provided for under the Convention. Upon completion of the investigation of an alleged breach, the Council has to consider the report of the inspectors and make recommendations to the Conference. In cases of particular gravity and urgency, the Executive Council may bring the issue directly to the UN General Assembly and the UN Security Council.

The Technical Secretariat consists of a Director-General appointed for a four-year term, inspectors and such scientific, technical and other personnel as may be required. It carries out the verification measures provided for in the Convention and performs functions delegated to it by the Conference and the Council. The Secretariat prepares a draft programme of the OPCW, as well as a draft report on the implementation of the Convention; provides administrative and technical support to the Conference, the Council and subsidiary bodies; provides technical assistance to parties in the implementation of the provisions of the Convention, including evaluation of scheduled and unscheduled chemicals; and negotiates agreements or arrangements relating to verification, subject to approval by the Council. The Secretariat has the duty to inform the Council of any problems encountered in the discharge of its verification functions, including any doubts, ambiguities or uncertainties about compliance with the Convention, which it has been unable to clarify through consultations with the party concerned.

CTBTO. This Organization is to be established in accordance with the 1996 CTBT in order to ensure the implementation of the Treaty and, like the OPCW, provide a forum for consultation and cooperation. The CTBTO must conduct its verification activities in the least intrusive way, seek to utilize existing expertise and facilities, and maximize cost efficiencies through cooperative arrangements with other organizations. It may request only such information and data that are necessary to fulfil its responsibilities, and must take every precaution to protect the confidentiality of information on civil and military activities and facilities coming to its knowledge.

The Conference of all parties, the main organ of the CTBTO, oversees the activities of the Organization's Executive Council and Technical Secretariat. It meets in regular sessions, but special sessions may also be convened. Its voting procedure is the same as that of the Conference of the parties to the CW Convention.

The Executive Council of the CTBTO is to consist of 51 members. Taking into account the need for equitable geographical distribution, the Council must comprise: ten states from Africa, seven states from Eastern Europe, nine states from Latin America and the Caribbean, seven states from the Middle East and South Asia, ten states from North America and Western Europe, and eight states from South-East Asia, the Pacific and the Far East. All states in each of the above regions are listed in an annex to the CTBT, which may be updated. At least one-third of the seats allocated to each region shall be filled by parties in that region designated on the basis of the nuclear capabilities relevant to the Treaty, as determined by international data as well as all or any of the following criteria: the number of monitoring facilities of the IMS; expertise and experience in monitoring technology; and contribution to the budget of the CTBTO. One of the seats allocated to each region shall be filled on a rotational basis by the party that is first in the English alphabetical order among the parties in that region that have not served as members of the Executive Council for the longest period of time. The remaining seats allocated to each region are to be filled by parties designated from among all parties in that region by rotation or elections.

In addition to many other duties, the Executive Council must supervise the activities of the Technical Secretariat of the CTBTO; cooperate with the relevant national authority of each party; consider and submit to the Conference a draft annual programme and budget of the Organization; conclude agreements or arrangements with states parties, other states and international organizations, and supervise their implementation; and consider any concern raised by a party about possible non-compliance with the Treaty or abuse of the rights established by the Treaty. Decisions of the Executive Council on matters of substance are to be taken by a two-thirds majority of all its members.

The functions of the Technical Secretariat, which comprises, as an integral part, the International Data Centre, include: supervising and coordinating the operation of the IMS; routinely receiving, processing, analysing and reporting on IMS data; providing technical assistance in, and support for, the installation and operation of monitoring stations; and receiving requests for on-site inspections and processing them. The Technical Secretariat is to comprise a Director-General, appointed by the Conference, as well as scientific, technical and other personnel.

Consultative Committees

The 1985 Treaty of Rarotonga established a Consultative Committee, constituted of representatives of the parties, to be convened from time to time by the Director of the South Pacific Bureau for Economic Co-operation – the Depositary of the Treaty. The Committee is to be chaired at any given meeting by the representative of the party which last hosted the meeting of heads of government of members of the South Pacific Forum. The Director reports annually on the status of the Treaty and on matters arising in relation to it. The Consultative Committee decides whether a special inspection should be carried out. The costs of the Committee, including the costs of special inspections, are to be borne by the South Pacific Bureau for Economic Co-operation (since 1988 the South Pacific Forum Secretariat).

The Consultative Committee of Experts, provided for in the 1977 Convention on the Prohibition of Military or any other Hostile Use of Environmental Modification Techniques (Enmod Convention), has a more circumscribed authority. To solve problems that may arise relating to the objective of, or in the application of the provisions of, this Convention, the Depositary must convene, upon request from a state party, a Consultative Committee to which all parties may appoint an expert. According to the rules of procedure set out in the annex to the Enmod Convention, there is to be no voting in the Committee on matters of substance. A summary of the findings, incorporating all views and information presented to the Committee during its proceedings, must be transmitted to the Depositary, who is to distribute it to all parties. It would be up to the complainant to draw conclusions from the information received, and to decide upon further action, which may include the lodging of the complaint with the UN Security Council.

To attain the objectives of the 1997 Inter-American Convention against the Illicit Manufacturing of and Trafficking in Firearms, Ammunition, Explosives, and Other Related Materials, the parties to the Convention established a Consultative Committee responsible for promoting the exchange of relevant information; encouraging cooperation between national liaison authorities; promoting training and exchange of knowledge and experience, technical assistance between the parties and international organizations, as well as academic studies; and requesting from non-party states, when appropriate, information on the illicit manufacturing of and trafficking in the objects in question. The decisions of the Consultative Committee are recommendatory in nature. The Committee is to hold one regular meeting each year and convene a special meeting if necessary. The host country for each regular meeting is to serve as Secretariat *pro tempore* of the Committee until the next regular meeting. When a regular meeting is held at the headquarters of the General Secretariat of the Organization of American States, the state party that will serve as Secretariat *pro tempore* is to be elected at that meeting.

Fact-Finding Missions

In accordance with the 1977 Protocol I Additional to the Geneva Conventions of 1949 and Relating to the Protection of Victims of International Armed Conflicts, an International Fact-Finding Commission was established in 1991, after the number of parties accepting the relevant optional provision of the Protocol had reached 20. The Commission consists of 15 members considered to be of high moral standing and acknowledged impartiality. The parties may at any time declare that they recognize, in relation to any other party accepting the same obligation, the competence of the Commission to enquire into allegations by such other party. The Commission is mandated to enquire into any facts alleged to be a grave breach, as defined in the Conventions and the Protocol, or other serious violation of these treaties, and to facilitate, through its good offices, the restoration of respect for the Conventions and the Protocol. In other situations, the Commission may institute an enquiry at the request of a party to the conflict only with the consent of the other party or parties concerned.

The 1995 Treaty of Bangkok gives each party the right to request the Executive Committee of the Commission for the Southeast Asia Nuclear Weapon-Free Zone to send a fact-finding mission to another state party to clarify and resolve a situation which may be considered ambiguous. Once the Executive Committee decides that the request is not frivolous, abusive or clearly beyond the scope of the Treaty, it must forward the request to the receiving state, indicating the proposed date for sending the fact-finding mission. The mission would consist of three inspectors from the IAEA who are not nationals of either the requesting or the receiving state. The inspectors must be provided unimpeded access to the location in which the situation giving rise to doubts about compliance with the Treaty has occurred. The receiving state would be allowed to take measures to prevent disclosures of confidential information and data not related to the Treaty.

Under the 1997 APM Convention, the Meeting of the States Parties or the Special Meeting of the States Parties may authorize a fact-finding mission and decide on its mandate by a majority of parties present and voting to clarify a question relating to compliance with the provisions of the Convention. The mission, consisting of up to nine experts, may collect information on the spot or in other places directly related to the compliance issue under the jurisdiction or control of the requested party. Upon receiving a request for a fact-finding mission, the UN Secretary-General, as the Depositary of the APM Convention, shall, after consultations with the requested party, appoint the members of the mission, including its leader. Nationals of states requesting the fact-finding mission or directly affected by it may not be appointed. The requested party must make all efforts to ensure that the fact-finding mission is given the opportunity to speak with all persons who may be able to provide information related to the issue. The requested party must grant access for the fact-finding mission to all areas and installations under its control where facts relevant to the compliance issue could be expected to be collected, subject to arrangements that the requested party may consider necessary for the protection of sensitive equipment, information and areas. All information provided in confidence and not related to the subject matter of the fact-finding mission shall be treated on a confidential basis.

The UN Secretary-General, in his capacity of the chief administrative officer of the United Nations, may also resort to fact-finding missions. Such missions could be particularly useful when generally recognized principles of international law, not yet fully codified in treaties, are alleged to have been violated.

Confidentiality

Several arms control agreements which envisage on-site inspections contain clauses requiring confidentiality of information acquired by the inspectors. (The need for such clauses became particularly evident when, in the 1990s, it was revealed that certain UN inspectors had been engaged in espionage activities against Iraq for the benefit of their government.) The most developed provisions to this effect appear in the 1993 CW Convention. According to the so-called Confidentiality Annex to this Convention, information is considered confidential if it is so designated by the state party from which the information was obtained and to which the information refers or if, in the judgement of the Director-General of the OPCW, its unauthorized disclosure could reasonably be expected to cause damage to the party to which it refers or to the mechanisms for the implementation of the Convention.

The Director-General of the OPCW has the primary responsibility for ensuring the protection of confidential information. The staff must enter into individual secrecy agreements with the Technical Secretariat of the OPCW, covering the period of employment and a period of five years after the employment has terminated. If, in the opinion of the Director-General, there is sufficient indication that obligations concerning the protection of confidential information have been violated, he should promptly initiate an investigation. He should also promptly initiate an investigation if an allegation concerning a breach of confidentiality is made by a party. Punitive and disciplinary measures must be imposed on staff members who have violated their obligations to protect confidential information. In cases of serious breaches, the Director-General may waive the immunity from jurisdiction. States parties shall, to the extent possible, cooperate and support the Director-General in investigating any breach or alleged breach of confidentiality, and in taking appropriate action in case a breach has been established.

The OPCW cannot be held liable for any breach committed by members of the Technical Secretariat. For breaches involving both a state party and the OPCW, a commission for the settlement of disputes related to confidentiality, set up as a subsidiary organ of the OPCW Conference and appointed by the Conference, is to consider the case.

National Verification Bodies

Several arms control agreements require that parties should take measures, in accordance with their constitutional processes, to prevent on their territories all action contrary to the international obligations assumed by their governments. This provision has led to appropriate laws being adopted or decrees issued by state authorities when the agreements entered into force. In certain cases, special national organizations have been established to enforce the contracted prohibitions or limitations. However, since governments themselves are potential violators of interstate agreements, the credibility of such verification bodies depends primarily on their being independent of the authorities of the country within which they function.

Whatever their trustworthiness in ensuring compliance, specialized national organizations are indispensable for the exchange of data among parties, for submission and receipt of notifications or other information, as well as for hosting inspections. In the multilateral field, the best-established national verification organizations are the atomic energy commissions which, under different names, exist in countries conducting nuclear activities. These commissions work out and implement internal regulations for the management and control of nuclear material and nuclear facilities maintain contact with the IAEA and cooperate with it in applying international safeguards. According to the CW Convention, each party must designate or establish a national authority to serve as the national focal point for liaison with the OPCW and with other parties. The CTBT also requires the designation or setting up of such an authority.

Proposals for a Universal Verification Organization

In recent years proposals have been made for the establishment of a global verification agency covering all arms control and disarmament undertakings. The idea is not new. As early as in 1961, the need for such an agency was recognized in the McCloy–Zorin Statement of agreed principles for negotiations on general and complete disarmament (see Chapter 3). One of the principles included in this Statement stipulated that an international disarmament organization should be created within the framework of the United Nations to 'implement control over and inspection of disarmament'. Accordingly, the Soviet draft treaty on general and complete disarmament provided that an organization of the parties to the treaty would receive information, supplied by the parties, about the armed forces, armaments, military production and military appropriations, and that it would have its own staff, recruited internationally, to exercise control. The United States envisaged the establishment of an international organization to ensure that all obligations were observed during and after the implementation of general and complete disarmament; inspectors of the organization would have unrestricted access to all places necessary for the purpose of effective verification. Although different views were subsequently expressed about the composition and terms of reference of an international disarmament organization, the parties were agreed that complete universal disarmament would require a comprehensive treatment of verification on a global scale to guard against risks to the security interests of all states. However, the proposition was purely hypothetical: the talks on general and complete disarmament had no chance of succeeding because neither of the two superpowers really contemplated the complete renunciation of arms.

The question debated later concerned whether a global verification organization was needed to deal with arms limitation (as distinct from total disarmament) measures – both those already agreed upon and those under consideration. Advocates of centralized verification arrangements argued that compliance with arms control treaties is of concern to all states and that, consequently, verification must be an international responsibility. Indeed, the 1978 Special Session of the UN General Assembly emphasized the requirement for all parties to participate in the verification process. In this context, France proposed the establishment of an International Satellite Monitoring Agency (ISMA). A group of UN experts which had studied the French proposal concluded that technical facilities for an ISMA could be acquired in stages, beginning with an image-processing and -interpretation centre, proceeding to stations which would receive appropriate data from observation satellites of various states, and ending with the ISMA having its own space segment (in addition to the ground segment), to consist of a certain number of satellites. The idea did not materialize because of the negative attitudes of the countries already possessing reconnaissance satellites. In any event, international monitoring by satellites – although valuable – could not replace other means of verification.

To sceptics, the prospect of creating an omnibus verification organization seems unrealistic and even undesirable, for the following reasons. Those multilateral arms control treaties that are already in force do not need a centralized organizational framework to strengthen their verification provisions. For significant new treaties, special expert bodies will be needed, comprising staffs of inspectors with special skills to handle verification. Control methods and procedures will differ from treaty to treaty, and different factors may have to be taken into account in initiating and carrying out investigations. It would therefore seem sensible to make use, where possible, of UN-affiliated and other authoritative international institutions (both intergovernmental and non-governmental) that deal with related peaceful matters, rather than to subsume all verification activities under one umbrella organization. In this respect, the IAEA has set a good precedent: although created only to promote the peaceful uses of nuclear energy, it now also renders services in verifying arms control obligations under the NPT and the nuclear-weapon-free zone agreements. The Agency could be of further use in checking a cut-off of production of fission-able material for weapon purposes, should such a measure be internationally agreed. At the regional level, states are likely to rely on regional rather than world-wide verification arrangements.

It thus appears necessary for multilateral arms control agreements to provide for specialized verification mechanisms. Through such mechanisms, information about compliance – particularly that collected by sophisticated technical methods – could be shared by all parties. Some have suggested delegating these responsibilities to the United Nations, but that would inflate the bureaucracy of that world body, which has no experience in the practical handling of arms control matters, and would most probably diminish the effectiveness of verification. Moreover, serious legal complications could arise because not all UN members would necessarily become parties to the same treaty, and the applicability of UN procedures (including UN scale of financial contributions) to treaties concluded and operated outside the United Nations could be questioned. Nonetheless, the United Nations is in a position to help in working out general guidelines for verification; to develop a database on all aspects of verification and compliance; and to facilitate exchanges between experts and diplomats, as suggested in the 1990 UN *Study on the Role of the United Nations in the Field of Verification*. The United Nations may also undertake investigations upon recommendation of the Security Council, if such a procedure is envisaged in arms control agreements. In certain cases, it could do so at the initiative of the Secretary-General himself.

19.6 Compliance

However well intentioned governments are at the time of signing a treaty, they may at a later stage be unable to resist incentives to acquire clandestinely the weapons they have renounced or to engage in other outlawed activities. The more comprehensive the arms restraints, the greater may be the incentive to cheat.

The vast majority of charges of violation have come from the United States and have been directed against the Soviet Union (now Russia), which has frequently responded with counter-charges. The most important allegations are listed below.

US Allegations

Regarding the 1972 ABM Treaty. In the early 1980s, the Soviet Union was accused of acting in violation of the ban on developing a territorial anti-ballistic missile defence by constructing a large (that is, having a high potential) phased-array radar (LPAR) near Krasnoyarsk in Siberia. The only LPARs allowed under the Treaty are those installed in ABM deployment areas and at test ranges, those serving early warning against ballistic missiles at locations along the periphery of the party's national territory and oriented outwards, as well as those used for tracking satellites and other objects in space. Since the Krasnoyarsk radar did not belong to any of these categories, the United States demanded that it be destroyed. After a long period of denial, the Soviet Union admitted that it had disregarded the provisions of the Treaty and pledged to take corrective action. In 1992, the United States agreed to a Russian proposal to convert the Krasnoyarsk radar into a furniture factory.

Regarding the 1972 BW Convention. After the outbreak of an anthrax epidemic in 1979 in the Soviet city of Sverdlovsk, the Soviet Union (and subsequently Russia) was accused of maintaining an offensive biological weapons programme. The programme allegedly included production, weaponization and stockpiling of biological weapons. The accusation was based on the suspected airborne release of anthrax spores from a Soviet biological facility which caused an outbreak of anthrax in April and May 1979, infecting dozens of people as well as livestock. The issue was the subject of bilateral US–Soviet consultations and, in 1992, the Russian authorities admitted that a breach of the Convention had been committed. They undertook – under a decree issued by the Russian President – to open secret military biological research centres to international inspection and convert them to civilian use.

In 1981 the United States accused the Soviet Union of being involved in the production, transfer and use of trichothecene mycotoxins in Laos, Kampuchea (Cambodia) and Afghanistan. The Soviet Union categorically rejected the allegation. US charges were based on reports by alleged victims and eyewitnesses who stated that, since the latter part of the 1970s, enemy aircraft had been producing 'yellow rain' by spraying a toxic yellow material. Chemical analyses of samples of this material and medical checks of the affected persons were conducted to substantiate the case. However, as the investigation proceeded with the involvement of laboratories in different countries and a careful scrutiny of the eyewitnesses' reports, the reliability of the evidence was increasingly questioned. Some authoritative scientists found that the yellow substance consisted to a large extent of excrement of wild honeybees. The United States did not formally retract its allegation.

Regarding the 1963 PTBT and the 1974 TTBT. For many years, the Soviet Union was charged with conducting its underground nuclear-weapon tests in a manner incompatible with the PTBT: on numerous occasions radioactive debris from these tests was found outside the Soviet territorial limits. It was also alleged to have conducted nuclear tests in excess of the threshold agreed under the TTBT. These allegations were denied.

Soviet Allegations

Regarding the 1972 ABM Treaty. The United States was accused of violating the Treaty prohibition against establishing an ABM defence of the territory of the parties and, in particular, the ban on the development, testing and deployment of space-based ABM systems or components. The Soviet Union considered the construction of a 'new' LPAR at Thule, Greenland, and of a similar radar near Fylingdales, England, contrary to the provisions of the Treaty. It alleged that new, so-called Pave Paws LPARs operational on the US Atlantic and Pacific coasts and the radars deployed in the southern United States had capabilities to provide a base for ABM radar coverage of a significant portion of US territory. The United States was also accused of having deployed a radar with ABM capabilities on Shemya Island in the Aleutians, of having undertaken to develop mobile ABM radars, of testing Minuteman ICBMs to provide them with ABM capabilities, and of developing multiple warheads for ABM interceptor missiles – all in conflict with the Treaty. The United States denied these charges.

Regarding the 1963 PTBT and the 1974 TTBT. According to Soviet allegations, radioactive debris from US underground nuclear explosions spread beyond national boundaries, in violation of the PTBT, and US nuclear tests exceeded the yield set by the TTBT. The United States admitted that it had had some difficulty in totally containing underground nuclear explosions and that there had been a few incidents of local seepage of radioactive gases at the Nevada Test Site. It asserted, however, that this venting had not resulted in a spread of radioactivity beyond US national borders, and it rejected the charge regarding the explosion yields.

Other Allegations

Regarding the 1925 Geneva Protocol. In 1951 and 1952, during the Korean War, North Korea, China and the Soviet Union accused the US forces of using chemical and biological warfare agents, mainly the latter. The United States denied the charges. No independent investigation took place.

In the course of the 1980–88 Iraq–Iran War, Iran complained that Iraqi forces were using chemical warfare agents against combatants and civilians, in violation of the 1925 Geneva Protocol. These allegations were found justified. Several teams of experts dispatched by the UN Secretary-General to conduct on-the-spot investigations confirmed that on several occasions recourse had been made to these prohibited means of warfare. The chemical agents used included nerve gas, probably never before employed in military operations. However, for political reasons, unrelated to arms control, the United Nations took no action in response to these violations other than stating its disapproval.

Regarding the 1968 NPT. Iraq, a party to the NPT, committed a breach of the Treaty by failing to declare all its nuclear activities and submit them to international control, and by engaging in illicit production of weapon-grade fissile material as well as of other nuclear-weapon components. In the aftermath of the 1991 Gulf War, the United Nations imposed a series of coercive measures on Iraq to abolish its nascent nuclear potential.

In 1992, North Korea was found in breach of its safeguards obligations, undertaken in accordance with the NPT, when it refused an IAEA special inspection. The matter was brought to the attention of the UN Security Council. However, no coercive measures against North Korea – as proposed by certain states – could be taken, mainly because there was no agreement among the permanent members of the Council.

To negotiate a resolution of the above dispute, the delegations of the United States and North Korea held a series of talks, as a result of which, in October 1994, the following measures were incorporated in the so-called Agreed Framework. North Korea's graphite-moderated reactors and related facilities, which had produced nuclear-weapon-grade plutonium, were to be replaced with two light-water reactors (LWR) which do not produce such plutonium. The latter reactors, to be financed and supplied by an international consortium called the Korean Peninsula Energy Development Organization (KEDO), should reach a total generating capacity of approximately 2,000 MW(e) by the year 2003, whereas the former reactors should be completely dismantled upon completion of the second LWR. To offset the energy forgone by North Korea because of the agreed freeze of its graphite-moderated reactors, the United States undertook, pending completion of the first LWR unit, to provide free of charge heavy oil, up to 500,000 tons annually, for heating and electricity production. Spent fuel from the North Korean 5 MW(e) experimental reactor was to be stored safely during the construction of the LWR project and disposed of in a manner that did not involve reprocessing in North Korea.

The IAEA was allowed to monitor the freeze of the North Korean reactors and, upon conclusion of the contract for the supply of new reactors, to resume routine inspections, under its safeguards agreement with North Korea, with respect to the facilities not subject to the freeze. Only when a significant portion of the LWR project is completed, but before the delivery of key nuclear components, will North Korea come into full compliance with its safeguards agreement with the IAEA. This includes taking steps that might be deemed necessary by the IAEA to verify the accuracy and completeness of North Korea's initial report on nuclear material in the country, and that might require some three to four years. Until then, it will not be clear how much plutonium North Korea has actually reprocessed.

Regarding the 1972 BW Convention. Cuba has, at various times, accused the United States of conducting biological warfare against it. In 1997 Cuba claimed that an aircraft operated by the US Department of State sprayed, during its flight over Cuban territory in October 1996, an insect pest called *Thrips palmi*. The pest feeds on a wide range of plants and has the ability to develop large populations in a short time, causing severe damage to vegetable and other agricultural crops. The United States denied the allegation and suggested that a failure of the Cuban border quarantine may have been responsible for the infestation.

Typology of Allegations

Instances in which a material breach of an arms control agreement was deliberately committed to significantly increase a state's military capability in a way prohibited by the treaty have not been very frequent. The most important among them are described above. Other allegations or admitted transgressions were of lesser military importance and were often due to insufficiently precise definitions of the banned items or activities. In fact, in the course of drafting a treaty, ambiguous formulas leading to divergent interpretations are at times consciously resorted to in order to resolve impasses in the negotiating process. Some problems of compliance may have been intentionally magnified during periods of political tension.

In certain cases, suspicions arose because the relevant treaties had not yet entered into force. Thus, for example, before the TTBT had become formally effective, the signatories accused each other of exceeding the agreed yield threshold for nuclear explosions, whereas the exchange of data necessary to establish a correlation between explosion yields and the seismic signals produced by explosions was held up, pending ratification of the Treaty. Some authoritative expert reports suggested that it was the lack of adequate information about the geological features of the Soviet nuclear test sites that may have led to equivocal seismological evidence of non-compliance. In fact, the parties themselves recognized the difficulty of predicting precise yields of nuclear explosions. They therefore reached an understanding that one or two breaches per year would not be considered a 'violation'. Similarly, because underground nuclear explosions cannot be completely contained under the earth's surface, whatever the precautions taken, the United States and the Soviet Union decided, in the first few years after the conclusion of the PTBT, that instances in which radioactive substances from such explosions spread outside the territory of the testing state would be treated as mere technical breaches that could be tolerated, as distinct from militarily or politically significant cheating that could not. Only in the mid-1980s did the superpowers include these occurrences in the lists of grievances against each other, without even mentioning whether any adverse environmental consequence had ensued. Many other allegations were based on evidence admittedly not sufficient to pass a definitive judgement; several presumed violations were characterized as 'probable' or 'likely'.

Agreements that have inadequate verification provisions generate charges of violation. Although the accused party always has a possibility to demonstrate its innocence, regardless of the letter of the treaty, there is a need for dedicated mechanisms – both to clarify suspicions to the satisfaction of the suspecting party and to protect parties against malevolent allegations.

19.7 Responses to Violations

Responses to established breaches of arms control agreements may differ depending on the extent to which a breach is considered serious by those affected by it. They may range from deliberately overlooking certain occurrences for overriding political or security reasons (for example, unwillingness to reveal the source of information) to abrogation of the treaty followed by some punitive action. Between these extremes there exists a possibility of using diplomacy to effect a change in the behaviour of the guilty party. In some cases this has proved useful.

Many multilateral arms control treaties provide for formal notification of a suspected or committed violation to the United Nations and/or another international organization, thus making the event public. As no government likes to be pilloried as a violator of legal obligations, publicity may be helpful as an instrument of sanction, especially in democratic countries, which are sensitive to public disapproval. A reported violation may lead some states to take such action as the recall of ambassadors, the reduction of embassy staffs and even the severance of diplomatic relations. International organizations may pass condemnatory resolutions. However, to make the violating state rectify its behaviour, stronger measures of enforcement may be needed.

UN Action

After a competent body has made a definitive finding that a state has violated an arms control agreement, the UN Security Council may, if so requested, consider the matter. The Council is not authorized by the UN Charter to take action against violators of arms control agreements, but if it finds that the situation brought about by the violation could lead to international friction it may, under Chapter VI of the Charter, recommend to the state or states concerned 'appropriate procedures or methods of adjustment'. The Council may also decide that a specific violation or a certain type of violation constitutes a 'threat to the peace'. It could then, under Chapter VII of the UN Charter, call on UN members to apply sanctions – complete or partial interruption of economic relations and of rail, sea, air, postal, telegraphic, radio and other means of communication. It could also recommend to the UN General Assembly the suspension of the rights and privileges of UN membership or even expulsion from the Organization. Finally, the Council may decide that military sanctions should be employed, including demonstrations, blockade and other operations by the air, sea or land forces of UN members.

Thus, in a formal sense, the Council possesses the means necessary to restore international peace that has been broken as a result of arms control violations. The determination to resort to these means was expressed in the 1992 statement by the President of the Security Council, on behalf of the members of the Council, to the effect that the proliferation of weapons of mass destruction would constitute a threat to international peace and security and that appropriate action would be taken to prevent it. Significantly, such action would affect all states breaking the rule of non-proliferation – not only parties to the relevant agreements – even though the ban on the proliferation of either nuclear, or chemical, or biological weapons is not yet a rule of customary international law binding on all states alike. However, a statement by the President of the Security Council does not have a binding legal effect. To have such effect, it would need to be converted into a formal decision of the Council. In addition, the term 'proliferation', which lends itself to different interpretations, would have to be unambiguously defined. Only then would the Council be entitled to take coercive measures against the violators of the non-proliferation norms.

In practice, it is difficult to obtain agreement on the application of drastic measures from the UN members not directly concerned. Even with the requisite two-thirds majority, the Council may prove unable to act if any one of its permanent members decides to exercise the right of veto – as specified in the UN Charter – to protect its own interests or the interests of its allies, or if it is opposed to the treaty in question. The problem of reconciling the right of veto with the proper functioning of treaties restricting armaments was recognized as early as in 1946, when the United States put forward the Baruch Plan for the creation of an international atomic development authority. At that time, the US government stressed the importance of immediate punishment for infringements, maintaining that there must be no veto to protect violators of international agreements – a proposition that the Soviet Union categorically rejected.

In connection with several arms control agreements, the UN Security Council has been granted functions which have the appearances of sanctions. Thus, according to Security Council Resolutions 255 of 1968 and 984 of 1995, parties to the 1968 NPT received a pledge of assistance (technical, medical, scientific or humanitarian) in the event they were aggressed or threatened to be aggressed with nuclear weapons. Under the 1972 BW Convention and the 1977 Enmod Convention, states have undertaken to provide or support assistance to any requesting party if the UN Security Council determines that such party has been harmed (or is likely to be harmed) or exposed to danger as a result of a violation. These assurances simply reaffirm the existing obligation of the United Nations to provide assistance to a country attacked or threatened with an attack, whatever the weapon used.

The 1993 CW Convention and the 1996 CTBT go somewhat further in enforcing compliance. They stipulate that some (unspecified) collective measures may be taken by the parties without reference to the UN Security Council if one of them engages in prohibited activities that can damage the object and purposes of the agreements. Urgent cases of non-compliance with the CW Convention or the CTBT may be brought to the attention of the United Nations if the required majority of parties decide to do so. However, the relevant provisions of the UN Charter would then apply (as they would in the case of the BW Convention and the Enmod Convention), and these provisions, as pointed out above, may prove inoperative. It is true that Iraq, which had committed a breach of the NPT, was forced under the 1991 Security Council Resolutions 687 and 715 to dismantle or destroy the key elements of its nuclear weapon development programme under the supervision of the UN Special Commission on Iraq (UNSCOM). However, these sanctions were imposed not because of the breach of the NPT, but because of Iraq's aggression against Kuwait in violation of the UN Charter.

The General Assembly is another principal organ of the United Nations to which complaints of treaty violations can be addressed. Its actions are not subject to veto; only a two-thirds majority is required for a recommendation concerning international peace and security. However, with the present composition of the Assembly of nearly 190 states, obtaining such a majority may not be easy. Even when it is duly adopted, a resolution of the Assembly – unlike a decision of the Security Council – is not binding on UN members.

IAEA Action

Another intergovernmental organization capable of dealing with breaches of arms control obligations is the IAEA. As envisaged in Article XII of its Statute, cases of non-compliance with nuclear safeguards agreements are to be reported to the UN Security Council and the General Assembly. If corrective action is not taken within a reasonable time, the IAEA Board of Governors may direct curtailment or suspension of assistance provided by the Agency or a member-state and call for the return of materials and equipment made available to the transgressing member. A non-complying state may also be suspended from exercising the privileges and rights of

IAEA membership. Since no country enjoys the right of veto in the IAEA Board, adoption of decisions to apply such sanctions cannot be ruled out, but their effectiveness is doubtful.

The IAEA provides very little direct assistance to states – and certainly not for their nuclear power programmes. As regards possible curtailment of assistance provided by states, such a decision may be adopted by the Board, but it is not as unambiguously mandatory under the IAEA Statute as are decisions of the UN Security Council. Even if all the deliveries of nuclear items were actually cut off to penalize the offending state, that state might not feel significantly disadvantaged in a world where no country is exclusively dependent on nuclear power and where the supply of nuclear materials and equipment exceeds demand. Withdrawal of materials and equipment already supplied is not a realistic measure, because it would require voluntary cooperation of the state being penalized, which is unlikely. Moreover, return of nuclear supplies may be both exceedingly expensive and dangerous, and the supplier may be unwilling to take them back. Suspension of IAEA membership does not seem to be an effective measure either. In concrete terms, it would involve: withdrawing the right to receive Agency assistance which, as explained above, is not an important sanction; barring access to information possessed by the Agency which is available to non-members as well; and exclusion from Agency meetings, which cannot be particularly hurtful. Expulsion from the Agency is not provided for. The weakness of the IAEA enforcement mechanism has been best illustrated by the case of North Korea, which refused international inspection of some suspect facilities without provoking immediate and effective sanctions.

Other Collective Action

Collective sanctions against a violator of a multilateral agreement may be taken even in the absence of an enforcement provision. Such sanctions, when applied, are usually related to the nature of the particular offence. Thus, in the 1970s, the breach by India of its undertaking under international cooperation agreements to use nuclear energy exclusively for peaceful purposes prompted a number of countries to restrict their supplies of nuclear materials and equipment to India. Iraq's use of chemical weapons against Iran, in violation of the 1925 Geneva Protocol, went unpunished but, even before the 1991 Gulf War, certain industrialized states decided to ban exports to Iraq of chemicals which could be used in the manufacture of chemical warfare agents. However, to produce the desired effect, 'in-kind' sanctions would have to be complemented by such measures as cancellation of economic assistance, imposition of trade restrictions, and even suspension or termination of vitally needed supplies unrelated to the breach.

Abrogation

All major arms control agreements contain a clause permitting a party to withdraw from the agreement if it decides that extraordinary events have jeopardized its supreme interests. Violation could be considered as an 'extraordinary' event justifying withdrawal, and the requirement to give advance notification and to explain the reasons would not prevent states from taking this step. However, even in the absence of a withdrawal clause, a material breach of a treaty by one party makes it possible for another party to denounce the agreement in accordance with general international law.

In bilateral relations, the threat of abrogation is the primary means of enforcing a treaty, for it may deprive the violating nation of the advantages it has gained from entering it. Alternatively, the party injured by a violation may respond by taking the same prohibited action as the offender, without repudiating the agreement as a whole. Such a tit-for-tat interplay – which would be equivalent to informally modifying the terms of the treaty – is conceivable only as long as the main purpose of the treaty has not been perverted. In multilateral relations, abrogation or retaliation with a similar violation could lead to the collapse of the treaty to the detriment of the complying parties.

An Alternative Approach

The traditional responses to established violations of multilateral arms control agreements encounter a number of obstacles which are difficult to overcome. Removing these obstacles would require, among other things, radical changes in the structure and working of the main organs of the United Nations as well as of other international organizations. In particular, the force of the UN General Assembly resolutions would have to be enhanced, the Security Council permanent members' veto would have to be restricted or ended, and the prerogatives of the executive bodies of the arms control implementing organizations would have to be widened and their decisions made mandatory. Such changes, the implications of which would go well beyond the field of arms control, would certainly be regarded by many states as politically undesirable and therefore not feasible in the foreseeable future.

If a response to a violation of a multilateral obligation is to be effective, all or most parties must act in solidarity with the state or states hurt by the violation. However, solidary action is not always possible, because non-compliant behaviour by some states may pose little or no security threat to others and because many countries are opposed to applying sanctions that have not been decided upon by competent international bodies. If collective enforcement measures against a culprit state were to be applied without the requirement that an international decision must be taken in each individual case, such measures would have to be formally agreed in advance.

In devising possible responses, a distinction must be made between different types of violation. Violations can vary from technical to material breaches, that is, from inaccurate or incomplete reporting to non-observance of procedural clauses, to offences resulting from misunderstanding, to violations of provisions essential to the accomplishment of the object or purpose of the treaty, including obstruction of the control system. Violations can be committed by governmental authorities, by non-governmental institutions or even by individuals (with or without the consent or knowledge of the authorities). Further differentiation is necessary between intentional and unintentional breaches; the latter – usually easier to remedy – may result from sheer negligence. Some breaches may be reversible, while others may not be.

The most appropriate approach would be to make responses to violations part and parcel of the complex of obligations contracted by the parties. The agreed responses, different for different treaties but proportionate to the offences, could be listed in the treaty itself, or in a protocol attached to it, or in a protocol added to the treaty already in force. They might include the measures mentioned in the preceding sections, with the exception of the use of armed force. The UN Security Council alone may decide military sanctions.

The responses would have to be graduated from mild to severe, so as to increase pressure on the violator over time and force him finally to change his behaviour. The conditions for transition from one response to another would have to be clearly spelled out. The mere existence of a list of predetermined sanctions could fulfil the function of deterrence and reduce the probability of violation. A government declining to react to violations and abstaining from efforts to uphold the validity of an arms control agreement would be in breach of its treaty obligation and would expose itself to both international and domestic criticism.

It is obvious that evasion cannot be prevented and compliance cannot be restored by the threat or use of sanctions alone, even those enjoying broad international support; political and economic inducements are equally important. It is also obvious that the stronger and the richer the country, the easier it may be for it to withstand outside pressure. Nonetheless, it is essential that violations of arms control treaties not be ignored and that no country, large or small, developed or undeveloped, be immune from deserved penalties. The general public tends to equate arms control violations with immediate threats to national security. Reactions to violations should, therefore, be predictable. Violators must apprehend detection.

20

Concluding Remarks

Arms control is not an aim in itself. It forms an integral part of human endeavours to bring about a safe and peaceful world. It is therefore affected by the vicissitudes of the political and economic relations among states, as well as by the changing global security environment.

The achievements and shortcomings of the arms control agreements described in detail in the preceding chapters can be summarized and assessed as follows.

20.1 Nuclear Arms Control

The Comprehensive Nuclear Test-Ban Treaty, signed in 1996, was meant to stop the substantial qualitative improvements of nuclear weapons. However, its entry into force is highly problematic.

In the course of the last decade of the 20th century, the numbers of strategic nuclear offensive arms were reduced, the intermediate-range nuclear forces of Russia and the United States were eliminated, and the deployment of tactical nuclear weapons was restricted. However, the nuclear warheads retained would still suffice to devastate the entire planet.

The danger of a deliberate, full-scale exchange of nuclear missiles between Russia and the United States has receded to the point of seeming unthinkable. However, the probability of a nuclear war between these powers has not vanished, in spite of the confidence-building measures taken to attenuate the risks of an unauthorized or accidental use of nuclear weapons and in spite of the apparent political rapprochement between the two powers.

The abrogation of the 1972 ABM Treaty, prohibiting the deployment of a nation-wide missile defence system, may give rise to a new race in strategic arms, both defensive and offensive, among all the powers possessing nuclear weapons, including India and Pakistan, states involved in political, territorial and religious conflicts with each other.

The 1968 Non-Proliferation Treaty has attracted a record number of adherents and has been extended for an indefinite period of time. Safeguards against the diversion of nuclear energy from peaceful to non-peaceful purposes have been strengthened. However, the disarmament obligations of the nuclear-weapon powers contracted under the Treaty have not been implemented. The opening of negotiations for a cut-off of production of fissile material for nuclear explosive devices – an important measure which had been expected to follow the cessation of nuclear test explosions – encountered serious problems. Large quantities of nuclear weapon-grade material accumulated by the nuclear-weapon powers are not under international safeguards.

Agreements setting up nuclear-weapon-free zones in different parts of the world have been concluded. They are an asset for the cause of nuclear non-proliferation, but the treaties which established them suffer from several shortcomings. Proposals for the denuclearization of the Middle East and South Asia – two regions of tension – have not resulted in treaties.

The so-called negative security assurances notwithstanding, most great powers envisage the use of nuclear weapons in response to attacks on them or their allies, whatever the weapon used by the attacker. It is this military doctrine that is a major obstacle to the abolition of nuclear weapons.

20.2 Chemical and Biological Arms Control

The establishment by the 1993 Chemical Weapons Convention of an international, legally binding norm against the possession of chemical weapons was an important achievement. However, for different reasons, mainly of an economic and technical nature, the destruction of chemical weapon stocks is not likely to be completed within the time frame prescribed by the Convention.

The 1972 Biological Weapons Convention reinforced the ban on the use of bacteriological means of warfare, which – like the ban on the use of asphyxiating, poisonous or other gases – is embodied in the 1925 Geneva Protocol. However, unlike the CW Convention, the BW Convention does not provide for measures to verify compliance with the obligations assumed by the parties. Efforts to fill this serious gap by adopting a verification protocol have proved unsuccessful.

20.3 Conventional Arms Control

Owing to the significant, verified cuts in the arsenals of European countries under the 1990 Treaty on Conventional Armed Forces in Europe, no state or group of states has today the capability to launch a surprise armed attack in Europe. However, for the post-Cold War situation the ceilings for both armaments and military personnel are still exceedingly high.

The Anti-Personnel Mines Convention, concluded in 1997, is to do away with a type of weapon that is in widespread use in both international and internal conflicts. However, the provisions of the Convention dealing with definitions and exceptions may give rise to divergent interpretations affecting the implementation of the parties' obligations; no organization was set up to oversee the Convention's operation on a continuous basis.

Guidelines, principles and codes of conduct regarding other categories of conventional armament, including small arms and light weapons, have been adopted as a result of global and regional initiatives. However, none of these documents is legally binding. The recommended measures are meant chiefly to control arms trading activities. They do not prohibit or restrict either the possession or the manufacture of the arms in question.

20.4 Prospects

As seen from the above balance sheet of achievements and shortcomings, considerably less has been achieved in the field of arms control than was expected by world

opinion after the end of the Cold War. To consolidate the achievements, it is imperative to carry into effect the agreements signed but not ratified, attract new adherents
to agreements already in force and strengthen them with tightened measures of verification and, if necessary, supplementary formal arrangements or informal understandings. To this end, a variety of incentives will probably have to be resorted to.
As argued in chapter 1.2, unilateral undertakings assumed without treaties cannot be
a substitute for treaties.

The multilateral arms control negotiating machinery must be improved and made
more efficient. It would, however, be inopportune to insist that all states, both those
that are militarily significant and those that are not, should be engaged in *all* arms
control negotiations on an equal footing. Such 'globalization' is not necessary; in
any case, it is hardly achievable. Non-participation in negotiations does not prevent
states from joining agreements relevant to their interests. On the other hand, regional
arms control requires the direct involvement of most, if not all, states in the region
in negotiating and drafting a treaty, especially when the treaty is linked to the resolution of a regional conflict. In all circumstances, when sovereign states enter into
agreements, whether of universal or regional application, it is essential that these
agreements provide for equal rights and obligations of the parties. Enforcement
mechanisms, set up within or outside the United Nations, must be capable of providing adequate responses to any party that violates its treaty obligations.

The choice of arms control measures to be negotiated and agreed upon in the
future will depend on whether states will decide to satisfy their security requirements through arms acquisition or through arms limitation and reduction within the
framework of cooperative security arrangements. The road to total disarmament –
the objective set by the United Nations several decades ago – remains long and
uncertain.

Appendix

Table A1 Parties to the Major Arms Control Agreements[a]
Status as of 1 January 2002

State	Geneva Protocol 1925	Antarctic Treaty 1959	PTBT 1963	Outer Space Treaty 1967	NPT 1968	Seabed Treaty 1971	BW Conv. 1972	Enmod Conv. 1977	CW Conv. 1993	APM Conv. 1997
Afghanistan	1986		1964	1988	1970	1971	1975	1985		
Albania	1989				1990		1992		1994	2000
Algeria	1992			1992	1995	1992	2001	1991	1995	2001
Andorra					1996					1998
Angola	1990				1996					
Antigua and Barbuda	1988		1988	1988	1985	1988		1988		1999
Argentina	1969	1961	1986	1969	1995	1983	1979	1987	1995	1999
Armenia			1994		1993		1994		1995	
Australia	1930	1961	1963	1967	1973	1973	1977	1984	1994	1999
Austria	1928	1987	1964	1968	1969	1972	1973	1990	1995	1998
Azerbaijan					1992				2000	
Bahamas			1976	1976	1976	1989	1986			1998
Bahrain	1988				1988		1988		1997	
Bangladesh	1989		1985	1986	1979		1985	1979	1997	2000
Barbados	1976			1968	1980		1973			1999
Belarus	1970		1963	1967	1993	1971	1975	1978	1996	
Belgium	1928	1960	1966	1973	1975	1972	1979	1982	1997	1998
Belize					1985		1986			1998

Country									
Benin	1986	1964	1986	1972	1986	1975	1986	1998	1998
Bhutan	1979	1978		1985		1978			
Bolivia	1985	1965		1970		1975		1998	1998
Bosnia and Herzegovina		1994			1994	1994		1997	1998
Botswana		1968		1969	1972	1992		1998	2000
Brazil	1970	1964	1969	1998	1988	1973	1984	1996	1999
Brunei Darussalam				1985		1991		1997	
Bulgaria	1934	1963	1967	1969	1971	1972	1978	1994	1998
Burkina Faso	1971		1968	1970		1991		1997	1998
Burundi				1971				1998	
Cambodia	1983			1972		1983			1999
Cameroon	1989			1969				1996	
Canada	1930	1964	1967	1969	1972	1972	1981	1995	1997
Cape Verde	1992	1979		1979	1979	1977	1979		2001
Central African Republic	1970	1964		1970	1981				
Chad		1965		1971					1999
Chile	1935	1965	1981	1995		1980	1994	1996	2001
China	1952		1983	1992	1991	1984		1997	
Colombia	1989	1985		1986		1983		2000	2000
Comoros				1995					
Congo, Dem. Rep. of		1965		1970		1975			
Congo, Republic of				1978	1978	1978			2001
Cook Islands								1994	
Costa Rica		1967		1970	1978	1973	1996	1996	1999

State	Geneva Protocol 1925	Antarctic Treaty 1959	PTBT 1963	Outer Space Treaty 1967	NPT 1968	Seabed Treaty 1971	BW Conv. 1972	Enmod Conv. 1977	CW Conv. 1993	APM Conv. 1997
Côte d'Ivoire	1970		1965		1973	1972			1995	2000
Croatia			1991		1992	1993	1993		1995	1998
Cuba	1966	1984		1977		1977	1976	1978	1997	
Cyprus	1966		1965	1972	1970	1971	1973	1978	1998	
Czech Republic	1993	1993	1993	1993	1993	1993	1993	1993	1996	1999
Denmark	1930	1965	1964	1967	1969	1971	1973	1978	1995	1998
Djibouti					1996					1998
Dominica					1984		1978	1992	2001	1999
Dominican Republic	1970		1964	1968	1971	1972	1973			2000
Ecuador	1970	1987	1964	1969	1969		1975		1995	1999
Egypt	1928		1964	1967	1981			1982		
El Salvador			1964	1969	1972		1991		1995	1999
Equatorial Guinea	1989		1989	1989	1984		1989		1997	1998
Eritrea					1995				2000	2001
Estonia	1931	2001			1992		1993		1999	
Ethiopia	1935				1970	1977	1975		1996	
Fiji	1973		1972	1972	1972		1973		1993	1998
Finland	1929	1984	1964	1967	1969	1971	1974	1978	1995	
France	1926	1960		1970	1992		1984		1995	1998
Gabon			1964		1974				2000	2000
Gambia	1966		1965		1975		1991		1998	
Georgia					1994		1996		1995	

Germany	1929	1979	1964	1971	1975	1975	1983	1983	1994	1998
Ghana	1967		1963		1970	1972	1975	1978	1997	2000
Greece	1931	1987	1963	1971	1970	1985	1975	1983	1994	
Grenada	1989				1975		1986			1998
Guatemala	1983	1991	1964		1970	1996	1973	1988		1999
Guinea					1985				1997	1998
Guinea-Bissau	1989		1976	1976	1976	1976	1976			2001
Guyana					1993				1997	
Haiti					1970					
Holy See	1966				1971				1999	1998
Honduras			1964		1973		1979			1998
Hungary	1952	1984	1963	1967	1969	1971	1972	1978	1996	1998
Iceland	1967		1964	1968	1969	1972	1973		1997	1999
India	1930	1983	1963	1982		1973	1974	1978	1996	
Indonesia	1971		1964		1979		1992		1998	
Iran	1929		1964		1970	1971	1973		1997	
Iraq	1931		1964	1968	1969	1972	1991			
Ireland	1930		1963	1968	1968	1971	1972	1982	1996	1997
Israel	1969		1964	1977	1975					
Italy	1928	1981	1964	1972	1975	1974	1975	1981	1995	1999
Jamaica	1970		1991	1970	1970	1986	1975		2000	1998
Japan	1970	1960	1964	1967	1976	1971	1982	1982	1995	1998
Jordan	1977		1964		1970	1971	1975		1997	1998
Kazakhstan				1998	1994				2000	

State	Geneva Protocol 1925	Antarctic Treaty 1959	PTBT 1963	Outer Space Treaty 1967	NPT 1968	Seabed Treaty 1971	BW Conv. 1972	Enmod Conv. 1977	CW Conv. 1993	APM Conv. 1997
Kenya	1970		1965	1984	1970		1976		1997	2001
Kiribati					1985				2000	2000
Korea, North	1989	1987			1985		1987	1984		
Korea, South	1989	1986	1964	1967	1975	1987	1986	1986	1997	
Kuwait	1971		1965	1972	1989		1972	1980	1997	
Kyrgyzstan					1994					
Laos	1989		1965	1972	1970	1971	1973	1978	1997	
Latvia	1931				1992	1992	1997		1996	
Lebanon	1969		1965	1969	1970		1975			
Lesotho	1972		1964		1970	1973	1977		1994	1998
Liberia	1927				1970					1999
Libya	1971		1968	1968	1975	1990	1982			
Liechtenstein	1991				1978	1991	1991		1999	1999
Lithuania	1932				1991		1998		1998	
Luxembourg	1936		1965		1975	1982	1976		1997	1999
Macedonia (FYROM)				1968	1995		1996		1997	1998
Madagascar	1967		1965	1968	1970			1978		1999
Malawi	1970		1964		1986				1998	1998
Malaysia	1970		1964		1970	1972	1991		2000	1999
Maldives	1966				1970		1993		1994	2000
Mali				1968	1970				1997	1998
Malta	1964		1964		1970	1971	1975		1997	2001

Country											
Marshall Islands											
Mauritania	1970		1964		1993					1998	2000
Mauritius	1932		1969	1969			1971	1972	1992	1993	1997
Mexico			1963	1968	1969		1984	1974		1994	1998
Micronesia					1995					1999	
Moldova					1994					1996	2000
Monaco	1967				1995			1999		1995	1998
Mongolia	1968		1963	1967	1969		1971	1972	1978	1995	
Morocco	1970		1966	1967	1970		1971			1995	
Mozambique					1990					2000	1998
Myanmar (Burma)			1963	1970	1992						
Namibia					1992					1995	1998
Nauru					1982					2001	2000
Nepal	1969		1964	1967	1970		1971			1997	
Netherlands	1930	1967	1964	1969	1975		1976	1981	1983	1995	1999
New Zealand	1930	1960	1963	1968	1969		1972	1972	1984	1996	1999
Nicaragua	1990		1965		1973		1973	1975		1999	1998
Niger	1967		1964	1967	1992		1971	1972	1993	1997	1999
Nigeria	1968		1967	1967	1968		1971	1973		1999	2001
Niue											1998
Norway	1932	1960	1963	1969	1969		1971	1973	1979	1994	1998
Oman					1997			1992		1995	
Pakistan	1960		1988	1968				1974	1986	1997	
Palau					1995						

State	Geneva Protocol 1925	Antarctic Treaty 1959	PTBT 1963	Outer Space Treaty 1967	NPT 1968	Seabed Treaty 1971	BW Conv. 1972	Enmod Conv. 1977	CW Conv. 1993	APM Conv. 1997
Panama	1970		1966		1977	1974	1974		1998	1998
Papua New Guinea	1981	1981	1980	1980	1982		1980	1980	1996	
Paraguay	1933				1970		1976		1994	1998
Peru	1985	1981	1964	1979	1970		1985		1995	1998
Philippines	1973		1965		1972	1993	1973		1996	2000
Poland	1929	1961	1963	1968	1969	1971	1973	1978	1995	
Portugal	1930			1996	1977	1975	1975		1996	1999
Qatar	1976				1989	1974	1975		1997	1998
Romania	1929	1971	1963	1968	1970	1972	1979	1983	1995	2000
Russia	1928	1960	1963	1967	1970	1972	1975	1978	1997	
Rwanda	1964		1963		1975	1975	1975			2000
Saint Kitts and Nevis	1989				1993		1991			1998
Saint Lucia	1988				1979		1986	1993	1997	1999
Saint Vincent & the Gren.				1999	1984	1999	1999	1999		2001
Samoa, Western			1965		1975					1998
San Marino			1964	1968	1970		1975		1999	1998
Sao Tome and Principe					1983	1979	1979	1979		
Saudi Arabia	1971			1976	1988	1972	1972		1996	
Senegal	1977		1964	1978	1970		1975		1998	1998
Seychelles			1985		1985	1985	1979		1993	2000
Sierra Leone	1967		1964	1967	1975		1976			2001
Singapore			1968	1976	1976	1976	1975		1997	

Slovakia	1993	1993	1993	1993	1993	1993	1993		1995	1999
Slovenia		1993	1992		1992	1992	1992		1997	1998
Solomon Islands	1981				1981	1981	1981	1981		1999
Somalia					1970					
South Africa	1930	1960	1963	1968	1991	1973	1975		1995	1998
Spain	1929	1982	1964	1968	1987	1987	1979	1978	1994	1999
Sri Lanka	1954		1964	1986	1979		1986	1978	1994	
Sudan	1980		1966		1973				1999	
Suriname			1993		1976		1993		1997	
Swaziland	1991		1969		1969	1971	1991	1984	1996	1998
Sweden	1930	1984	1963	1967	1970	1972	1976		1993	1998
Switzerland	1932	1990	1964	1969	1977	1976	1976	1988	1995	1998
Syria	1968		1964	1968	1969	1972	1973			
Taiwan			1964	1970	1970			1999		
Tajikistan	1963				1995				1995	1999
Tanzania			1964		1991				1998	2000
Thailand	1931		1963	1968	1972		1975			1998
Togo	1971		1964	1989	1970	1971	1976		1997	2000
Tonga	1971		1971	1971	1971		1976			
Trinidad and Tobago	1962		1964		1986				1997	1998
Tunisia	1967		1965	1968	1970	1971	1973	1978	1997	1999
Turkey	1929		1965	1968	1980	1972	1974		1997	
Turkmenistan					1994		1996		1994	1998
Tuvalu					1979					

State	Geneva Protocol 1925	Antarctic Treaty 1959	PTBT 1963	Outer Space Treaty 1967	NPT 1968	Seabed Treaty 1971	BW Conv. 1972	Enmod Conv. 1977	CW Conv. 1993	APM Conv. 1997
Uganda	1965		1964	1968	1982		1992		2001	1999
UK	1930	1960	1963	1967	1968	1972	1975	1978	1996	1998
Ukraine		1992	1963	1967	1994	1971	1975	1978	1998	
United Arab Emirates					1995				2000	
Uruguay	1977	1980	1969	1970	1970		1981	1993	1994	2001
USA	1975	1960	1963	1967	1970	1972	1975	1980	1997	
Uzbekistan					1992		1996	1993	1996	
Vanuatu					1995		1990			
Venezuela	1928	1999	1965	1970	1975		1978		1997	1999
Viet Nam	1980			1980	1982	1980	1980	1980	1998	
Yemen	1971		1979	1979	1979	1979	1979	1977	2000	1998
Yugoslavia	1929		1964		1970	1973	1973		2000	
Zambia			1965	1973	1991	1972			2001	2001
Zimbabwe					1991		1990		1997	1998

Geneva Protocol   1925 Protocol for the Prohibition of the Use in War of Asphyxiating, Poisonous or Other Gases, and of Bacteriological Methods of Warfare

Note: Although the Geneva Protocol is considered part of the humanitarian law of armed conflict, it is included in this table to illustrate the process of transition from the ban on the use of biological and chemical weapons under the Protocol to the ban on the possession of these weapons under the BW and CW Conventions.

Antarctic Treaty   1959 Treaty ensuring the use of Antarctica for peaceful purposes only

PTBT   1963 Treaty Banning Nuclear Weapon Tests in the Atmosphere, in Outer Space and Under Water

Outer Space Treaty   1967 Treaty on Principles Governing the Activities of States in the Exploration and Use of Outer Space, Including the Moon and Other Celestial Bodies

NPT   1968 Treaty on the Non-Proliferation of Nuclear Weapons (Non-Proliferation Treaty)

Seabed Treaty   1971 Treaty on the Prohibition of the Emplacement of Nuclear Weapons and other Weapons of Mass Destruction on the Seabed
                and the Ocean Floor and in the Subsoil thereof

BW Conv.   1972 Convention on the Prohibition of the Development, Production and Stockpiling of Bacteriological (Biological) and Toxin
                Weapons and on their Destruction

Enmod Conv.   1977 Convention on the Prohibition of Military or Any Other Hostile Use of Environmental Modification Techniques

CW Conv.   1993 Convention on the Prohibition of the Development, Production, Stockpiling and Use of Chemical Weapons and on their
                Destruction

APM Conv.   1997 Convention on the Prohibition of the Use, Stockpiling, Production and Transfer of Anti-Personnel Mines and on their
                Destruction

a The years indicated are those of the ratification, accession, succession, acceptance or approval by individual countries (see Chapter 1.2).

Table A2 *Parties to the Major Agreements on Humanitarian Law of Armed Conflict[a]*
Status as of 1 January 2002

State	Genocide Conv. 1948	Geneva Conv. IV 1949	Prot. I to 1949 Geneva Convs 1977	Prot. II to 1949 Geneva Convs 1977	Prot. I to 1981 CCW Conv.[b] 1981	Prot. II to 1981 CCW Conv.[b] 1981	Amended Prot. II to 1981 CCW Conv.[b] 1996	Prot. III to 1981 CCW Conv.[b] 1981	Prot. IV to 1981 CCW Conv.[b] 1995
Afghanistan	1956	1956							
Albania	1955	1957	1993	1993					
Algeria	1963	1962	1989	1989					
Andorra		1993							
Angola	1984	1984	1984						
Antigua and Barbuda	1988	1986	1986	1986					
Argentina	1956	1956	1986	1986	1995	1995	1998	1995	1998
Armenia	1993	1993	1993	1993					
Australia	1949	1958	1991	1991	1983	1983	1997	1983	1997
Austria	1958	1953	1982	1982	1983	1983	1998	1983	1998
Azerbaijan	1996	1993							
Bahamas	1975	1975	1980	1980					
Bahrain	1990	1971	1986	1986					
Bangladesh	1998	1972	1980	1980	2000	2000	2000	2000	2000
Barbados	1980	1968	1990	1990					
Belarus	1954	1954	1989	1989	1982	1982		1982	2000
Belgium	1951	1952	1986	1986	1995	1995	1999	1995	1999

Belize	1998	1984	1984	1984					
Benin		1961	1986	1986	1989			1989	1989
Bhutan		1991							
Bolivia	1992	1976	1983	1983	2001	2001	2001	2001	2001
Bosnia and Herzegovina		1992	1992	1992	1993	1993	2000	1993	2001
Botswana		1968	1979	1979					
Brazil	1952	1957	1992	1992	1995	1995	1999	1995	1999
Brunei Darussalam		1991	1991	1991					
Bulgaria	1950	1954	1989	1989	1982	1982	1998	1982	1998
Burkina Faso	1965	1961	1987	1987					
Burundi	1997	1971	1993	1993					
Cambodia	1950	1958	1998	1998	1997	1997	1997	1997	1997
Cameroon		1963	1984	1984					
Canada	1952	1965	1990	1990	1994	1994	1998	1994	1998
Cape Verde		1984	1995	1995	1997	1997	1997	1997	1997
Central African Republic		1966	1984	1984					
Chad		1970	1997	1997					
Chile	1953	1950	1991	1991					
China	1983	1956	1983	1983	1982	1982	1998	1982	1998
Colombia	1959	1961	1993	1995	2000	2000	2000	2000	2000
Comoros		1985	1985	1985					
Congo, Dem. Rep. of	1962	1961	1982						

State	Genocide Conv. 1948	Geneva Conv. IV 1949	Prot. I to 1949 Geneva Convs 1977	Prot. II to 1949 Geneva Convs 1977	Prot. I to 1981 CCW Conv.[b] 1981	Prot. II to 1981 CCW Conv.[b] 1981	Amended Prot. II to 1981 CCW Conv.[b] 1996	Prot. III to 1981 CCW Conv.[b] 1981	Prot. IV to 1981 CCW Conv.[b] 1995
Congo, Republic of									
Costa Rica	1950	1969	1983	1983	1998	1998	1998	1998	1998
Côte d'Ivoire	1995	1961	1989	1989					
Croatia	1992	1992	1992	1992	1993	1993		1993	
Cuba	1953	1954	1982	1999	1987	1987		1987	
Cyprus	1982	1962	1979	1996	1988	1988		1988	
Czech Republic	1993	1993	1993	1993	1993	1993	1998	1993	1998
Denmark	1951	1951	1982	1982	1982	1982	1997	1982	1997
Djibouti		1978	1991	1991	1996	1996		1996	
Dominica		1981	1996	1996					
Dominican Republic		1958	1994	1994					
Ecuador	1949	1954	1979	1979	1982	1982	2000	1982	
Egypt	1952	1952	1992	1992					
El Salvador	1950	1953	1978	1978	2000	2000	2000	2000	2000
Equatorial Guinea		1986	1986	1986					
Eritrea		2000							
Estonia	1991	1993	1993	1993	2000		2000	2000	2000
Ethiopia	1949	1969	1994	1994					
Fiji	1973	1971							
Finland	1959	1955	1980	1980	1982	1982	1998	1982	1996

France	1950	1951	2001	1984	1988	1988	1998		1998
Gabon	1983	1965	1980	1980					
Gambia	1978	1966	1989	1989					
Georgia	1993	1993	1993	1993	1996	1996	1997	1996	1997
Germany	1954	1954	1991	1991	1992	1992	1999	1992	1997
Ghana	1958	1958	1978	1978					
Greece	1954	1956	1989	1993	1992	1992		1992	1997
Grenada		1981	1998	1998					
Guatemala	1950	1952	1987	1987	1983	1983	2001	1983	
Guinea	2000	1984	1984	1984					
Guinea-Bissau		1974	1986	1986					
Guyana		1968	1988	1988					
Haiti	1950	1957							
Holy See		1951	1985	1985	1997	1997	1997	1997	1997
Honduras	1952	1965	1995	1995					
Hungary	1952	1954	1989	1989	1982	1982	1998	1982	1998
Iceland	1949	1965	1987	1987					1999
India	1959	1950			1984	1984	1999	1984	
Indonesia		1958							
Iran	1956	1957							
Iraq	1959	1956							
Ireland	1976	1962	1999	1999	1995	1995	1997	1995	1997

State	Genocide Conv. 1948	Geneva Conv. IV 1949	Prot. I to 1949 Geneva Convs 1977	Prot. II to 1949 Geneva Convs 1977	Prot. I to 1981 CCW Conv.b 1981	Prot. II to 1981 CCW Conv.b 1981	Amended Prot. II to 1981 CCW Conv.b 1996	Prot. III to 1981 CCW Conv.b 1981	Prot. IV to 1981 CCW Conv.b 1995
Israel	1950	1951			1995	1995	2000		2000
Italy	1952	1951	1986	1986	1995	1995	1999	1995	1999
Jamaica	1968	1964	1986	1986					
Japan		1953			1982	1982	1997	1982	1997
Jordan	1950	1951	1979	1979	1995		2000	1995	
Kazakhstan	1998	1992	1992	1992					
Kenya		1966	1999	1999					
Kiribati		1989							
Korea, North	1989	1957	1988						
Korea, South	1950	1966	1982	1982	2001		2001		
Kuwait	1995	1967	1985	1985					
Kyrgyzstan	1997	1992	1992	1992					
Laos	1950	1956	1980	1980	1983	1983		1983	
Latvia	1992	1991	1991	1991	1993	1993		1993	1998
Lebanon	1953	1951	1997	1997					
Lesotho	1974	1968	1994	1994	2000	2000		2000	
Liberia	1950	1954	1988	1988					
Libya	1989	1956	1978	1978					
Liechtenstein	1994	1950	1989	1989	1989	1989	1997	1989	1997
Lithuania	1996	1996	2000	2000	1998		1998	1998	1998

Country									
Luxembourg	1981	1953	1989	1989	1996	1996	1999	1996	1999
Macedonia (FYROM)	1994	1993	1993	1993	1996	1996		1996	1996
Madagascar		1963	1992	1992					
Malawi		1968	1991	1991					
Malaysia	1994	1962							
Maldives	1984	1991	1991	1991	2000		2000	2000	2000
Mali	1974	1965	1989	1989	2001		2001		2001
Malta		1968	1989	1989	1995	1995		1995	
Mauritania		1962	1980	1980					
Mauritius		1970	1982	1982	1996	1996		1996	1998
Mexico	1952	1952	1983	1982	1982	1982		1982	
Micronesia		1995	1995	1995					
Moldova	1993	1993	1993	1993	2000	2000	2001	2000	2000
Monaco	1950	1950	2000	2000	1997		1997		
Mongolia	1967	1958	1995	1995	1982	1982		1982	1999
Morocco	1958	1956							
Mozambique	1983	1983	1983						
Myanmar (Burma)	1956	1992	1994	1994					
Namibia	1994	1991							
Nauru					2001		2001		2001
Nepal	1969	1964							
Netherlands	1966	1954	1987	1987	1987	1987	1999	1987	1999

State	Genocide Conv. 1948	Geneva Conv. IV 1949	Prot. I to 1949 Geneva Convs 1977	Prot. II to 1949 Geneva Convs 1977	Prot. I to 1981 CCW Conv.[b] 1981	Prot. II to 1981 CCW Conv.[b] 1981	Amended Prot. II to 1981 CCW Conv.[b] 1996	Prot. III to 1981 CCW Conv.[b] 1981	Prot. IV to 1981 CCW Conv.[b] 1995
New Zealand	1978	1959	1988	1988	1993	1993	1998	1993	1998
Nicaragua	1952	1953	1999	1999	2000		2000	2000	2000
Niger		1964	1979	1979	1992	1992		1992	
Nigeria		1961	1988	1988					
Norway	1949	1951	1981	1981	1983	1983	1998	1983	1998
Oman		1974	1984	1984					
Pakistan	1957	1951			1985	1985	1999	1985	2000
Palau		1996	1996	1996					
Panama	1950	1956	1995	1995	1997	1997	1999	1997	1997
Papua New Guinea	1982	1976							
Paraguay	2001	1961	1990	1990					
Peru	1960	1956	1989	1989	1997		1997	1997	1997
Philippines	1950	1952		1986	1996	1996	1997	1996	1997
Poland	1950	1954	1991	1991	1983	1983		1983	
Portugal	1999	1961	1992	1992	1997	1997	1999	1997	
Qatar		1975	1988						
Romania	1950	1954	1990	1990	1995	1995		1995	
Russia	1954	1954	1989	1989	1982	1982		1982	1999
Rwanda	1975	1964	1984	1984					
Saint Kitts and Nevis		1986	1986	1986					

Country									
Saint Lucia		1981	1982	1982					
Saint Vincent & the Gren.	1981	1981	1983	1983					
Samoa, Western		1984	1984	1984					
San Marino		1953	1994	1994					
Sao Tome and Principe		1976	1996	1996					
Saudi Arabia	1950	1963	1987						
Senegal	1983	1963	1985	1985			1999	1999	
Seychelles	1992	1984	1984	1984	2000	2000	2000	2000	2000
Sierra Leone		1965	1986	1986					
Singapore	1995	1973							
Slovakia	1993	1993	1993	1993	1993	1993	1999	1993	1999
Slovenia	1992	1992	1992	1992	1992	1992		1992	
Solomon Islands		1981	1988	1988					
Somalia		1962							
South Africa	1998	1952	1995	1995	1995	1995	1998	1995	1998
Spain	1968	1952	1989	1989	1993	1993	1998	1993	1998
Sri Lanka	1950	1959							
Sudan		1957							
Suriname		1976	1985	1985					
Swaziland		1973	1995	1995					
Sweden	1952	1953	1979	1979	1982	1982	1997	1982	1997
Switzerland	2000	1950	1982	1982	1982	1982	1998	1982	1998

State	Genocide Conv. 1948	Geneva Conv. IV 1949	Prot. I to 1949 Geneva Convs 1977	Prot. II to 1949 Geneva Convs 1977	Prot. I to 1981 CCW Conv.b 1981	Prot. II to 1981 CCW Conv.b 1981	Amended Prot. II to 1981 CCW Conv.b 1996	Prot. III to 1981 CCW Conv.b 1981	Prot. IV to 1981 CCW Conv.b 1995
Syria	1955	1953	1983						
Tajikistan		1993	1993	1993	1999	1999	1999	1999	1999
Tanzania	1984	1962	1983	1983					
Thailand		1954							
Togo	1984	1962	1984	1984	1995	1995		1995	
Tonga	1972	1978							
Trinidad and Tobago		1963	2001	2001					
Tunisia	1956	1957	1979	1979	1987	1987		1987	
Turkey	1950	1954							
Turkmenistan		1992	1992	1992					
Tuvalu		1981							
Uganda	1995	1964	1991	1991	1995	1995		1995	
UK	1970	1957	1998	1998	1995	1995	1999	1995	1999
Ukraine	1954	1954	1990	1990	1982	1982	1999	1982	
United Arab Emirates		1972	1983	1983					
Uruguay	1967	1969	1985	1985	1994	1994	1998	1994	1998
USA	1988	1955			1995	1995	1999		
Uzbekistan	1999	1993	1993	1993	1997	1997		1997	1997
Vanuatu		1982	1985	1985					
Venezuela	1960	1956	1998	1998					

	Genocide Conv.	Geneva Conv. IV	Prot. I to 1949 Geneva Convs	Prot. II to 1949 Geneva Convs	CCW Conv.	Prot. I to 1981 CCW Conv.	Prot. II to 1981 CCW Conv.	Amended Prot. II to 1981 CCW Conv.	Prot. III to 1981 CCW Conv.	Prot. IV to 1981 CCW Conv.
Viet Nam	1981	1957	1981							
Yemen	1987	1970	1990	1990						
Yugoslavia	2001	2001	2001	2001	2001	2001	2001			2001
Zambia	1966		1995	1995	2001		2001			
Zimbabwe	1991	1983	1992	1992						

Genocide Conv. — 1948 Convention on the Prevention and Punishment of the Crime of Genocide

Geneva Conv. IV — 1949 Geneva Convention (IV) relative to the Protection of Civilian Persons in Time of War

Prot. I to 1949 Geneva Convs — 1977 Protocol I Additional to the 1949 Geneva Conventions, and Relating to the Protection of Victims of International Armed Conflicts

Prot. II to 1949 Geneva Convs — 1977 Protocol II Additional to the 1949 Geneva Conventions, and Relating to the Protection of Victims of Non-International Armed Conflicts

CCW Conv. — 1981 Convention on Prohibitions or Restrictions on the Use of Certain Conventional Weapons which may be Deemed to be Excessively Injurious or to have Indiscriminate Effects

Prot. I to 1981 CCW Conv. — 1981 Protocol I to the 1981 CCW Convention on Non-Detectable Fragments

Prot. II to 1981 CCW Conv. — 1981 Protocol II to the 1981 CCW Convention on Prohibitions or Restrictions on the Use of Mines, Booby-Traps and Other Devices

Amended Prot. II to 1981 CCW Conv. — Protocol II to the 1981 CCW Convention, as amended in 1996

Prot. III to 1981 CCW Conv. — 1981 Protocol III to the 1981 CCW Convention on Prohibitions or Restrictions on the Use of Incendiary Weapons

Prot. IV to 1981 CCW Conv. — 1995 Protocol IV to the 1981 CCW Convention on Blinding Laser Weapons

[a] The years indicated are those of the ratification, accession, succession, acceptance or approval by individual countries (see Chapter 1.2).

[b] The CCW Convention is a framework agreement, under which specific agreements are subsumed in the form of protocols. At the time of the deposit of its instrument of ratification, acceptance or approval of the Convention or of accession thereto, each state must notify the depository of its consent to be bound by any two or more of the protocols.

Table A3   *Membership of Multilateral Weapon and Technology Export Control Regimes*
Status as of 1 January 2002

State	Zangger Committee	Nuclear Suppliers Group	Australia Group	Missile Technology Control Regime	Wassenaar Arrangement
Argentina	•	•	•	•	•
Australia	•	•	•	•	•
Austria	•	•	•	•	•
Belarus		•			
Belgium	•	•	•	•	•
Brazil		•		•	
Bulgaria	•	•	•		•
Canada	•	•	•	•	•
China	•				
Cyprus		•	•		
Czech Republic	•	•	•	•	•
Denmark	•	•	•	•	•
Finland	•	•	•	•	•
France	•	•	•	•	•
Germany	•	•	•	•	•
Greece	•	•	•	•	•
Hungary	•	•	•	•	•
Iceland			•	•	
Ireland	•	•	•	•	•
Italy	•	•	•	•	•
Japan	•	•	•	•	•

Country					
Korea, South	•	•	•	•	•
Latvia	•	•			
Luxembourg	•	•	•	•	•
Netherlands	•	•	•	•	•
New Zealand	•	•	•	•	•
Norway	•	•	•	•	•
Poland	•	•	•	•	•
Portugal	•	•	•	•	•
Romania	•	•	•		•
Russia	•	•	•	•	•
Slovakia	•	•	•		
Slovenia	•	•			
South Africa	•	•	•	•	•
Spain	•	•	•	•	•
Sweden	•	•	•	•	•
Switzerland	•	•	•	•	•
Turkey	•	•	•	•	•
UK	•	•	•	•	•
Ukraine	•	•		•	•
USA	•	•		•	•
Total number	**35**	**39**	**33**	**33**	**33**

Source: Stockholm International Peace Research Institute (SIPRI).

Select Bibliography

Albright, D., Berkhout, F. and Walker W., 1997. *Plutonium and Highly Enriched Uranium 1996: World Inventories, Capabilities and Policies.* Oxford: Stockholm International Peace Research Institute (SIPRI) and Oxford University Press.

Aron, R., 1963. *Le grand débat, initiation à la stratégie atomique* (The Great Debate, Introduction to Atomic Strategy). Paris: Calmann-Levy.

Anthony, I. & Rotfeld, A. D., eds, 2001. *A Future Arms Control Agenda: Proceedings of Nobel Symposium 118, 1999.* Oxford: Stockholm International Peace Research Institute (SIPRI) and Oxford University Press.

Bailey, S. D., 1972. *Prohibitions and Restraints in War.* London, Oxford, New York: Oxford University Press.

Barnaby, F., Goldblat, J., Jasani, B. & Rotblat, J., eds, 1979. *Nuclear Energy and Nuclear Weapon Proliferation.* London: Stockholm International Peace Research Institute (SIPRI) and Taylor & Francis.

Beaton, L. & Maddox, J., 1962. *The Spread of Nuclear Weapons.* London: Chatto and Windus.

Bernauer, T., 1990. *The Projected Chemical Weapons Convention: A Guide to the Negotiations in the Conference on Disarmament.* New York: United Nations.

Bernauer, T. & Ruloff, D., eds, 1999. *The Politics of Positive Incentives in Arms Control.* Columbia, South Carolina: University of South Carolina Press.

Blackaby, F., Goldblat, J. & Lodgaard, S., eds, 1984. *No-First-Use.* London, Philadelphia: Stockholm International Peace Research Institute (SIPRI) and Taylor & Francis.

Blackaby, F. & Milne, T., eds, 2000. *A Nuclear-Weapon-Free World: Steps Along the Way.* Houndmills, Basingstoke, Hampshire and London: Macmillan.

Blacker, C. D. & Duffy, G., 1984. *International Arms Control.* Stanford, California: Stanford University Press.

Blair, B. G., 1995. *Global Zero Alert for Nuclear Forces.* Washington, DC: Brookings Institution.

Bogdanov, O. V., 1972. *Razoruzhenie Garantia Mira* (Disarmament – Guarantee of Peace). Moscow: Izdatelstvo Mezhdunarodnye Otnoshenia.

Brauch, H. G., ed., 1986. *Vertrauens-Bildende Massnahmen und Europäische Abrüstungskonferenz* (Confidence-building Measures and the European Disarmament Conference). Gerlingen: Bleicher Verlag.

Bring, O., 1987. *Nedrustningens folkrätt* (International Law of Disarmament). Stockholm: Norstedts.

Bull, H., 1965. *The Control of the Arms Race.* New York: Praeger.

Burns, E. L. M., 1972. *A Seat at the Table.* Toronto: Clarke, Irwin & Company.

Caflisch, L. & Tanner F., eds, 1989. *The Polar Regions and Their Strategic Significance*. Geneva: Graduate Institute of International Studies.

Carlton, D. & Shaerf, C., eds, 1975. *The Dynamics of the Arms Race*. London: Croom Helm.

Catrina, C., 1988. *Arms Transfers and Dependence*. New York, Philadelphia, Washington, DC, London: UN Institute for Disarmament Research (UNIDIR) and Taylor & Francis.

Charpak, G. & Garwin, R. L., *Feux follets et champignons nucléaires* (Will-o'-the Wisp and Nuclear Mushrooms). Paris: Editions Odile Jacob.

Cowen Karp, R., ed., 1991. *Security with Nuclear Weapons? Different Perspectives on National Security*. Oxford: Stockholm International Peace Research Institute (SIPRI) and Oxford University Press.

Dahlitz, J., 1983. *Nuclear Arms Control with Effective International Agreements*. Melbourne: McPhee Gribble Publishers.

Dahlman, O. & Israelson, H., 1977. *Monitoring Underground Nuclear Explosions*. Amsterdam: Elsevier Scientific Publishing Company.

Davidov, V. F., 1980. *Nerasprostranenie Yadernogo Oruzhiya i Politika S. Sh. A.* (Non-Proliferation of Nuclear Weapons and US Policy). Moscow: Izdatelstvo Nauka.

Dhanapala, J., ed., 1993. *Regional Approaches to Disarmament: Security and Stability*. Aldershot, Brookfield USA, Hong Kong, Singapore, Sydney: UN Institute for Disarmament Research (UNIDIR) and Dartmouth.

Eisenbart, C. & Daase, C., eds, 2000. *Nuklearwaffenfreie Zonen: Neue Aktualität eines alten Konzeptes* (Nuclear-Weapon-Free Zones: The Topicality of an Old Concept). Heidelberg: FEST.

Epstein, W., 1976. *The Last Chance, Nuclear Proliferation and Arms Control*. New York: The Free Press.

Etzioni, A., 1962. *The Hard Way to Peace: A New Strategy*. New York: Collier.

Fieldhouse, R., ed., 1990. *Security at Sea: Naval Forces and Arms Control*. Oxford: Stockholm International Peace Research Institute (SIPRI) and Oxford University Press.

Fischer, D. & Szasz, P., 1985. *Safeguarding the Atom: A Critical Appraisal*. London, Philadelphia: Stockholm International Peace Research Institute (SIPRI) and Taylor & Francis.

Fischer, D., 1993. *Towards 1995: The Prospects for Ending the Proliferation of Nuclear Weapons*. Aldershot, Brookfield USA, Hong Kong, Singapore, Sydney: UN Institute for Disarmament Research (UNIDIR) and Dartmouth.

Fischer, D., 1997. *History of the International Atomic Energy Agency: The First Forty Years*. Vienna: International Atomic Energy Agency (IAEA).

Fischer, G., 1969. *La non-prolifération des armes nucléaires* (The Non-Proliferation of Nuclear Weapons). Paris: Pichon et Durant-Augias.

Fry, M., Keatinge, N. P. & Rotblat, J., eds, 1990. *Nuclear Non-Proliferation and the Non-Proliferation Treaty*. Berlin, Heidelberg, New York, London, Paris, Tokyo, Hong Kong: Springer-Verlag.

Gasparini Alves, P., 1991. *Prevention of an Arms Race in Outer Space: A Guide to the Discussions in the Conference on Disarmament*. New York: United Nations.

Gasteyger, C., 1985. *Searching for World Security: Understanding Global Armament and Disarmament*. London: Frances Pinter.

Geissler, E., ed., 1986. *Biological and Toxin Weapons Today*. Oxford: Stockholm International Peace Research Institute (SIPRI) and Oxford University Press.

Ghebali, V-Y., 1989. *Confidence-Building Measures within the CSCE Process: Paragraph-by-Paragraph Analysis of the Helsinki and Stockholm Régimes*. New York: United Nations.

Ghebali, V-Y., 2000. *L'OSCE dans l'Europe post-communiste, 1990–1996: Vers une identité paneuropéenne de sécurité* (The OSCE in Post-Communist Europe, 1990–1996: Towards a Pan-European Security Identity). Brussels: Etablissements Emile Bruyant.

Gjelstad, J. & Njolstad, O., eds, 1996. *Nuclear Rivalry and International Order*. Oslo, London: International Peace Research Institute, Oslo (PRIO) and SAGE Publications.

Glasstone, S. & Dolan, P. J., eds, 1977. *The Effects of Nuclear Weapons*. Washington, DC: US Department of Defense and US Department of Energy.

Goldblat, J., 1971. *CB Disarmament Negotiations 1920–1970*. Stockholm: Stockholm International Peace Research Institute (SIPRI) and Almqvist & Wiksell.

Goldblat, J. & Vinas, A., 1985. *La no proliferacion de armas nucleares* (The Nonproliferation of Nuclear Weapons). Madrid: Fundación de Estudios sobre la Paz y las Relaciones Internacionales (FEPRI)

Goldblat, J., ed., 1985. *Non-Proliferation: The Why and the Wherefore*. Stockholm, London: Stockholm International Peace Research Institute (SIPRI) and Taylor & Francis.

Goldblat, J. & Cox, D., eds, 1988. *Nuclear Weapon Tests: Prohibition or Limitation?* Stockholm, Ottawa, Oxford: Stockholm International Peace Research Institute (SIPRI), Canadian Institute for International Peace and Security (CIIPS) and Oxford University Press.

Goldblat, J., 1990. *Twenty Years of the Non-Proliferation Treaty: Implementation and Prospects*. Oslo: International Peace Research Institute, Oslo (PRIO).

Goldblat, J., 1992. *The Non-Proliferation Treaty: How to Remove the Residual Threats?* New York: United Nations.

Goldblat, J., ed., 1992. *Maritime Security: The Building of Confidence*. New York: United Nations.

Goldblat, J., 1994. *Arms Control: A Guide to Negotiations and Agreements*. London, Thousand Oaks, New Delhi: SAGE Publications.

Goldblat, J., 1997. *The Nuclear Non-Proliferation Regime: Assessment and Prospects*. The Hague/Boston/London: Martinus Nijhoff.

Goldblat, J., ed., 2000. *Nuclear Disarmament: Obstacles to Banishing the Bomb.* London and New York: I. B. Tauris.

Goldschmidt, B., 1980. *Le complexe atomique* (The Atomic Complex). Paris: Fayard.

Graduate Institute of International Studies (Geneva), 2001. *Small Arms Survey 2001: Profiling the Problem.* Oxford: Oxford University Press.

Grin, J. & Graaf, H. V. D., eds, 1990. *Unconventional Approaches to Conventional Arms Control Verification: An Exploratory Assessment.* Amsterdam: VU University Press.

Haar, B., ter, 1991. *The Future of Biological Weapons.* New York, London: Praeger.

Hamel-Green, M., 1990. *The South Pacific Nuclear Free Zone Treaty: A Critical Assessment.* Canberra: Australian National University.

Herby, P., 1992. *The Chemical Weapons Convention and Arms Control in the Middle East.* Oslo: International Peace Research Institute, Oslo (PRIO).

Independent Commission on Disarmament and Security Issues, 1982. *Common Security: A Blueprint for Survival.* New York: Simon & Schuster.

Jacobson, H. & Stein, E., 1966. *Diplomats, Scientists and Politicians: The United States and the Nuclear Test Ban Negotiations.* Ann Arbour, Michigan: University of Michigan Press.

Jasani, B., ed., 1987. *Space Weapons and International Security.* Oxford: Stockholm International Peace Research Institute (SIPRI) and Oxford University Press.

Jasani, B., ed., 1991. *Peaceful and Non-Peaceful Uses of Outer Space: Problems of Definition for the Prevention of an Arms Race.* New York, Philadelphia, Washington, DC, London: UN Institute for Disarmament Research (UNIDIR) and Taylor & Francis.

Joyce, J. A., 1980. *The War Machine: The Case Against the Arms Race.* London, Melbourne, New York: Quartet Books.

Kalinowski, M. B., ed., 2000. *Global Elimination of Nuclear Weapons.* Baden-Baden: Nomos.

Kalshoven, F., 1973. *The Law of Warfare, A Summary of its Recent History and Trends in Development.* Leiden, Geneva: Sijthoff and Henry Dunant Institute.

Karem, M., 1988. *A Nuclear Weapon Free Zone in the Middle East: Problems and Prospects.* New York: Greenwood Press.

Karp, A., 1996. *Ballistic Missile Proliferation: The Politics and Technics.* Oxford: Stockholm International Peace Research Institute (SIPRI) and Oxford University Press.

Khan, S. A., 1990. *Non-Proliferation in a Disarming World: Prospects for the 1990s.* Geneva: Bellerive Foundation.

Klein, J., 1964. *L'Entreprise du désarmement depuis 1945* (The Business of Disarmament since 1945). Paris: Editions Cujas.

Klein, J., 1987. *Sécurité et désarmement en Europe* (Security and Disarmament in Europe). Paris: Institut français des relations internationales (IFRI).

Kokoski, R., 1995. *Technology and the Proliferation of Nuclear Weapons.* Oxford: Stockholm International Peace Research Institute (SIPRI) and Oxford University Press.

Krass, A. S., 1985. *Verification: How Much Is Enough?* London, Philadelphia: Stockholm International Peace Research Institute (SIPRI) and Taylor & Francis.

Lachs, M., 1972. *The Law of Outer Space, An Experience in Contemporary Lawmaking.* Leiden: Sijthoff.

Leventhal, P. & Alexander, Y., 1987. *Preventing Nuclear Terrorism.* Lexington, Massachusetts: Lexington Books.

Lifton, R. & Falk, R., 1982. *Indefensible Weapons: The Political and Psychological Case Against Nuclearism.* New York: Basic Books.

Lodgaard, S., ed., 1990. *Naval Arms Control.* Oslo, London, Newbury Park, New Delhi: International Peace Research Institute, Oslo (PRIO) and SAGE Publications.

Lodgaard, S. & Thee, M., eds, 1983. *Nuclear Disengagement in Europe.* London, New York: Stockholm International Peace Research Institute (SIPRI), Pugwash and Taylor & Francis.

Lumsden, M., 1975. *Incendiary Weapons.* London, Stockholm: Stockholm International Peace Research Institute (SIPRI), MIT Press and Almqvist & Wiksell.

Lumsden, M., 1978. *Anti-Personnel Weapons.* London: Stockholm International Peace Research Institute (SIPRI) and Taylor & Francis.

Marks, A. W., ed., 1975. *NPT: Paradoxes and Problems.* Washington, DC: Arms Control Association and Carnegie Endowment for International Peace.

McPhee, J., 1974. *The Curve of Binding Energy.* New York: Farrar, Straus and Giroux.

Melman, S., ed., 1958. *Inspection for Disarmament.* New York: Columbia University Press.

Moch, J., 1969. *Destin de la paix* (The Future of Peace). Paris: Mercure de France.

Müller, H., ed., 1989. *A Survey of European Nuclear Policy, 1985–87.* London: Macmillan.

Myrdal, A., 1976. *The Game of Disarmament.* New York: Pantheon.

Newhouse, J., 1973. *Cold Dawn. The Story of SALT.* New York: Holt, Rinehart and Winston.

Noel-Baker, P., 1958. *The Arms Race, A Programme for World Disarmament.* London: John Calder.

Norwegian Ministry of Foreign Affairs, 1992. *Towards a Comprehensive Test Ban Treaty: Expert Study on Questions Related to a Comprehensive Test Ban Treaty.* Oslo: Utenriksdepartementet.

Petrovski, V. F., 1982. *Razoruzhenie: kontseptsii, problemy, mekhanism* (Disarmament: Concepts, Problems and Mechanisms). Moscow: Politizdat.

Potter, W. C., ed., 1990. *International Nuclear Trade and Nonproliferation: The Challenge of the Emerging Suppliers*. Lexington, Massachusetts/Toronto: Lexington Books.

Prawitz, J., 1995. *From Nuclear Option to Non-Nuclear Promotion: The Sweden Case*. Stockholm: Utrikespolitiska Institutet.

Primicerio, M., ed., 1995. *Controllo o Disordine: Il Futuro della Proliferazione Nucleare* (In Control or Out of Control: The Future of Nuclear Proliferation). Milan: Franco Angeli.

Prokosch, E., 1995. *The Technology of Killing: A Military and Political History of Antipersonnel Weapons*. London & New Jersey: ZED BOOKS.

Roberts, A. & Guelff, R., eds, 1989. *Documents on the Laws of War*. Oxford: Clarendon Press.

Roberts, S. & Williams, J., 1995. *After the Guns Fall Silent: The Enduring Legacy of Landmines*. Washington, DC: Vietnam Veterans of America Foundation.

Robinson, J. P., 1971. *The Rise of CB Weapons*. Stockholm: Stockholm International Peace Research Institute (SIPRI) and Almqvist & Wiksell.

Robinson, J. P., 1973. *CB Weapons Today*. Stockholm: Stockholm International Peace Research Institute (SIPRI) and Almqvist & Wiksell.

Robles, A. G., 1967. *El Tratado de Tlatelolco, Génesis, Alcance y Propositos de la Proscripcion de las Armas Nucleares en la America Latina* (The Treaty of Tlatelolco: Origin, Scope and Purposes of the Ban on Nuclear Weapons in Latin America). Mexico, DF: El Colegio de México.

Robles, A. G., 1979. *La Asamblea General del Desarme* (The General Assembly on Disarmament). Mexico, DF: Editorial de el Colegio Nacional.

Röling, B. V. A. & Sukovic, O., 1976. *The Law of War and Dubious Weapons*. Stockholm: Stockholm International Peace Research Institute (SIPRI) and Almqvist & Wiksell.

Roskill, S., 1968. *Naval Policy between the Wars*. New York: Walker & Company.

Rotblat, J., 1981. *Nuclear Radiation in Warfare*. London: Stockholm International Peace Research Institute (SIPRI) and Taylor & Francis.

Rotblat, J., 1998. *Nuclear Weapons: The Road to Zero*. Boulder, Colorado: Westview Press.

Rotblat, J., Steinberger, J. & Udgaonkar, B., eds, 1993. *A Nuclear-Weapon-Free World: Desirable? Feasible?* Boulder, San Francisco, Oxford: Westview Press.

Rotfeld, A. D. & Stützle, W., eds, 1991. *Germany and Europe in Transition*. Oxford: Stockholm International Peace Research Institute (SIPRI) and Oxford University Press.

Sanders, B., 1975. *Safeguards Against Nuclear Proliferation*. Cambridge, Massachusetts, London, Stockholm: Stockholm International Peace Research Institute (SIPRI), The MIT Press and Almqvist & Wiksell.

Schindler, D. & Toman, J., eds, 1988. *The Laws of Armed Conflicts: A Collection of Conventions, Resolutions and Other Documents*. Dordrecht, Geneva: Martinus Nijhoff Publishers and Henry Dunant Institute.

Schoettle, E. C. B., 1979. *Postures for Non-Proliferation: Arms Limitation and Security Policies to Minimize Nuclear Proliferation.* London: Stockholm International Peace Research Institute (SIPRI) and Taylor & Francis.

Scott, G., 1973. *The Rise and Fall of the League of Nations.* London: Hutchinson.

Shaker, M. I., 1980. *The Nuclear Non-Proliferation Treaty: Origin and Implementation. 1959–1979.* London/New York: Oceana.

Sharp, J. M. O., ed., 1990. *Europe After an American Withdrawal: Economic and Military Issues.* Oxford: Stockholm International Peace Research Institute (SIPRI) and Oxford University Press.

Shohno, N., 1986. *The Legacy of Hiroshima: Its Past, Our Future.* Tokyo: Kosei Publishing Co.

Simpson, J. & Howlett, D., eds, 1995. *The Future of the Non-Proliferation Treaty.* Houndmills, Basingstoke, Hampshire and London: Macmillan.

Simpson, J., ed., 1987. *Nuclear Non-Proliferation: An Agenda for the 1990s.* Cambridge, New York, New Rochelle, Melbourne, Sydney: Cambridge University Press.

Sims, N. A., 1988. *The Diplomacy of Biological Disarmament: Vicissitudes of a Treaty in Force, 1975–85.* London: Macmillan and London School of Economics and Political Science.

SIPRI Yearbooks. *Armaments, Disarmament and International Security.* Stockholm, London, Oxford: Stockholm International Peace Research Institute (SIPRI), Almqvist & Wiksell, Taylor & Francis and Oxford University Press.

Slack, M. & Chestnutt, H., 1990. *Open Skies: Technical, Organizational, Operational, Legal and Political Aspects.* Toronto: York University.

Spector, L. S., 1988. *The Undeclared Bomb.* Cambridge, Massachusetts: Ballinger.

Spector, L. S. & Smith, J. R., 1990. *Nuclear Ambitions. The Spread of Nuclear Weapons 1989–1990.* Boulder, San Francisco, Oxford: Westview Press.

Stützle, W., Jasani, B. & Cowen, R., eds, 1987. *The ABM Treaty: To Defend or Not to Defend?* Oxford, New York: Stockholm International Peace Research Institute (SIPRI) and Oxford University Press.

Sur, S., 1988. *Verification Problems of the Washington Treaty on the Elimination of Intermediate-Range Missiles.* New York: United Nations.

Sur, S., ed., 1991. *Verification of Current Disarmament and Arms Limitation Agreements: Ways, Means and Practices.* Aldershot, Brookfield USA, Hong Kong, Singapore, Sydney: UN Institute for Disarmament Research (UNIDIR) and Dartmouth College.

Sur, S., ed., 1992. *Verification of Disarmament or Limitation of Armaments: Instruments, Negotiations, Proposals.* New York: United Nations.

Tanner, F., ed., 1992. *From Versailles to Baghdad: Post-War Armament Control of Defeated States.* New York: United Nations.

Thee, M., 1986. *Military Technology, Military Strategy and the Arms Race.* London, New York: Croom Helm and St. Martins Press.

Timerbayev, R. M., 1983. *Kontrol za Ogranicheniem Vooruzheniy i Razoruzheniem* (Verification of Arms Limitation and Disarmament). Moscow: Mezhdunarodnye Otnoshenia.

Timerbayev, R. M., 1999. *Rossiya i Yadernoye Nerasprostranenye, 1945–1968* (Russia and Nuclear Non-Proliferation, 1945–1968). Moscow: Nauka.

Towpik, A., 1970. *Bezpieczenstwo Miedzynarodowe a Rozbrojenie* (International Security and Disarmament). Warsaw: Polski Instytut Spraw Miedzynarodowych.

UNITAR (UN Institute for Training and Research), 1987. *The United Nations and the Maintenance of International Peace and Security.* Dordrecht/Boston/Lancaster: Martinus Nijhoff.

Verification Yearbooks. London: Verification, Research, Training and Information Centre (VERTIC).

Veuthey, M., 1983. *Guérilla et Droit Humanitaire* (Guerrilla and Humanitarian Law). Geneva: Le Comité International de la Croix Rouge.

Wedar, C., Intriligator, M. & Vares, P., 1992. *Implications of the Dissolution of the Soviet Union for Accidental/Inadvertent Use of Weapons of Mass Destruction.* Tallinn: A/S MULTIPRESS.

Westing, A. H., ed., 1984. *Environmental Warfare: A Technical, Legal and Policy Appraisal.* London, Philadelphia: Stockholm International Peace Research Institute (SIPRI) and Taylor & Francis.

Westing, A. H., ed., 1984. *Herbicides in War: The Long-term Ecological and Human Consequences.* London, Philadelphia: Stockholm International Peace Research Institute (SIPRI) and Taylor & Francis.

Westing, A. H., ed., 1985. *Explosive Remnants of War: Mitigating the Environmental Effects.* London, Philadelphia: Stockholm International Peace Research Institute (SIPRI), UN Environment Programme (UNEP) and Taylor & Francis.

Westing, A. H., ed., 1990. *Environmental Hazards of War: Releasing Dangerous Forces in an Industrialized World.* Oslo, London, Newbury Park, New Delhi: International Peace Research Institute, Oslo (PRIO), UN Environment Programme (UNEP) and SAGE Publications.

Willot, A., 1964. *Le désarmement général et complet, Une approche* (An Approach to General and Complete Disarmament). Brussels: Editions de l'institut de sociologie, Université Libre.

Winkler, T., 1981. *Kernenergie und Aussenpolitik* (Nuclear Energy and Foreign Policy). Berlin: Berlin Verlag.

Wright, M., 1964. *Disarm and Verify: An Explanation of the Central Difficulties and of National Policies.* London: Chatto and Windus.

York, H. F., 1971. *Race to Oblivion: A Participant's View of the Arms Race.* New York: Simon & Schuster.

York, H., F., 1987. *Making Weapons, Talking Peace: A Physicist's Odyssey from Hiroshima to Geneva.* New York: Basic Books.

About the Author

Jozef Goldblat holds university degrees in international relations, law, economics and linguistics. He has been studying the problems of arms control since the late 1950s and has been involved in disarmament negotiations in various capacities, including service for the United Nations. He was active in international commissions responsible for verification of compliance with armistice agreements, has written reports, articles and books on truce supervision, the arms race and disarmament problems, which have appeared in several languages, and has lectured at several universities. From 1969 to 1989 he directed the arms control and disarmament research programme at the Stockholm International Peace Research Institute (SIPRI). In 1980 he assisted the UN Secretary-General in preparing a report on a comprehensive nuclear test ban. He is Vice-President of the Geneva International Peace Research Institute (GIPRI). He is also Associate Editor of *Security Dialogue,* a journal published by the International Peace Research Institute, Oslo (PRIO), and Resident Senior Fellow at the United Nations Institute for Disarmament Research (UNIDIR) in Geneva. Among the recent publications, he is the author of *Arms Control: A Guide to Negotiations and Agreements* (SAGE, 1994) and *The Nuclear Non-Proliferation Régime: Assessment and Prospects* (The Hague Academy of International Law, 1997) and the editor of *Nuclear Disarmament: Obstacles to Banishing the Bomb* (I.B. Tauris, 2000). His book *Arms Control Agreements: A Handbook* (Taylor & Francis, 1983) received a CHOICE Award, and in 1984 Jozef Goldblat was awarded the Pomerance Award by the NGO Committee on Disarmament in recognition of his scholarship in the field of disarmament and arms control.

Index

The index includes entries for Part I of this volume. The agreements reproduced in Part II, on CD-ROM, are given in bold print.

Index compiled by Peter Rea.